P9-DMF-275

To my wife, Edna, whose loving support
sustained me while writing this book.

To my parents, without whose collaboration
this book could not have been written.

To David, Clare, and Robert.

Brooks/Cole Publishing Company
A Division of Wadsworth, Inc.

Printed in the United States of America
10 9 8 7 6 5 4 3 2 1

Library of Congress Cataloging-in-Publication Data

Steward, Donald V.
Software engineering with systems analysis and design.

Includes index.
1. Computer software. 2. System analysis.
3. System design. I. Title.
QA76.754.S74 1987 005.1 86-26808
ISBN 0-534-07506-1

Sponsoring Editor: *Cynthia C. Stormer*
Project Development Editors: *John Bergez and Liz J. Currie*
Marketing Representative: *Jill McGillen*
Editorial Associate: *Corinne Kibbe*
Production Editor: *Ellen Brownstein*
Manuscript Editor: *Jonas Weisel*
Permissions Editor: *Carline Haga*
Interior and Cover Design: *Sharon L. Kinghan*
Cover Image: *Lisa Thompson*
Cover Photo: *Lee Hocker*
Art Coordinator: *Sue C. Howard*
Interior Illustration: *Maggie Stevens-Huft*
Typesetting: *Graphic Typesetting Service*
Printing/Binding: *The Maple-Vail Book Manufacturing Group*

Software Engineering
with Systems Analysis and Design

Donald V. Steward

California State University, Sacramento

Brooks/Cole Publishing Company
Monterey, California

Software Engineering
with Systems Analysis and Design

Preface

Software Engineering with Systems Analysis and Design develops the natural integration between software engineering and systems analysis and design. The field of software engineering has become increasingly complex. A bewildering variety of methodologies are currently in use and new methodologies are continually being developed. This book presents the classical approaches to software engineering—data flow diagrams, structure charts, Warnier-Orr Diagrams—and explores many of the newer techniques, such as Trees and higher order software. Furthermore, it presents an integrated approach that serves to illustrate basic principles while solving many of the problems intrinsic to the classical methods. It can be used alone to develop systems or in conjunction with readily-available computer aids.

Software Engineering with Systems Analysis and Design is written with both the student and the practitioner in mind. The book is appropriate for classes in software engineering or systems analysis, as well as in the spectrum of computer science and management information systems classes. The software engineering practitioner will find the text a valuable tool for improving his or her current methods.

The impetus for this book comes from two sources, one in industry and one in academic life. My twenty years of experience in industry (General Electric and Burroughs Research Laboratories) and my observations of senior projects at California State University, Sacramento, convinced me that students are frequently ill-prepared for the real-world problem solving they encounter when they graduate. This realization was the main motivation for my developing the course in software engineering that is now required of all computer science majors at CSUS. It has also guided my approach in the text.

Organization and Features

Software Engineering with Systems Analysis and Design is divided into five parts:

Part 1 introduces the principles used throughout the rest of the book;
Part 2 develops techniques for describing the product and working with its requirements and design;
Part 3 describes how the project is managed;
Part 4 examines information-systems technologies; and
Part 5 applies the techniques developed in the previous parts to each phase of the project.

Part 1 should be read first. Parts 2, 3, and 4 may be read in any order. Part 5 builds on the principles developed in Parts 1 through 4 and should be read in conjunction with the case study in the appendix.

The integrated approach presented here is supported by such unifying themes as the tree, which is used as a hierarchical representation of systems or programs; matching requirements and means; and the role of expectations in project management. The reader is introduced to needs and feasibility analyses, Two-Entity Data Flow Diagrams, Trees, and structured design, and then learns how these techniques can be extended and integrated throughout the entire software development life cycle. This approach shows the reader how to simplify transitions between phases in the life cycle, improve traceability, provide progress reports to managers, and estimate, plan, schedule, and control projects. Principles and general methods are presented first, followed by step-by-step applications of these methods through the phases of the project. The need for up-front planning is emphasized and methods of representation and progress measurement are provided. Also included in this book are chapters on problem definition, cost, and feasibility, topics that are often omitted or underplayed in other software engineering books. A case study at the end of the book provides a practical model for the concepts presented in the text.

To the Software Engineering Practitioner

There are currently many software engineering tools in use (data flow diagrams, structure charts, Warnier-Orr Diagrams, entity relation diagrams, state transition diagrams), each with its own application to a specific aspect of the system. If we use structure charts, we do structured design but, if we use Warnier-Orr Diagrams, we probably don't. We may use data flow diagrams and structured English for requirements, structure charts for design, and pseudocode and a high-level programming language for implementation. With all these methods in use on the same project and different methods in use during each phase, it is very difficult to provide consistency of representation, to ensure that ideas and work are not lost when converting from one method to another, and to track the progress of the entire project throughout its life cycle. It is also difficult to ensure that the requirements developed at the beginning of the project have been preserved in the end product, harder still to collect data that will aid in estimating future projects, and very difficult to use computer aids.

This book provides a means of using both the conventional and newer, more powerful methods of software engineering more effectively. It presents an integrated approach, based on the fundamental concepts of the classical approach, that can greatly increase productivity. The integrated approach provides the following advantages:

1. a consistent measure of progress throughout the entire cycle, from requirements to code generation
2. a means of tracking requirements throughout the cycle, ensuring that no requirements are lost from one phase to the next
3. a method of representation (tree editors) that does not have to be modified from one phase to another
4. a means of calculating metrics using computers
5. a method of collecting data automatically as the system develops
6. a method that can be implemented with computer aids.

In short, the integrated approach presented here will help the software engineering practitioner incorporate the methods he or she already uses with those that are newly developed and those that undoubtedly will be developed as the field of software engineering continues to grow.

Acknowledgments

There are many people to whom I owe a debt of gratitude for their valuable suggestions and contributions along the way. On our faculty at CSUS, I wish to thank: Richard Hill, Robert Buckley, Martin Meyers, Ronald Ernst, Nancy Minor, and Joan Al-Kazily. At Brooks/Cole, Ellen Brownstein, Mike Needham, Cindy Stormer, and Neil Oatley have lent that vital assistance needed to turn these ideas into a book.

Thanks are also due to the reviewers of this book: Meledath Damodaran, University of Bridgeport; Richard Hill, California State University, Sacramento; Gregory Jones, Utah State University, Logan; William Junk, University of Idaho, Moscow; Richard LeBlanc, Georgia Institute of Technology, Atlanta; Christopher Pidgeon, University of California, Irvine; Vaclav Rajlich, Wayne State University, Detroit; K.V. K. Reddy, McNeese State University, Lake Charles, Louisiana; Paul Ross, Millersville State College, Millersville, Pennsylvania; and Stephen Thebaut, University of Florida, Gainesville.

Donald V. Steward

Contents

PART 3

1 PART

Introducing Software Engineering

1

What Is Software Engineering?

- The field of software engineering is concerned with all of the activities involved in the solution of problems through the development of computer systems.
- Today careful planning and coordination are necessary to produce software because the problems they solve and the programs themselves are complex, the development must be done by teams, software is expensive, and mistakes are costly and difficult to correct.
- Software engineering is the management of expectations, computer technology, human skills, time, and money in order to create a software product that meets the expectations of the client with a satisfactory return to the producer.
- Software projects should be planned at the start rather than ad hoc as they proceed. They require communication between the two cultures of producer and client, require a commitment of resources before many important questions can be resolved, and culminate in an entirely new product.
- Software engineering continues to evolve, finding better methods for producing today's increasingly complex software.

What Does Software Engineering Involve?

The first point to be made about software engineering has to do with its scope of activities. Software engineering is the management of the entire process of developing computer systems to solve problems. We'll expand this definition as we proceed, but here we want to emphasize the inclusiveness of the field.

As software systems have grown more sophisticated and complex, software developers have sought new methods for their development. Software engineering is a response to that need. A comparatively new field, it is still rapidly changing and maturing. In this chapter we will define software engineering and describe the circumstances under which it has become so important.

Typically, when someone is told to write a program, someone else has told him or her *what* the program is to do and *why*; the programmer is concerned only with *how* to write it. As software engineers, however, we will be the ones concerned with the *what* and *why*. Software engineering includes the whole range of activities having to do with problem solving—from helping the client define the problem or opportunity, to evaluating the client's satisfaction with the solution.

Developing a software system may require writing a whole collection of programs to tell machines what to do, writing procedures to tell people what to do, and providing training so the people understand how to do it. We may need to convert data from an old system so it can be run by a new system, hire people and acquire machines to run the programs, and obtain space in which the machines and people can do their work. What we are building is a system of many parts working together. Such an endeavor requires patience and flexibility. We will have to fix the system when it does not do what we expected, or the client may ask us to change the system because he or she wants it to do something not previously intended.

Programming may be no more than 20 percent of the total scope of software engineering, and the fraction of effort involved with programming can be expected to drop as improved methods are used for developing software systems. As time goes on, more of our effort will go into managing the overall process and less into programming.

What Is a Successful Software Development Project?

The goal of software engineering is to carry out successful software development projects. The goal of a successful project is to produce an acceptable product. Let us define some terms that will lead us to understand what this means.

A **software product** is a system of computer programs and everything needed to run them not already available to the client. The product would thus include the programs themselves and generally the documentation needed to use and change them. It could also include training and data conversion to help the client make the change from the old system to the new, and maintenance and consulting services to help him or her continue to run the system successfully.

A **software development project** is the effort resulting from the producer's promise to create a software product in exchange for payment by the client. The client and producer agree on expectations of what that software product will do, how well it will do it, what the client will need to do to use it, what it will cost, and when it will be available to be used.

The producer begins with the tools of the trade: a knowledge of how similar systems have been done before, his or her personal capabilities and confidence, an image of what is to be produced, and a plan showing how to do it.

A project is normally accomplished by a team that comes into existence just to produce this particular product and then goes out of existence when the project has been finished or is shown not to be worth continuing. This project team usually exists as part of a larger ongoing organization, such as a company or a department of government. Within this larger organization, as some projects fold, other projects are being formed. People move from one project to another.

A successful software development project, therefore, delivers a software product within limits on time and cost in a way that meets the expectations of both the client and the producer.

Why Do We Need Software Engineering?

To understand the necessity for software engineering, we must pause briefly to look back at the recent history of computing. This history will help us to understand the problems that started to become obvious in the late sixties and early seventies, and the solutions that have led to the creation of the field of software engineering. These problems were referred to by some as "The Software Crisis," so named for the symptoms of the problem. The situation might also have been called "The Complexity Barrier," so named for the primary cause of the problems. Some refer to the software crisis in the past tense. The crisis is far from over, but thanks to the development of many new techniques that are now included under the title of software engineering, we have made and are continuing to make progress.

In the early days of computing the primary concern was with building or acquiring the hardware. Software was almost expected to take care of itself. The consensus held that "hardware" is "hard" to change, while "software" is "soft," or easy to change. Accordingly, most people in the industry carefully planned hardware development but gave considerably less forethought to the software. If the software didn't work, they believed it would be easy enough to change it until it did work. In that case, why make the effort to plan?

The cost of the software amounted to such a small fraction of the cost of the hardware that no one considered it very important to manage its development.

Everyone, however, saw the importance of producing programs that were efficient and ran fast because this saved time on the expensive hardware. People time was consumed to save machine time. Making the people-process efficient received little priority.

This approach proved satisfactory in the early days of computing, when the software was simple. However, as computing matured, programs became more complex and projects grew larger. Whereas programs had once been routinely specified, written, operated, and maintained all by the same person, programs began to be developed by teams of programmers to meet someone else's expectations.

Individual effort gave way to team effort. Communication and coordination that once went on within the head of one person had to occur between the heads of many persons, making the whole process very much more complicated. As a result, communication, management, planning, and documentation became critical.

Consider this analogy: a carpenter might work alone to build a simple house for himself or herself without more than a general concept of a plan. He or she could work things out or make adjustments as the work progressed. That's how early programs were written. But if the house is more elaborate, or if it is built for someone else, the carpenter has to plan more carefully how the house is to be built. Plans need to be reviewed with the future owner before construction starts. And if the house is to be built by several carpenters, the entire project certainly has to be planned before work starts so that as one carpenter builds one side of the house, another is not building the other side of a different house. Scheduling becomes a key element so that cement contractors pour the basement walls before the carpenters start the framing. As the house becomes more complex and more people's work has to be coordinated, blueprints and management plans are required.

As programs became more complex, the early methods used to make blueprints (flowcharts) were no longer satisfactory to represent this greater complexity. And thus it became difficult for one person who needed a program written to convey to another person, the programmer, just what was wanted, or for programmers to convey to each other what they were doing. In fact, without better methods of representation it became difficult for even one programmer to keep track of what he or she was doing. We'll come back to this problem of communication in a moment.

The times required to write programs and their costs began to exceed all estimates. It was not unusual for systems to cost more than twice what had been estimated and to take months or years longer than expected to complete. The systems turned over to the client frequently did not work correctly because the money or time had run out before the programs could be made to work as originally intended. Or the program was so complex that every attempt to fix a problem produced more problems than it fixed. As clients finally saw what they were getting, they often changed their minds about what they wanted. At least one very large military software systems project costing several hundred million dollars was abandoned because it could never be made to work properly.

The quality of programs also became a big concern. As computers and their programs were used for more vital tasks, like monitoring life support equipment, program quality took on new meaning. Since we had increased our dependence on computers and in many cases could no longer get along without them, we discovered how important it is that they work correctly.

Making a change within a complex program turned out to be very expensive. Often even to get the program to do something slightly different was so hard that it was easier to throw out the old program and start over. This, of course, was costly. Part of the evolution in the software engineering approach was learning to develop systems that are built well enough the first time so that simple changes can be made easily

At the same time, hardware was growing ever less expensive. Tubes were replaced by transistors and transistors replaced by integrated circuits until microcomputers costing less than three thousand dollars have become more powerful than the older mainframe machines costing several million dollars. As an indication of how fast change was occurring, the cost of a given amount of computing decreases by one half every two years. Given this realignment, the times and costs to develop the software were no longer so small, compared to the hardware, that they could be ignored.

As the cost of hardware plummeted, software continued to be written by humans, whose wages were rising. The savings from productivity improvements in software development from the use of assemblers, compilers, and data base management systems did not proceed as rapidly as the savings in hardware costs. Indeed, today software costs not only can no longer be ignored; they have become larger than the hardware costs. Some current developments, such as nonprocedural (fourth generation) languages and the use of artificial intelligence (fifth generation), show promise of increasing software development productivity, but we are only beginning to see their potential.

Another problem was that in the past programs were often written before it was fully understood what the program needed to do. Once the program had been written, the client began to express dissatisfaction. And if the client was dissatisfied, ultimately the producer, too, was unhappy. As time went by software developers learned to lay out with paper and pencil exactly what they intended to do before starting. Then they could review the plans with the client to see if they met the client's expectations. It is simpler and less expensive to make changes to this paper-and-pencil version than to make them after the system has been built. Using good planning makes it less likely that changes will have to be made once the program is finished.

Unfortunately, until several years ago no good method of representation existed to describe satisfactorily systems as complex as those that are being developed today. The only good representation of what the product would look like was the finished product itself. Developers could not show clients what they were planning. And clients could not see whether the software was what they wanted until it was finally built. Then it was too expensive to change.

Again, consider the analogy of building construction. An architect can draw a floor plan. The client can usually gain some understanding of what the architect has planned and give feedback as to whether it is appropriate. Floor plans are reasonably easy for the layperson to understand because most people are familiar with drawings representing geometrical objects. The architect and the client share common concepts about space and geometry. But the software engineer must represent for the client a system involving logic and information processing. Since they do not already have a language of common concepts, the software engineer must teach a new language to the client before they can communicate. Moreover, it is important that this language be simple so it can be learned quickly.

How Do We Define Software Engineering?

Having considered the problems software engineering is intended to solve, we can define software engineering as follows: **Software engineering** is the management of expectations, computer technologies, people and their skills, time, cost, and other resources to create a product that meets the expectations of the client with a process that meets the expectations of the producer.

Or we may take the approach of defining software engineering by describing in a few sentences what a software engineer does. A **software engineer** integrates the various computer science technologies and people skills to build a computing system to satisfy defined expectations for how well the product meets the client's needs, how much it costs, and when it is delivered. This implies producing a high-quality product using a high-productivity process. If the project is to be a success, the client must be satisfied with the product, and the producer must be satisfied that he or she was so compensated that the effort put into the process was worth it.

We might further refine our definition by considering what we will assume software engineering includes and excludes. Software engineering includes a number of topics such as systems analysis, requirements specification, system design, and project management. Project management, in turn, includes such concerns as negotiating expectations, estimating, budgeting and scheduling, organizing, directing, measuring, planning, and controlling the project. Software engineering is also concerned with metrics to measure the performance of the product and of the process to produce it. A central consideration of software engineering is quality assurance, which checks that the right product is being produced and that it is being produced as planned.

Our study of software engineering excludes the computer science technologies themselves such as programming and programming languages, data base management systems and data base design, communications, systems enhancement software such as operating systems and compilers, computer science theory, analysis of algorithms, or computer architecture. But software engineers manage the use of all these techniques. We will discuss the technology we apply in Part 4 of this book. Software engineering also excludes the specific applications areas such as data processing or scientific calculations. However, since the software engineer is responsible for bringing together these technologies and applications to solve a problem, he or she needs to be familiar with all of them.

We also hear the terms **systems engineer** and **systems analyst**. A systems engineer may be responsible for a whole system, including hardware, software, and people, with the responsibility for the software part of the system delegated to a software engineer. If the major part of the new system is software, there may not be a systems engineer as such, and the responsibility for the whole system may rest with a software engineer. The title *systems analyst* may apply to the person responsible for just the initial analysis phase of the project. However, the title has been used for what we would today call a software engineer.

In some circles the term *software engineering* refers to writing software for engineering applications, or writing drivers that interface directly with the hardware. We would prefer to call this field engineering software, not software engineering.

In systems involving hardware, software, and people, they all have to be considered together. Once it was thought that one built the hardware first, then wrote the software, and finally trained the people how to run it. Now it is recognized that it may not be that simple. For example, the Burroughs 5500 was designed as a machine to process the type of code seen by compilers and operating systems. The hardware design was strongly influenced by the software to be run on it. As interfaces with people become more important, people have to be considered in the design of the hardware and software.

The importance of software engineering was first recognized in the development of large, complex projects, where the problems of the software crisis have been most critical. But the principles are important in the development of smaller software systems as well.

The job of a software engineer can be particularly appealing to someone who likes the challenge of going beyond programming and would like to grapple with the harder challenge of solving problems and defining what the programs are to do. Compared with the programmer, the software engineer deals more directly with the people who have the problem, and he or she deals more with the programmers than with the programming itself. The software engineer may deal more with abstract representations of hardware and software than with the actual hardware and software itself. However, although the software engineer may not actually program, he or she must know how to program in order to deal intelligently with programmers and retain their respect. Software engineering tends to be a people- and problem-oriented job. The software engineer must be continually aware of the relationships between software, hardware, and people.

Software engineers should also be comfortable with a great deal of ambiguity because often the problem may not be well formulated when first encountered.

A programmer usually has little or no contact with the system until the software engineer has formulated the problem and has resolved this ambiguity.

Finally, we can define this field by saying that software engineers are typically concerned with such problems as the following:

- What are the expectations of the current system? Are they being met and, if not, in what way and why?
- How might these or other expectations be met by a new system, and what cost and time might be required before they can be met?
- How might the present system be changed, or another system built to meet new expectations?
- What outputs are required, and in what format should they be? Whom should the output allow to make what decisions?
- What input is required to produce this output? Where can it be obtained and in what form? How is it entered into the system?
- What records need to be kept by the system? How fast must they be retrieved and what descriptors must be used to retrieve them? How long must the records be kept, how are they to be stored, and how must they be protected?
- What processes transform this data, and what algorithms do these processes use?
- How should data entry, output, processing, and storage be distributed in time and location?

Software engineers' concerns are, of course, not confined to this list. They are concerned with anything and everything necessary to achieve a successful software product.

What Are the Characteristics of Software Engineering Projects?

To clarify and expand on the definition of the preceding section, let's consider the range of qualities that might typify work in this field. A software engineering project has the following characteristics:

- The project is concerned with making a change to an existing system or building a new system. In either case we will talk about the result as the *product*.
- The change is not easy to undo or modify once it is made. The cost of fixing errors is high. Thus one should visualize the change carefully before it is made. Ultimately less expense is involved in planning the project carefully and doing it right in the first place rather than building the product carelessly and then trying to change it until it is right.
- The project cannot be done by one person, and usually must be done with a team of people with different skills. Thus, we must be concerned with the management of the people and the communication among them.

- The product must work with other systems, or the work to produce the product must be coordinated with the work of other people. Thus, how and when the project is done must be planned to fit constraints.
- Projects involve two cultures: the client who wants the change, and the people who possess the technology to make the change. They may have different backgrounds and may talk different technical languages. Thus communication between these two sets of people can be difficult, but nevertheless it is very important.
- Resources must be committed to the project, and providers or clients may express concern as to whether the value of the product is worth the cost of committing these resources.
- The product is new. It has never been made before. Thus, not until the end of the project does one know exactly what the product is, how it behaves, what it is worth, and what it cost to make. Software engineers and their clients would like to have all this information before they decide to begin the project. But they never do, so they must offer estimates and use them to make decisions before this information becomes available.
- Clients commit themselves to buying a system they have never seen. Thus they want a good visualization or description of what they will get before they obligate themselves.
- To undertake a project is a gamble. It may fail.

How Is Software Engineering Evolving?

Software engineering began with the recognition and understanding of the problems that led to the software crisis. While there has been progress that has significantly improved the quality of software and the productivity of those developing it, much is still left to be done. We are not completely out of the software crisis yet.

Many of the published software engineering techniques are more acknowledgments that solutions are still actively being sought than solutions themselves. The field is still in flux. There is a need to look at software engineering methods critically, choosing what makes sense and works, building on the foundations, and recognizing what still needs to be done. Anyone pursuing software engineering as a career should plan to keep up with the literature in order to follow the latest developments, continually evaluating which new methods he or she can use.

There are now available several major software engineering methodologies, each solving problems the others do not address, or solving the same problems differently. Several of these methods will be discussed in Chapters 3 and 4.

In addition to presenting the classical methods, this book will put these various methods together into one unified system. This unified approach can be used as a reference from which to understand any of the other methodologies that one might be asked to use, or this unified approach can be used as a methodology in its own right.

Such an integrated approach is important if we are to be able to use computer aids effectively in software engineering. Without an integrated approach we would find ourselves spending too much time reentering the same information in a different form each time we move into a different phase of the project. We would be working for the computer instead of having the computer work for us. We will show in Chapters 3 and 4 how the methods developed in this book can be put on the computer.

Summary

Software engineering has developed as a field to help manage the process of creating today's more complicated software. It is an inclusive discipline, comprising all the problem-solving tasks of software development.

The actual work of software engineering can be thought of as an organized sequence of activities, or a project. A software product includes the programs themselves, as well as the documentation, training, maintenance, and consultation necessary to run the programs. A software development project delivers the software product to the client according to the agreed-on time and cost restraints and the expectations of the client and producer.

Forethought and organization are necessary today in developing software because, over the years, the programs have grown in complexity and the production costs have risen significantly. Software was once thought to be simple enough and cheap enough to be created in a rather haphazard fashion. Contemporary software is developed by a team of skilled professionals. Advanced planning, communication, and clarification of expectations have all become crucial to the process.

Software engineering encompasses the administration of client and producer expectations, computer technology, people's talents, schedules, and budgets to create a product that satisfies both client and producer. Systems analysis and design, requirements specification, and project management are part of the field.

Software engineering projects share a number of characteristics. The product is new and therefore presents a set of unknowns. Errors are difficult, time consuming, and expensive to fix. A team, instead of a single person, is required. Two cultures, the producer and the client, must learn to communicate with one another. Resources must be committed to the project before many important questions can be resolved, and the project's success is a gamble.

The field is a young, dynamic discipline. As software engineering evolves, it can be expected to offer new solutions and methods to the approaches now available.

Exercises

1. List all the activities you can think of that must be done from the time one conceives of a possible need for a software system to replace an existing system until that new system finally goes out of use.
2. Consider your education in computing. To what extent has it emphasized programming, and to what extent has it emphasized the other aspects of software systems development?
3. What forces contribute to making the software systems we are now developing increasingly complicated? How has this affected the methods we use in developing those systems?
4. Name some of the changes that you have been able to observe in how software systems are developed or in how their development is taught.

References

Fairley, Richard E. *Software Engineering Concepts*. New York: McGraw Hill, 1985.

Jensen, Randall W., and Charles C. Tonies. *Software Engineering*. Englewood Cliffs, N.J.: Prentice-Hall, 1979.

Pressman, Roger S. *Software Engineering: A Practitioner's Approach*. New York: McGraw Hill, 1979.

Shooman, Martin L. *Software Engineering*. New York: McGraw Hill, 1983.

Turner, Ray. *Software Engineering Methodology*. Reston, Va.: Reston, 1984.

Software Engineering Principles

- At the start of a project, the client and system developer must define expectations, matching the anticipated needs to the means available to meet them.
- Once the client and system developer have agreed on the project requirements, the developer can begin to design a proposed system, and then analyze that system to see if it satisfies the requirements.
- To understand and communicate our work with complex systems, we look first at the broad picture, ignoring details, then look at successively smaller parts in greater detail.
- Given a system's complexity, we cannot test every possible contingency to ensure complete accuracy; rather we must depend on very careful development to ensure a quality product.
- Tactics to avoid errors at a later stage in the process include describing the system in terms of how it transforms information before describing the people or devices that make the transformations, assisting clients in visualizing the project and offering feedback, and reviewing work at regular intervals.
- Work on a project moves through a series of phases, each consisting of specifications, performance, and evaluation. Phases include problem definition, requirements analysis, feasibility evaluation, system and components design, components and system building, and product evaluation.
- Managing a software development project involves estimating the time and cost of producing the system, planning the work, controlling the work to conform to the plan, replanning, and tracking changes.
- The final quality of a product can be judged by such factors as completeness, reliability, efficiency, integrity, and expandability. Everyone on the project team should take responsibility for the product's quality.

This chapter presents an overview of several fundamental principles of software engineering that will be used throughout the rest of the book. These principles offer ways to understand complex systems and to communicate that understanding effectively with the client. Among the topics covered are defining expectations of what the project will accomplish, matching the client's needs to the means to satisfy them, planning the project to accommodate risks, and setting quality standards.

The Definition of Expectations

Expectations are what we believe should be or will be. The client has expectations about the product he or she will receive to meet his or her needs, and the developers have expectations about how they will be compensated for producing that product. Both clients and developers must try to be as descriptive and precise as possible when they discuss these expectations.

Matching Needs and Means

A software engineering project begins with expectations that are not now being satisfied. An expectation involves a perceived need and a perceived means that may satisfy that need. The purpose of a project is to satisfy expectations. Sometimes this may involve changing the expectations.

For example, we may have an order entry system that is not working as it was intended to. Or maybe it worked well when it was first built but, now that the business has grown, it can no longer handle the increased order volume. This is a perceived need. We believe we may be able to develop a new order entry system that would handle the order volume and not cost too much or take too long to develop. This is a perceived means to satisfy the need.

Before we could actually seek a solution in this example we would have to define the need and the means more specifically. As far as the need is concerned, exactly how many orders a day must we process now, and how many may we expect in the future? The means must be described as a particular piece of software running on a well-defined machine, operated by specific people using explicit procedures. A need seeking a means is called a problem.

Let us consider another type of situation. Here we have an idea for a means and a suspicion that someone has a need for it. For example, assume that we have an idea of how we can develop a program to coordinate the activities of people working in an electronic office. We have a proposed means. We suspect that there are people working in electronic offices that might want to use our means. We would like to develop and sell our program and make money doing so. But first we must confirm our suspicion that there are enough people who would recognize the need for our system and be willing to buy it from us so that it would be worth our while to pursue this venture. A means seeking a need is called an opportunity.

In both the cases just described, before we spend a lot of time and money, we would like to have some confidence that once we finished our work we would have a well-defined need that corresponds to a well-defined means. If we conclude that there may be a need, but not a practical means to satisfy it, we drop the project. Similarly, if we believe we have a means but no one recognizes that they have a need that our means will satisfy, we also drop the project. Figure 2.1 illustrates this concept.

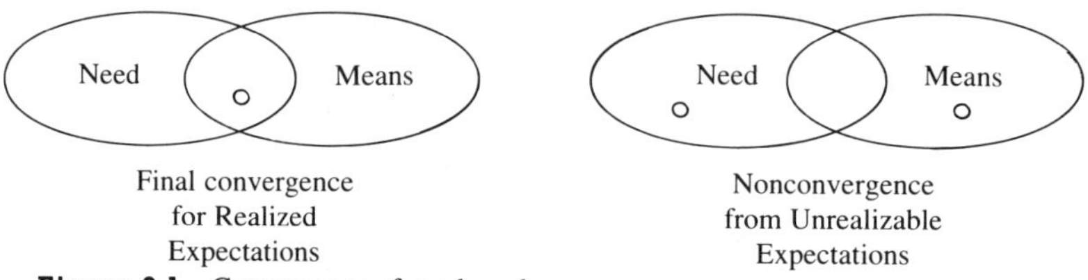

Figure 2.1 Convergence of needs and means.

Doing What Has Not Been Done Before

When we agree to supply a product to a client, we are making a match between a need and a means that has never been made before. If the match had been made before, there would be no need for our doing it again.

Salespeople can also say they make a match between a client's need and a means that satisfies that need. What distinguishes software engineers from salespeople is that salespeople are usually selling something that already exists. They can show the client the product right on the shelf, or at least they can show a picture in a catalog. But software engineers promise to deliver something that does not yet exist. Con artists also make a match between a recognized need and a means that does not exist. What distinguishes software engineers from con artists, hopefully, is that software engineers intend to deliver and believe they can. But software engineers must be careful that they do not get intoxicated with what they think they can do, only to find they are not able to deliver.

Making Changes

Another way of looking at a project is from the point of view of making a change. Either we change an existing system, or we produce an entirely new system. In fact, the latter can also be considered to be making a change—from the situation of having no system to the situation of having a system.

A useful metaphor for making a change is making a move to a new city and job. We are someplace. We think we would like to be someplace else. Before we make that change, though, we want to answer some questions:

1. Where are we now?
   a. Why are we not happy here?
   b. What expectations do we have that are not being met?
2. Where do we want to be?
   a. Why might we expect we would be happy there?
   b. How would it meet our expectations?
3. How would we get there?
   a. What would it cost and how long would it take?
4. Would it be worth the move?

We make changes because some desirable expectations are not now being met and we believe that they could be. Clearly, before we proceed to spend any time and money, we want to know what those expectations are. Fundamental to both of these points of view—matching needs and means, and making a move—is that we must define expectations, what we might do to satisfy those expectations, and, if necessary, how we might change the expectations.

As the project unfolds, new information becomes available that can cause our expectations to change. Perhaps we find that what we had expected to do is not possible or not affordable. Perhaps in the process we find other desirable results we can reasonably expect. For this reason we have included in the definition of software engineering the concept of managing expectations.

Interface between Client and Designer

The people who will use the system and those who will develop it are usually different people. The client may be a business executive who talks about the discounted value of money, and the developer may be a computer person who talks about disk access times. Their backgrounds are dissimilar, and what they understand and do not understand is different. This makes it very hard for them to talk intelligently to each other about what is wanted and what will be produced. Neither has the time to understand completely the whole domain of the other's business. They need to concentrate their communication efforts on a carefully defined interface that they both can understand. This interface is the **requirements specification.** Clients specify what they want so the system fits into their world, while developers put together the technology from their world so that it does what the clients specified.

We may think of a system as a box with the controls on the outside and the workings on the inside. The requirements specification is what clients say they want and software engineers agree to produce as seen from the outside. The design is how designers put the system together on the inside so that what is on the outside is what clients want.

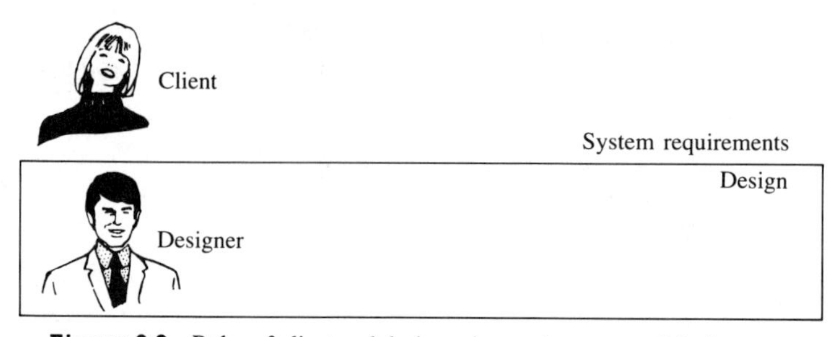

Figure 2.2 Roles of client and designer in requirements and design.

Design is how the parts are made to work inside the box so that the system does what the requirements specification call for on the outside (see Figure 2.2).

Specifying the Expectations

The specification game says: "I will specify what I want and a price and delivery time. If you believe you can profit by satisfying that need, price, and schedule, you may accept the contract. You have the freedom to provide it any way you wish. If you can demonstrate that what you give me meets the specifications I wrote, I must pay you. If I am not happy, it is because I did not write the specifications correctly to tell you what I wanted."

A facetious story illustrates the idea. A beachcomber found a bottle with a genie. The genie offered him two wishes. First he wished for a new car with a radio. Zap! He was driving down the road listening to the radio and began to sing along with a commercial, "I wish I was an Oscar Meyer wiener." Zap! He became the victim of a requirements specification error.

The requirements specification is the interface between the client and the developer. Both must understand that specification if they are to be satisfied. But how to specify exactly what is wanted and separate it from how it is to be built is very tricky. A lot of work has gone into methods for defining the *what* without the *how.*

How do we specify what we want without specifying how to build it? For software systems we usually make the following types of specifications:

1. *Environment and interfaces*—what larger system is this system part of, and how does it fit into or relate to other systems?
2. *Functions*—what must the system do, and how must the user utilize it to get it to do that?
3. *Performance*—how fast must the system perform, and what resources (memory, disk space, and so on) does it use?
4. *Logistics*—what must the user do to be able to utilize and maintain the system?
5. *Quality*—how well must the system work and last?

Requirements will be discussed in Chapter 13.

From Requirements to Design and Analysis

As the project gets under way, expectations are translated into the technical details that constitute the system. Here the process moves through three general stages: description of the requirements, design, and a check that the design fits the requirements.

Recursive Character of Requirements and Design

Designers search for a way of putting together components to build a system to meet the requirements specification. This putting together components is called **synthesis.** In the first step of the process designers may draw up the system using components that they do not yet have but that they assume they can somehow obtain or develop later. They become, therefore, clients for these components. They write a set of requirements for the components. These components may be designed of subcomponents for which requirements must be stated, and so on. This process continues until either one reaches a component that can be obtained, or one concludes such a component is not obtainable. In the latter case one has to go back to a previous step to compose the system with different components. Figure 2.3 illustrates the process.

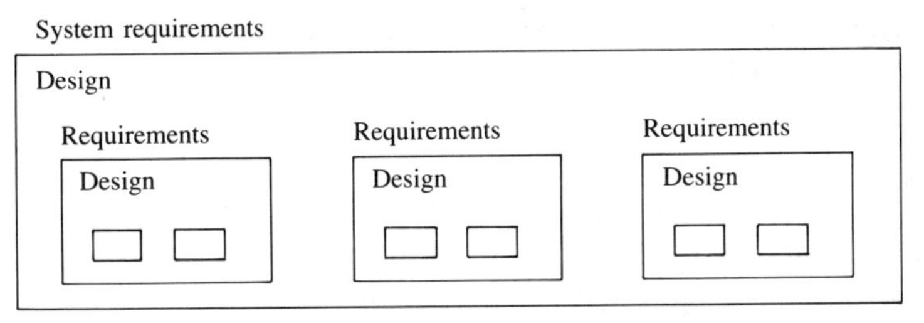

Figure 2.3 Recursive requirements and design.

When a process contains another process like itself, it is called **recursive.** This idea of boxes within boxes is also related to the concept of **abstraction.** Composing data structures from other data structures yet to be defined is called building abstract data structures. Building a program from program structures yet to be defined is known as designing a program using abstractions. The yet-to-be-defined structures are called abstract structures.

Analysis

Once we have requirements and a design, we need a step to check how they match up. **Analysis** is the study of a system to determine what it will do under various circumstances and the consequences of what it does. Initially we analyze the existing system to understand whether it is meeting current expectations, and if not, why not. Analysis comes up again when we design a new system. We write requirements specifications to describe what we want the system to do, and we will synthesize the design of a system, that is, we will consider what components we will use and how we will put them together into a system. Then we analyze the proposed design to see what it would do and whether it would meet the requirements specification. If it won't meet the specification, we synthesize another design and analyze it, and so on.

We can state this in another way. When we design a system, it is a candidate to replace an existing system. Just as we analyzed the existing system to see if it was meeting its expectations, now we analyze the proposed design to see if it would meet the expectations we have for it as given in the requirements specifications (see Figure 2.4).

Requirements

Design
  Synthesis
  Analysis

Figure 2.4 Relation of requirements, design, synthesis, and analysis.

Dealing with Complexity

In Chapter 1 we blamed much of the software crisis on the complexity of systems. But what is a system, and what is complexity? More importantly, how do we come to grips with complexity in order to plan a project?

Systems and Complexity

A **system** is a set of parts that interact in such a way that the behavior of the whole depends on the interactions among the parts as well as on the behaviors of the parts. Another approach, which emphasizes the purpose of the system, would be to say that a system is a set of parts that are made to relate in such a way as to achieve some objective.

Given the parts and their behaviors, **complexity** is the difficulty involved in using the relations between the parts to infer the behavior of the whole. Or, if you wish, complexity is the extent by which a system is greater than the sum of its parts.

As computer systems were developed to perform more complex functions, the systems became more complex. And as the systems became more complex, they became harder to change. Because of the interaction of the parts, a change to fix a problem or make an improvement in one part would often have undesirable effects on other parts. From the definition of complexity, we can already see that one approach to avoiding complexity is to build the software from parts such that there is as little interaction among the parts as possible. Later in this chapter we will discuss the top-down approach, which addresses this problem.

Software development is not the only area where the problems of complexity have caused difficulty. When everything relates to everything else in a system, a simple change can have effects that are easily overlooked. One example can be seen in the development of the Bay Area Rapid Transit system (BART), a high-speed commuter rail network in the San Francisco Bay area. The original plans called for running trains often enough that no one would be expected to stand and the trains would never be more than six cars long. It was later decided they would have to run longer trains less frequently. Now standing passengers sometimes miss their stops because the station signs were positioned to be read only by seated passengers, and some passengers get wet because the roofs in the elevated stations were only long enough to cover six cars. In a similar way, a software engineer has to think of everything that may occur.

Fallacy of Relying on Testing

Given the complexity of most of the software systems with which we will be concerned, software quality cannot be guaranteed by testing. It can be obtained only by careful development. The concern for quality must underlie every step of the development.

This means careful attention all the way from defining the problem or opportunity, through developing the requirements specification, preparing the design, doing the implementation, and training the people who will use the system.

We could use testing to determine whether the software works as specified in the requirements if we could test every possible input and compare the output with an independently derived result. But no software that is of interest to software engineers is small enough that we can afford to do this.

In most practical software systems the number of test cases needed to test all possibilities could not be run on a computer in a lifetime. Even if the computer could generate and run all the needed test cases, we could never afford to compute and compare the answers manually to see if they were right.

Game playing is a similar exercise. We can compute all the outcomes for the game of tic-tac-toe. But almost any game of real interest, such as checkers, chess, or Go, is sufficiently complicated that not even the fastest of computers can compute all outcomes.

We can be clever about how we test. We can analyze the code to plan tests so that each test will confirm (or deny) the working of a large class of cases. By this means we can significantly reduce the number of test cases required. But tremendous effort may be required just in designing such tests. Computer programs have been developed that will analyze the code to design these tests. But computer-designed tests tend to prove that the code does what it does rather than show that it does what we intended.

Consider an engineer designing a hardware circuit to multiply two 24-bit numbers. (This might be the mantissa of a double-precision floating point number for a 16-bit machine, about the precision of seven decimal digits.) The number of possible outcomes is 2^{48}, or about 3×10^{14}. The engineer designing this hardware could not test to make sure it works by checking out all these combinations. At a million multiplications per second it would take the machine ten years just to compute these numbers, let alone print them. Even once the answers were printed out, think how long it would take to do the hand calculations for each of the 3×10^{14} cases and to compare them to see if these results were correct.

Instead, in a case like this, engineers design the multiplier carefully according to well-established principles so that if the multiplier is built as specified, they can be reasonably sure it will produce the right answers. Instead of testing all possible results, they need only test to ensure that it was built the way they designed it. Tests should be used to confirm that it is built correctly, not that it was designed correctly.

Today researchers are working on the concept that software should be designed as a mathematical proof. So far they have succeeded only with very simple programs. Although we usually cannot develop our software with the rigor of mathematical proofs, we can develop it using such care and well-defined construction principles that we have a high confidence it will do what it should do. We will discuss construction principles in Chapter 4.

Miller's Principle

To penetrate this "complexity barrier," new methods of representing systems and managing their development have had to be developed. Since this understanding of complex systems has to be conveyed to people, we might take a moment here to consider how people process information.

The psychologist George Miller (1956; and see Simon, 1981) did some experiments to develop a model of how people process information. He presented nonsense syllables to human subjects and asked them to recall these syllables in various orders. His observations can be explained by a model that includes a very small memory in which information can be processed randomly. If the subject is disturbed while processing information in this memory, information can be lost. This small dynamic memory is backed up by a very large memory, which holds its information indefinitely. Information can be stored and retrieved from the large memory, but it can be processed only by bringing it into the small memory (see Figure 2.5).

Since people can deal with only a limited amount of information at one time, a good method of representation would be one that would allow the individual to read through the whole system but only have to absorb a small amount of information at any time. Miller's model of human memory provided the basis

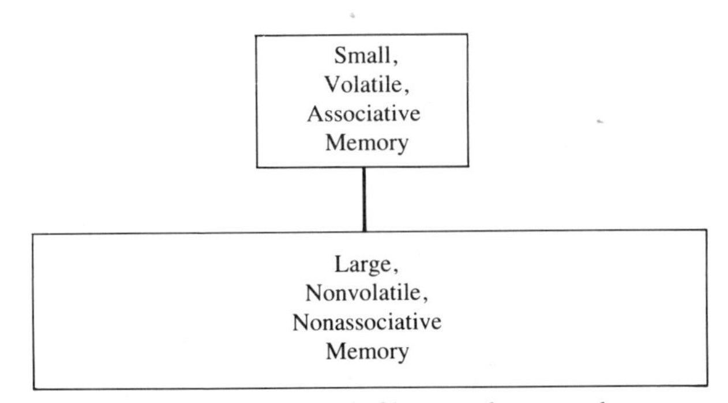

Figure 2.5 Miller's model of how people process data.

for a method of representing systems that would work with people's natural method of processing data and help them understand systems of greater complexity. We discuss that method in the next section.

Top-Down, or Hierarchical, Methods of Representation

The top-down approach provides us with a way of representing and understanding complex systems and this allows us to work within our human limitations as described by the Miller Principle.

Let us illustrate what we mean by the top-down approach. If we were to plan a trip from Coit Tower in San Francisco to the White House, we would first look at a road map of the United States to find a route from San Francisco to Washington, D.C. This is the top view. Then we would look at a street map of San Francisco to see how to get from Coit Tower to the cross-country route we found on the national map, and at a Washington street map to see how to get from the cross-country route to the White House.

Thus, when we take a top-down approach we look first at the broad picture, ignoring the details. Then we look at the details of just one part at a time, using the broad picture to tie each of these parts together. The value of a top-down approach is that it limits how much information we have to think about at one time, that is, the Miller Principle.

We will use the top-down approach to represent many types of systems and many aspects of those systems. But let us for the moment discuss the representation of just one of those types of systems, a computer program. Computer programs can become very complex. It certainly would be desirable to represent computer programs top-down. But is it possible? Fortunately Bohm and Jacopini (1966) were able to prove that by using just a few fundamental control constructs they could develop and describe complex programs in this way. These constructs became the basis for structured programming.

When discussing projects, we will focus on two distinct aspects: the *product* that is made, and the *process* we go through to produce it. The concepts involved in top-down and hierarchical methods are fundamental both to the methods for describing the product and to the methods for managing the process.

A **hierarchy** is like a set of boxes nested inside each other (see Figure 2.6). Each box has on its top an overall description of what is inside and what is to be done with the contents. The details inside can only be seen by opening the box. As you look into one box, you are concerned about the relationships among the other boxes you see there but not about the details within any of the other boxes.

Developing a system top-down involves defining the outer boxes first, then the inner boxes. The outer boxes describe a high-level view of what needs to be done. Opening a box reveals the details inside. One need look at only one box and its immediate contents at a time. The top-level view (that is, the outer box) is of interest to the client. The inner boxes detail what needs to be done to accomplish the high-level goals. The view inside the box is of interest to the developer. Note that this is the same pattern we saw in the recursive process of requirements and design in Figures 2.2 and 2.3. Top-down methodology conforms to the Miller Principle of requiring humans to deal with only a small amount of information at a time.

The usual form of flowcharts, on the other hand, was not very useful for this top-down thinking. Flowcharts cannot generally be understood in a succession of chunks, each small enough to be mentally digestible. GOTOs in programs can do great violence to this form of thinking by taking you from hither to yon without any underlying structure. This had led us to the use of structured programming.

How do we represent such a hierarchical structure? Figure 2.6 shows several methods: nested boxes, parentheses, BEGIN-END pairs (which are the form of hierarchy used in structured programming), indented outlines, and trees.

Requirements and Design without Seams

The similarity of the requirements-design structure in Figure 2.3 and the hierarchical structure of Figure 2.6 leads us to ask whether it may be possible to do requirements and design as a hierarchy (the answer is yes) and whether we can use one set of tools and methods for representation during the whole requirements and design process (the answer is also yes). This will allow us to develop a **seamless** approach to the process of requirements and design. This seamless approach means we do not have to change methods of representation when going from requirements to design and from design to code. This will help us not to lose track of whether we have met all the requirements as we develop the design and later as we write the code. Some of the current methods use different representations during requirements, during design, and during implementation. We call this a **patchwork quilt** approach. In Chapters 3 and 4, we will discuss both patchwork quilt and seamless approaches.

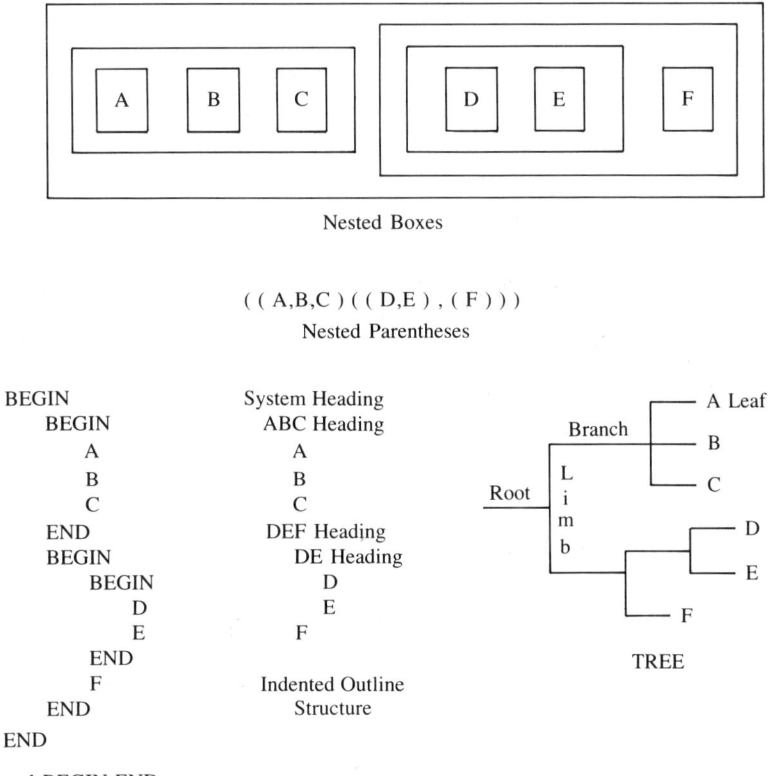

Figure 2.6 Equivalent representations for a nested structure.

Importance of Up-Front Planning

The top-down method allows us to formulate a careful, intelligible approach to a complicated project. But if something is overlooked or changes, the cost of mistakes goes up with each step. Thus, along the way, as we move from step to step, we can take advantage of certain up-front planning strategies before plunging ahead with the next task.

Rising Costs with Delays in Finding Mistakes

The cost of fixing an error rises as more work is built upon that error before it is found and fixed. The cost to catch a mistake and make a change at the time of writing the requirements specifications may be $10, and during the design $300. While the product is being built, the error may cost $3000; after the product has been delivered, the mistake could cost as much as $15,000 to fix, and possibly much more in losses to the client because the product didn't work. Thus, it pays to discover mistakes early in the project when the costs of changes are low.

With complex projects it is important to weed out errors early. Thus near the start of the project there should be:

1. a good *definition* of the problem or opportunity,
2. a good *description* of what the final product will look like so that clients can understand what they are receiving and producers understand what they are creating (this description is the requirements specification),
3. *negotiation* with clients so they know what to expect and so the product will be acceptable to them when completed, and,
4. a good *plan* to manage the process.

This planning becomes more important as the project gets larger and more complex. We must do a great deal of thinking on paper early in the project, long before we start to create the product itself. The paper is easier and less expensive to change than the final product. But it is harder to measure progress when you have only paper, not a product, to look at. We discuss how to measure progress during the paper phase in Chapters 4 and 6.

Logical Description before Physical Description

The reason for building an information system is to get information we need to make certain decisions. So when we specify a new system, we should start by considering what decisions must be made, then what information is required to make them, what information is available, how that available information can be transformed into the information we need, and what information must be retained in the system to make the transformation. This is called the **logical description** of the system. It concerns just the information, not the software, hardware, or peopleware who transform or retain that information.

Once we establish the logical description of the system we want, then we can assign the transformations and storage of information to the software, hardware, and peopleware who will do it. Only then do we write the software, acquire new hardware, and acquire and train new people. This is called the **physical description.**

Reviews

Another strategy to catch errors early is to schedule frequent reviews. A review involves people looking at specifications and plans to head off problems before they occur. Reviews can be made at several levels. One programmer can invite another to look over his or her program. In a more formal walkthrough a programmer's peers review the work and take notes on problems; and then a follow-up is made to ensure that these problems are resolved.

Compared to a test, a review is a vastly more powerful device for finding errors. First, a review brings human logic to bear so that problems can be found at a logical rather than a case level. Many problems found by human review might never be caught by a test case. Second, when a human catches a problem, he or she is usually looking right at the cause of it. This makes it easier to fix. But when a faulty case is found by testing, one still has to find the cause, which may not be easy. Third, a review can be made earlier in the development, when the product appears only on paper. At this stage it can be understood by humans but not yet by machines. Catching problems early means they are not built upon throughout the rest of the project to create greater problems. Reviews are discussed further in Chapter 7.

Thorough Requirements or Rapid Prototyping

The sooner we catch any deviation from what the client wants, the sooner we can get back on the right path. Visualization, or making a pictorial representation of the product before it exists, is how the client and provider express what they each see in their own mind's eye. Early visualization requires good methods of representation. This is discussed in Chapters 3 and 4. We must constantly ask our client, "Will this thing we are producing satisfy your expectations?"

There are two approaches to showing clients what they can expect to get before they make their commitment to the project.

1. Develop careful specifications and a good representation of the system on paper to help clients visualize what they will get. (A thorough requirements approach will be discussed in Chapter 13.)
2. Show the client a prototype—that is, a real system in hardware or software that behaves in some key way like the system or some part of the system you expect to provide. Prototyping depends upon ways to build cheap systems that look externally like the final system. Prototypes should, however, be easy to change, and they should not contain all the expensive internals needed to make the final system work under all circumstances. (The rapid prototyping approach will be discussed in Chapter 9.)

Managing with the Project Life Cycle

Another basic concept we will use is the **project life cycle**, so called because of its parallel to human lives. Projects are conceived when a need and a means join together. A product is born and takes on a life with its own needs, resulting in the conception of new projects. We will discuss two ways in which this life cycle may be broken up into manageable pieces: by stages and by phases.

Stepwise Commitment

After we have finished the project, we can ask several important questions:

1. Did it solve our problem?

2. How much was it worth to have solved the problem?
3. How much did it cost?
4. Was it worth doing?

It is one thing to answer these questions after we are finished. By then, though, the time, money, and resources have been spent. There is no way we can unspend them if we decide it was not worth doing.

We are faced then with a fundamental dilemma: We need to make the decisions now, but we won't have the information we need to make them with certainty until later.

At the beginning of the project	*At the end of the project*
The product is an abstract concept on paper.	The product is concrete, as visible and usable hardware and software.
It is not exactly clear what you will get.	You know exactly what you've got.
The representation of the product is easy and cheap to change.	The product is difficult and expensive to change.

So we begin by making guesses. We divide the project into review **stages.** At the end of each stage we have more information to improve upon our guesses than we had before. We have tried some things and seen whether they succeeded or failed, how much they cost, and how long they took. We can reevaluate where we are and where we are going. We are better able to estimate the times, costs, and benefits of finishing the project. This may cause us to replan how to do the rest of the project and may even cause us to have to redo some work. Or we may decide it is not worth continuing the project. Reevaluating these decisions after each stage may save us the cost and effort of continuing a project if new information shows the project is not worth continuing. Or the reevaluation may show that we must plan to do some of the work differently than we had originally planned. We do not wish to commit resources to the project faster than we build confidence in what payoffs will be produced.

At the end of each stage we make our final commitment of resources to the next stage. This allows us to limit our losses if the project fails. As resources are committed to each stage, the total commitment of resources should never be greater than the expected returns for the project. As more information develops and one has more confidence in these returns, a greater commitment of resources can be made (see Figure 2.7).

To make these guesses we need to use our own or someone else's experience with earlier projects. Thus, we will want to keep track of information about this project for its possible use in estimating future projects.

We must gamble. We must understand that we are making a decision to undertake this project using only guesses as to what it will cost and what it will be worth. We could be wrong. We sometimes are. But we hope to be right more often than wrong.

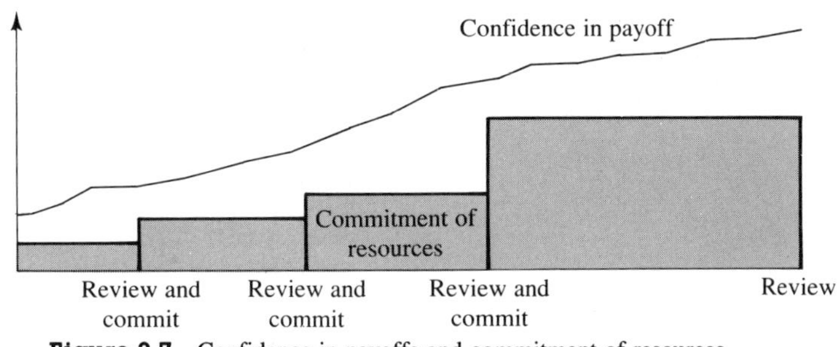

Figure 2.7 Confidence in payoffs and commitment of resources.

In the review at the end of each stage we must consider:

1. What has it cost so far in time, people, and other resources?
2. What is the current best estimate of the cost in time, people, and other resources to finish the project?
3. What is the current best estimate of the benefits to be derived from completing the project?
4. If the original benefits cannot be obtained, are there other benefits that could still be obtained that warrant continuing the project?
5. Should we discontinue the project? If so, should we document the work already done to retrieve some value that can be used for future projects?
6. If we continue the project, should we replan how we proceed with the rest of the project?
7. In view of this new information, do we have to redo some work already completed?
8. What is the plan and commitment of resources for the next stage? The plan for the next stage should be better than the plans for the stages beyond because we are closer to it.

We must make a plan for our work at the beginning of the project. As the work proceeds and we have more information, we must be prepared at the end of each stage to replan the work for the next stage and for the rest of the project. By the time the project is finished, each stage may have been replanned many times.

Parenthesis Structure of the Phases of a Project

The conventional approach to the project life cycle divides the project into phases, which correspond to the different types of activities that occur as the project matures. Each new phase builds upon the information from the earlier phases.

Each phase of the project can be represented by a parenthesis structure using the following format:

1. a planning process to determine what is to be done, followed by
2. a left parenthesis representing the specifications of what is to be done, followed by

3. doing it, which is enclosed by the parentheses, followed by
4. a right parenthesis representing an evaluation comparing how well it was done against the specifications. (See Figure 2.8.)

Planning to determine what you intend to do	(Specification of what to do	Do it	) Evaluate how well it was done against the specification

Figure 2.8 Parentheses used to define a phase.

Validation and Verification

The left and right parentheses correspond to two types of reviews we must make to ensure we are doing things correctly. The left parenthesis corresponds to validation, and the right to verification. When we review the specifications to make sure they call for what is really wanted before we do each phase of the project, we call that **validation.** When we review after we did each phase of the project to see if we did the work according to the specifications, we call that **verification.** According to Boehm (1981), one might envision the distinction as follows:

> Validation asks: "Are we doing or did we do the *right thing*?"
> Verification asks: "Are we doing or did we do the *thing right*?"

Validation is a check on the left parenthesis. It brings to bear an external view to ask whether the correct thing is being specified. The term *validation* can be used to describe the review made to determine that what is specified in this phase is consistent with what was called for by the previous phase. The term *validation* can also be used to describe the check made at the end of the project to see that the correct product was produced. Validate what you produced before you deliver.

Verification is a check on the right parenthesis. It is an internal check that the performance at the right parenthesis agrees with the specifications made at the left parenthesis. Figure 2.9 shows the basic structure of how validation and verification are associated with parentheses. This structure holds for each phase that we will describe in the next section.

Planning to determine what you intend to do	(Specifications of what to do	Do it	) Evaluate how well it was done against the specification
	Validation that what the specification called for is what is wanted		Verification that the specification was satisfied

Figure 2.9 Validation and verification with the parentheses.

Phase by Phase through a Project

Once each phase has been defined using this parenthesis structure, the whole project can be described by nesting these phases, like nesting parentheses. The individual project phases described using this parenthesis concept are shown in Figures 2.10 through 2.13.

We begin the project by defining and gaining an understanding of the problem or opportunity and making an analysis to determine whether that problem or opportunity really exists. This involves a study of the existing situation to determine why current expectations are not being met and what changes are wanted. We end the project with a product and make an analysis of whether the expectations for the product were met.

	(	)
Problem definition	Problem/opportunity statement and validation	Problem/opportunity verification

Figure 2.10 Parenthesis structure for problem definition.

We write a requirements specification document describing what the product needs to do if it is to solve the problem or realize the opportunity defined in the problem definition phase. Once the product is produced, we then verify the product against the requirements to determine whether the product does what the requirements said it was to do (see Figure 2.11).

	(	)
Requirements analysis	Requirements specification and validation	Requirements verification

Figure 2.11 Parenthesis structure for requirements analysis.

In the feasibility study phase we make an estimate of whether the system specified in the requirement specifications can be built, and if so, the cost and time required to build it and the benefits of having it. We must also define all the **constraints** that prevent us from doing some things, and the **capabilities** that allow us to do others. From this we can decide whether to undertake the project (see Figure 2.12).

	(	)
Feasibility study	Feasibility report	Evaluation

Figure 2.12 Parenthesis structure for feasibility.

Evaluation involves making these same determinations after the product has been built to learn whether we made a good decision to undertake this project and to improve our estimates and decisions in future projects. In the feasibility study we guess at what we will not know for certain until the evaluation at the end of the project.

In the design phase we establish how a system is to be built and define test specifications to determine whether the system was built correctly.

System design takes a top-level view to establish the parts (often called components or **modules**) of the system, their functions, and the interfaces between them, so that we will have a system that satisfies the requirements specifications. This phase also allocates these functions to be done by people, hardware, or software and determines whether the hardware and software will be built in house or purchased. It sets up test data for the interfaces to test the components. Just as the requirements analysis phase sets the requirements for the system, the system design phase sets the requirements for the components, as we discussed in conjunction with Figure 2.3.

The right parenthesis matching the systems design specification is the integration of the components into the system and the integration test to ensure that it works (see Figure 2.13).

Systems design	(	)
	Systems design specification	Systems integration and test
Component design	(	)
	Component design specification	Component test

Figure 2.13 Parenthesis structure for design.

Once the systems design is done, the design of the various components can proceed in parallel because the functions and interfaces between the components have been established by the system design. This arrangement of system and component design is an example of top-down system development. The right parenthesis matching the component design is the component test to check that the component works correctly.

Within the innermost parentheses the building of the components takes place. In a software system "building components" means writing the program modules. Verbs such as "build," "implement," or "construct" may be used interchangeably in this context.

Parenthesis Structure for the Whole Project

Putting these phases together in a nested form creates the structure of the whole project as in Figure 2.14.

In the conventional approach to the phases of the project life cycle, there is an implication that these phases progress in sequence, closing the door on one phase as the door is opened on the next. However, projects rarely proceed in such a simple pattern. As we have seen, requirements and design are recursively related, so we can't simply isolate one from the other. Sometimes we see projects laid out with the requirements phase shown before the feasibility phase, and other times with feasibility before requirements. In actuality, feasibility has to be considered while the requirements are being developed. Phases may overlap as different parts of the system go through their phases at different times. And problems may arise requiring some phases of the work on certain parts of the system to be redone. Although it has its limitations, this is still a useful conceptual model of how the work progresses with the maturity of the project.

```
Prob. def. (                                                        ) Prob. eval.
      Req. (                                                 ) Req. eval.
         Feas. (                                       ) Eval.
            Sys. design (                    ) Sys. build and test
                   Comp. design (Comp. build) Comp. test
                                     :
                               (Comp. build)
```

Figure 2.14 Parenthesis structure of the phases of a project.

Contracting

If we hand over the responsibility for some part of the work to someone else, we negotiate a contract with them. This process has a similar parenthesis structure (see Figure 2.15).

Negotiate	[Contract	Performance of contract	] Acceptance of product

Figure 2.15 Parenthesis structure for contracting.

The contract is an agreement between those who want the work done and those doing the work so that each knows what work is expected, to what quality, in what time, and for what cost.

It is useful to think of the structure of parentheses representing the phases of the project as though they too were contracts, each phase negotiating for the work of the next phase. Often the phases are done by different groups of people, and indeed this is a contractual process.

Controlling the Project

Part of the job of managing a software project is keeping changes and time and costs within bounds. If this control is not maintained, projects can turn into monsters and take on lives of their own.

Configuration and Change Control

However carefully we plan a project and attempt to follow that plan, we'll invariably face changes. One of the great problems of software development is evaluating proposed changes and keeping track of changes once they are made so that everyone has the same concept of what the latest version of the product is.

During the project much is learned that was not known when the project started. Many changes are made to the original concepts and plans. Some things cannot be done in the way they were planned. Clients may change their minds about the requirements, which necessitates the product being changed. Someone may have a new idea for how to do something to obtain a better product, less cost, or a quicker schedule. This leads to more changes. If changes are occurring all the time, project members cannot get their work done because no one knows what is going on. The resulting chaos is often the cause of project failure.

The **configuration** of the product is the current description of that product. **Change control** is a management process that includes evaluating proposed changes; determining their consequences on the product, cost, and schedule; approving appropriate changes; ensuring that documents and the schedule reflect the change; and notifying the people whose work is affected. This is an extremely important aspect of project management.

Estimation and Measurement

In addition to managing changes, another real challenge facing software engineers today is estimating the time and cost of producing a system, measuring the performance of the process, and controlling the work so that it might be done within or as close to its estimate as possible. There are several reasons for this current state of affairs in estimation, measurement, and control.

One problem is that we do not have models of how software is developed from which we can derive good measures. Without good models and measures we find it difficult to collect data from one project that we can apply effectively to estimating a different project. And by definition all projects are different. Why should we want to produce the same piece of software twice? In Chapter 6 we will consider models of software development that we can use to collect data on projects to be applied to estimating other projects.

Another problem is that during a busy project we do not have time to collect data that we might use to estimate later projects. The pressure of the here and now is more important than the future. We will consider automated tools in Chapter 3 and Chapter 4; these will not only help us during this project but will also collect much of the data we need for estimating future projects.

Assuming we do collect data on what the project cost and how long it took, when we try to compare this information to our original estimate to see how our estimate could have been improved, we find that what we estimated is not what we did. The project may have changed after our estimate was made, so the estimate and the actual project cannot be compared.

We are often pressured to make an estimate that someone wants to hear. We usually give in because, if we do not, we will not get a chance to work on what could be a very interesting project. Besides, we feel the pressure now, but the day of reckoning will not be until some time later. By then people hopefully will have forgotten what had been assumed when the estimates were made. Or we may be able to show that there were so many changes that what we estimated had little relation to what was done.

Software estimates are usually made by the people who will have to do the work. They tend to be optimistic about their own abilities. They do not want to admit that it may take as long as they really think it will. They know they will be rewarded or punished by how well their performance corresponds to their estimate. Thus, from a political point of view it is more useful to develop good excuses for wiggling out of an estimate later than it is to develop good ways of making better estimates now. All these reasons work against our improving our estimating ability.

If estimates are to improve, they should be made by people who specialize in estimating rather than those who will do the work (DeMarco, 1982). By specializing in estimating and thus estimating many projects, the professional estimators gain more experience. By not being responsible for doing the work themselves, they can be more objective in their estimates. They should be held responsible for how well they estimated, not for whether management liked the estimate. The estimators should be given the time and responsibility for collecting the data to improve their estimates. They should continually improve their estimates as the project proceeds.

Chapter 6 will discuss methods for making estimates for software development.

Concern for Quality

A project fails when we do not deliver a quality product. No matter how hard everyone worked, if the software does not perform as expected, the project cannot be considered a success. The more we know what makes up quality and how each team member contributes to it, the better able we'll be to create the best product possible.

Defining and Measuring Quality

If we are to specify the quality desired in the product, we must be able to measure it so we will know whether that quality is present when the product is delivered. There are a number of factors associated with the concept of quality. These factors should be specified when the client and developer contract to do the project. They should be considered during all steps of the project; and they should be evaluated when the product is delivered. But we need good ways of defining and measuring these quality factors.

Unfortunately, at this stage of the maturity of software engineering we are not yet able to define these factors in ways that can be agreed upon and measured easily and unambiguously. We can, however, give them verbal definitions, gain an understanding of why they are important, and pay them due attention throughout the process. As software engineering matures, we will look forward to having better definitions and measures for the quality characteristics we want in our products. Some of the factors of quality of software products are as follows:

- *Completeness*—does the product do all that the specifications say it should do?
- *Correctness*—does the product do its work correctly as specified?
- *Reliability*—how long will the product continue to do what the specifications say it should do, especially when it is exposed to new input and new environments that the specifications say the system must be able to handle?
- *Usability*—how easy is it for the user to operate the system without making mistakes, or alternatively, how well is the reliability of the system protected when the user is introduced?
- *Understandability/Simplicity*—how easily can someone understand what the product does, how it works, and how to control and use it?
- *Robustness*—how will the system survive bad inputs by identifying them and helping with their correction?
- *Efficiency*—how effectively does the system use resources to handle problems as large and as quickly as specified?
- *Survivability*—how well can the product perform its specified functions when its environment (for example, the hardware) has deteriorated?
- *Integrity*—how well does the system protect itself against hostile acts, or access by unauthorized persons?
- *Verifiability*—how easy is it to verify that the system is doing correctly what it is supposed to do?
- *Maintainability*—how inexpensive is the product to fix or change?
- *Versatility/Flexibility*—how inexpensive is it to change the system to do something not originally anticipated in the requirements?
- *Expandability*—how easy would it be to expand the system's capabilities?
- *Portability*—how easy would it be to get the system operating in another environment (for example, on different hardware or under a different operating system)?
- *Reusability*—how easy would it be to use the product or its parts in other systems?
- *Interoperability*—how easy is it to interface the product with other systems?

These factors are not independent. For example, if a system is maintainable, it is likely to be flexible and extendable. There are also likely to be trade-offs between factors. For example, what one does in the design to obtain efficiency might sacrifice maintainability.

We might assume that we need not develop the system to a higher standard of quality than what is specified in the requirements. To do so would add to the costs and time to develop the system, which would not be desirable to either the client or the developer. However, we must also be concerned that we meet our legal, moral, and ethical obligations, which may be beyond what is stated in the requirements specification.

At the present state of the art in defining and measuring quality, it is often easier to measure and control the process than to measure and control the product. Thus, quality assurance of software has often come to mean ensuring that the standards and procedures have been followed in the process of developing the product. If such an approach is taken, we must ensure that the process is so defined that there is a positive relation between doing the process right and creating the right product. But it is also important to ensure that quality assurance methods lead to a quality product and do not just generate unnecessary paperwork.

Attitude toward Quality

To obtain a high-quality software product we must rid ourselves of two dangerous attitudes:

1. Software always has defects, so why bother to avoid them.
2. Besides, we can always fix it later.

We must keep in mind that software is complex and as a consequence is very difficult and costly to change. Thus we can well afford extra effort so that we won't have to make changes.

In Chapter 17 we will discuss testing. The first sixteen chapters will be devoted to how we proceed so that we can be confident that we have a quality product before we test it. If we do not take this care as we proceed, it is likely to be too late, even before we start to test.

We must make sure that everyone understands clearly that the quality is built in; it cannot be tested in. This attitude must be in place before quality assurance measures will work.

The word *bug* goes back to Grace Hopper, who while looking for the source of a problem in an early Harvard relay machine, found a bug caught in one of the relay contacts. This word implies something that flew in from outside for which we are not responsible. We should begin by not calling the results of our faulty work "bugs." Before we can get rid of these mistakes we need to adopt the attitude that they are defects due to our own bad work for which we must take full responsibility.

DeMarco (1982, p. 197) tells the story of two engineers participating in one of his seminars. They had written only one significant piece of software between them, but this software had survived extensive use by other users over an extended time without any defects being found. Since these two had less experience in software development than had the other participants, yet apparently had more success in writing quality software, DeMarco asked them how they did it. Their answer was simple and to the point. "We're just engineers. We didn't know defects were allowed." We should develop software with the understanding that defects are not allowed.

One of the first places we should seek a change of attitude is with management. They need to understand that building software is little different from constructing buildings. Once the foundation is laid, it is almost impossible to change, and if the foundation is not laid correctly, you cannot build a good structure on it.

Management should ask for careful foundations to be laid and ask to see those foundations (problem/opportunity definition, requirements specification, design specifications) before they allow implementation to begin. Managers ought to require validation and verification reviews for each phase. Equally, management should not push their people into coding so they can count lines of code as a measure of progress. If these lines of code have to be changed or completely redone, does this represent progress? It costs many times as much to change a line of code as to write it in the first place.

Those directly involved in the work on the project are often the first to become aware of problems or potential problems. They must be listened to if these problems are to be headed off. Upward communication is essential for the creation of a quality product.

It is a blow to our ego to find a defect in what we have done. One way to protect ourselves is not to test carefully so we will not find the defects that embarrass us. This bit of deception is easy to fall into, whether consciously or unconsciously. Thus someone else must be assigned to test our work. Then we will be inclined to do our work carefully enough so we will not be embarrassed by their finding our errors.

The quality assurance group should not just test programs. They should be involved in defining the procedures used to develop the system. They should make sure that reviews are conducted and that recommendations are made and followed up. But the reviews themselves should be done by the development people, not the quality assurance people. The quality assurance team should be seen not as adversaries but as colleagues who help development people deliver a product that everyone can be proud of.

Summary

Software engineering projects are initiated because the client identifies an expectation that is not currently being satisfied. One important early step is to be sure that all parties agree on the exact nature of the need and that the means are available to meet the need. Since they are agreeing to do what hasn't been done before, software engineers must take care to be realistic about what they can achieve. A useful way of viewing a project is to imagine it as a move to a new place, which would involve considering why we want to move, what our lives will be like when we get there, and how we'll get there.

The client and provider must agree early in the project on the requirements specification—what the client wants the system to do and what the provider can offer by way of technology. The specifications describe the system's environment, functions, performance, logistics, and quality. The design process is recursive, because the designer works from requirements to determine the components and how they are put together, then he or she in turn writes requirements for these components. Putting the components together is known as synthesis; studying the system to see if it does what it's supposed to do is called analysis.

One of the greatest challenges in software engineering is coping with the system's complexity. A system is a group of parts that interact so that the behavior of the whole depends on the behaviors of the parts. Complexity is the difficulty in determining how the system behaves, given a knowledge of how its components behave. Testing a system's performance in all circumstances is usually not possible if the system is at all complex. Instead, systems must be designed carefully to ensure they work properly. The top-down, or hierarchical, method enables us to understand the designs of complex systems.

Complexity increases the chance of mistakes. Because the cost of an error rises as time increases before it is found and corrected, software engineers must do as much as they can to work out mistakes near the start. A number of tactics can help to avert costly trouble. One strategy is to describe the system logically, or in terms of how information is processed, before describing the devices or people that will do the processing. Clients must also be assisted in visualizing the completed project through a complete representation on paper or by way of a prototype. Frequent peer reviews can also help to spot possible problems.

In the life of each project there must be scheduled pauses to review work thus far and decide whether to continue, change the approach, or stop. Project development also follows a system life cycle, which is divided into phases. Each phase consists of a specification step, a performance step, and an evaluation step. Before work is done, its specifications are validated; after completion, the work is verified. The phases of a system life cycle proceed from problem definition, to requirements and feasibility analysis, to system and components design, to components and system building, and finally to system evaluation.

As the project proceeds changes are inevitable, and one great problem is keeping track of these changes. A configuration describes the current state of the product. Change control is the method of following changes and their consequences for product quality, cost, schedules, and tasks to be done. Accurate means are needed for estimating the time and cost of producing a system. Estimates will improve as we develop better models, use the experience of previous projects, and rely on people specially skilled in this type of estimating.

At the end of every project the final judgment depends on the quality of the product. Client and provider must decide in advance how they will gauge the product's quality. Among the factors of quality are completeness, correctness, reliability, usability, maintainability, and versatility. The product's quality inevitably relies on each team member's attitude about quality. Each person must take responsibility for the work he or she does and be willing to face a peer review.

Exercises

1. What similarities and differences can you see between a project to build a bridge and a project to build an information processing system?
2. Top-down requirements and design: This exercise will illustrate the concepts of requirements and design and show how design involves the processes of synthesis and analysis. It will also illustrate the use of abstraction in top-down design. We will do this in the context of designing a full adder. Obviously this is hardware, not software, but with a hardware example it is easier to make the requirements explicit and complete.

   We want to design a full adder circuit that will add one bit position of two binary numbers including the carry. We can represent this full adder as a box as follows:

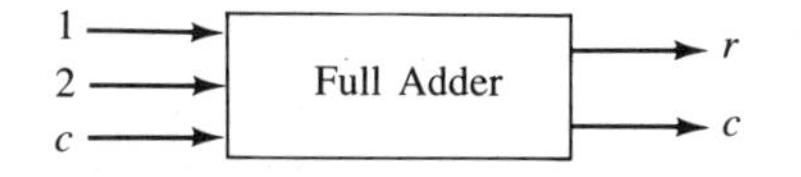

where:

1 and 2 are bits in the same position in the two numbers being added
r is the result in that bit position
c is the carry to or from an adjacent position.

We can define the functional requirements for this full adder by describing the output we get for each possible input as a logic table:

Input			Output	
c	1	2	c	r
0	0	0	0	0
0	0	1	0	1
0	1	0	0	1
0	1	1	1	0
1	0	0	0	1
1	0	1	1	0
1	1	0	1	0
1	1	1	1	1

We have available primitive circuits with which to build this adder. These primitives will perform the following functions, which we can also describe using logic tables:

AND

In 1	In 2	Out 0
0	0	0
0	1	0
1	0	0
1	1	1

OR

In 1	In 2	Out 0
0	0	0
0	1	1
1	0	1
1	1	1

NOT

In 1	Out 0
0	1
1	0

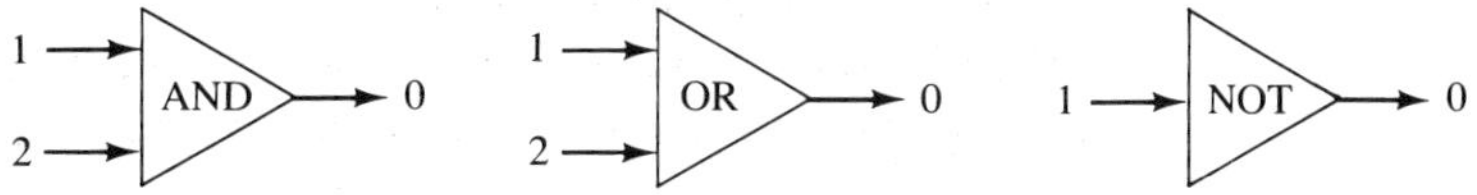

1 → AND → 0
2 →

1 → OR → 0
2 →

1 → NOT → 0

To make a design we will put together combinations of ANDs, ORs, and NOTs. This is *synthesis*. Then we will simulate how the design would work. Developing the implications of a proposed design is *analysis*.

Synthesis is a creative process. We usually do not have a recipe that we can use to grind out a design given the requirements we must satisfy. Design is an iterative, or repetitive, process. We synthesize and analyze iteratively until we are able to come up with a design that our analysis shows satisfies the requirements.

With a complex system it may be very difficult to synthesize a design that meets the requirements in one step. So we divide this process into several steps by doing the design top-down. We illustrate the top-down design of our full adder as follows: Assume that we believe we could build a device, which we will call here a half adder. We assume it satisfies the following requirements:

Half Adder

In 1	In 2	Out c	Out r
0	0	0	0
0	1	0	1
1	0	0	1
1	1	1	0

1 → Half Adder → r
2 → → c

This specification adds only two inputs and thus should be a simpler device to design. Since we have not yet designed or built such a device, it is at this time just an *abstraction*. What we are doing here is to use a classical problem solution method. We solve a complex problem in terms of simpler problems we assume we can solve later. This is the top-down approach.

The exercise here is to design a full adder by connecting the following half adders:

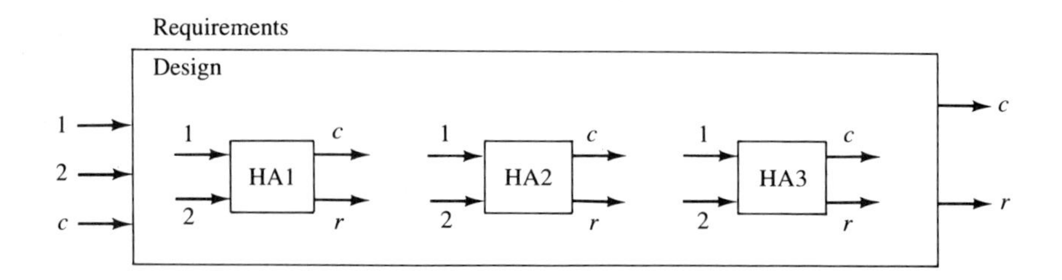

If we think we have a design, we can use the logic tables for the components to simulate the behavior of that design to see if the design satisfies the requirements. Fill in the following logic table for your proposed design and change the design as needed until the requirements are satisfied.

Input			HA1		HA2		HA3		Output	
c	1	2	*c*	*r*	*c*	*r*	*c*	*r*	*c*	*r*
0	0	0	—	—	—	—	—	—	—	—
0	0	1	—	—	—	—	—	—	—	—
0	1	0	—	—	—	—	—	—	—	—
0	1	1	—	—	—	—	—	—	—	—
1	0	0	—	—	—	—	—	—	—	—
1	0	1	—	—	—	—	—	—	—	—
1	1	0	—	—	—	—	—	—	—	—
1	1	1	—	—	—	—	—	—	—	—

Now we have designed a full adder using abstract components we have called half adders. Next we must design a half adder. If we are unable to design a half adder, we have to go back to the higher level and design the full adder again using a different abstraction. We leave it as an exercise to design a half adder by connecting the following ANDs, ORs, and NOTs.

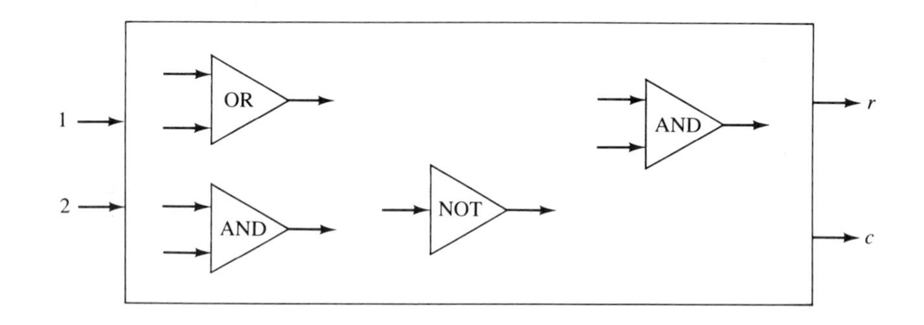

Fill in the following logic table for your proposed design and change the design as needed until the requirements are satisfied.

1	2	OR	AND	NOT	AND	Output c	r
0	0	—	—	—	—	—	—
0	1	—	—	—	—	—	—
1	0	—	—	—	—	—	—
1	1	—	—	—	—	—	—

Expand this hierarchical description of the design to show the design of the full adder directly in terms of ANDs, ORs, and NOTs. Can you see where you can delete some components without affecting the requirements?

3. The following is an interesting classical problem that shows why an important aspect of solving a problem is how you choose to represent it.

   A man wishes to ferry a fox, a goose, and a cabbage across a river. The boat must be rowed by the man and can carry one other item—the fox, goose, or cabbage—on each trip. The man must not leave the fox and goose together on one side unattended for the fox will eat the goose. He must not leave the goose and cabbage together on one side unattended for the goose will eat the cabbage.

   a. Show how he can get them all across the river without the goose or cabbage being eaten.
   b. How did you choose to represent the various states of what was on each side of the river and how transitions were made between those states?
   c. Could you have solved this problem without such a representation? If so, could you have shown someone else how you solved it so he or she could confirm the correctness of your solution without using such a representation?
   d. What sequence of decisions did you use? Did you work from the problem to the answer, from the answer to the problem, or a little of both?
   e. How does this apply to what was discussed in this chapter?

References

Alexander, Christopher. *Notes on the Synthesis of Form*. Cambridge, Mass.: Harvard University Press, 1964.

Baker, F. T. "Chief Programmer Team Management of Production Programming." *IBM Systems Journal* 11, no. 1 (1972): 56–73.

Boehm, Barry W. *Software Engineering Economics*. Englewood Cliffs, N.J.: Prentice-Hall, 1981.

Boehm, Barry W., et. al. *Characteristics of Software Quality*. Amsterdam: North-Holland, 1978.

Bohm, C., and G. Jacopini. "Flow Diagrams, Turing Machines and Languages with Only Two Formation Rules." *Communications of the ACM* 9, no. 5 (1966): 366–371.

Cougar, J. Daniel, and Robert W. Knapp, eds. *Systems Analysis Techniques*. New York: John Wiley & Sons, 1974.

Davis, C., and C. Vick. "The Software Development System." *IEEE Transactions on Software Engineering* 3, no. 1 (1977): 69–84.

DeMarco, Tom. *Controlling Software Projects: Management, Measurement and Estimation*. New York: Yourdon Press, 1982.

Deutsch, Michael S. "Verification and Validation," in Jensen, R. W., and Tonies, C. C., eds., *Software Engineering*. Englewood Cliffs, N.J.: Prentice-Hall, 1979, pp. 329–408.

Jensen, Randall W., and Charles C. Tonies. *Software Engineering*. Englewood Cliffs, N.J.: Prentice-Hall, 1979.

Katzan, H., Jr. *Systems Design and Documentation: An Introduction to the HIPO Method*. New York: Van Nostrand Reinhold, 1976.

Lundeberg, Mats, Goran Goldkuhl, and Anders Nilsson. *Information Systems Development: A Systematic Approach*. Englewood Cliffs, N.J.: Prentice-Hall, 1981.

Miller, G. A. "The Magical Number Seven, Plus or Minus Two: Some Limits on Our Capacity for Processing Information." *Psychological Review* 63 (1956): 81–97.

Myers, Glenford J. *Software Reliability: Principles and Practice*. New York: Wiley Interscience, 1976.

Myers, Glenford J. *The Art of Software Testing*. New York: Wiley Interscience, 1979.

Nassi, I., and B. Shneiderman. "Flowchart Techniques for Structured Programming." *ACM SIGPLAN Notices* 8, no. 8 (1973): 12–26.

Peters, L. J. *Software Design: Methods and Techniques*. New York: Yourdon Press, 1981.

Simon, Herbert A. *Sciences of the Artificial*. 2nd ed. Cambridge, Mass.: MIT Press, 1981.

Teichroew, D., and E. Hershey. "PSL/PSA: A Computer-Aided Technique for Structured Documentation and Analysis of Information Processing Systems." *IEEE Transactions on Software Engineering* 3, no. 1 (1977): 41–48.

Wirth, N. "Program Development by Stepwise Refinement." *Communications of the ACM* 14, no. 4 (1971): 221–27.

Yourdon, Edward N., ed. *Classics in Software Engineering*. New York: Yourdon Press, 1979.

PART 2

Describing the Product

The chapters in Part 2 show methods for representing systems. These methods are used when we describe and analyze the existing system to understand why it is not meeting expectations. They are also used in the communication with the client to establish that what we propose building will satisfy his or her needs. And they act as a tool in developing the design and describing how it is to be built. These methods of representation will help us understand what we are doing as we do it, coordinate our work with the work of others, and leave a record for anyone who needs to change or build upon what we do.

Part 2 includes chapters on data flow diagrams (Chapter 3) and tree structures (Chapter 4). In Chapter 3 we'll look at two types of data flow diagrams: the classical DeMarco Data Flow Diagram and the newer Two-Entity Data Flow Diagram. Chapter 4 will explore three types of tree structures: structure charts, Warnier-Orr diagrams, and Trees (with a capital T).

Data flow diagrams and tree structures stand for two different points of view in describing systems—each with a different emphasis, and each tending to appeal to a different type of person. Data flow diagrams focus on the data flow, while tree structures stress the hierarchy and control. Some developers feel more comfortable starting their clients with data flow diagrams. But this requires the developer to change to a tree structure method in order to continue into the system's design and implementation. Typically, with this approach, one starts with a DeMarco Data Flow Diagram and then transforms this information into a structure chart. In some circles it has also been popular to work initially with the client using a tree structure type of method known as Warnier-Orr Diagrams. We'll look at both these approaches.

The Two-Entity Data Flow Diagram and the Tree are extensions to the classical methods. They make it possible to start with a simple diagram that is understandable to the client and then to add successive levels of detail and sophistication to the same diagram as our work progresses. If we wish to start with the Two-Entity Data Flow Diagram, we can transform it into a Tree at the appropriate time. If we prefer to start with the Tree, we can avoid this transformation. In the latter case, the same basic method of representation can be used as one seamless description from the original concept, through the functional requirements, through design, and right into producing code. This seamless approach makes it less likely that we will lose information in making the transformations and makes it easier to use computer aids.

3

Data Flow Diagrams and Matrices

- Data flow diagrams, which are simplified depictions of the movement of data through a system, provide a common language for both clients and designers to describe a system.
- Rather than having many types of special symbols, data flow diagrams use distinct symbols only for processes, data flows, and data files.
- The data flow diagram acts as a table of contents, with labels within the symbols referring to more detailed information on attached documents called annotations.
- Data flow diagrams occupy only a single page each, but details of a process in one diagram can be expanded as another data flow diagram, providing a hierarchy of diagrams.
- The DeMarco version of the data flow diagram shows only the flow of information, and not other information such as control.
- Two-Entity Data Flow Diagrams can be extended as needed to show also control and triggers that initiate processes.
- Matrices can be used to represent in a computer data base the information in the data flow diagram and a dictionary of data elements.
- When changes are made the matrices can be used to identify the affected data elements, files, and processes.
- These matrices can also be used to compute sizes of record and files, frequencies with which processes are run and files passed, and other information of interest when designing information systems.

In this chapter we will be discussing two types of data flow diagrams, the DeMarco version (D-DFD) and the Two-Entity version (TE-DFD). Most of the techniques we will discuss will apply to both types. The DeMarco Data Flow Diagram was the original and is still the most widely used. However, we will focus most of our attention on the newer Two-Entity Data Flow Diagram because of its importance to the integrated approach we emphasize in this book.

In the latter part of the chapter we will discuss how to represent data flow diagrams in a computer data base so we can identity the effects of changes in a software system on other parts of the system.

We will use two examples in our discussions, a simple one to illustrate the concepts, and a more realistic-looking example involving an order processing system.

Data Flow Diagrams—Basic Concepts

A **data flow** is information in motion. It may cause something to happen when it is received. A **process** transforms input data flows into output data flows. When a data flow stops, waiting to be used at another time, it becomes a **data store,** which may also be called a **file.** A data flow diagram, be it the DeMarco or Two-Entity version, shows the connections between processes, data flows, and files. Rather than using a different symbol for each of the various types of processes, just three types of symbols are used: process, data flow, and file. A label in the symbol refers the reader to additional information elsewhere.

To illustrate these concepts consider this example. An order for merchandise is sent to a mail-order house. The order is transformed into a record of the order to be stored and a sheet of paper showing the pickers what items are to be picked from the warehouse. The incoming order, the order as it goes to the order file, and the paper showing the pickers what to pick are data flows. The record of the order is retained in a file. The transformation is done by a process, which might be called Process Order.

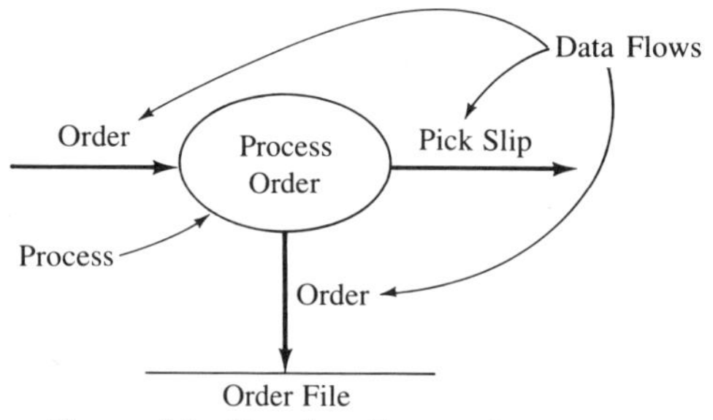

Figure 3.1 Data flow diagramming concepts.

DeMarco Data Flow Diagrams

When Tom DeMarco first presented his method in 1978 (DeMarco, 1978), it was a great triumph because it created an approach by which a developer and someone outside the field could communicate, solving the biggest and most fundamental problem in software systems development.

Previously software developers had described systems using system flowcharts. We will not consider system flowcharts here except to say that systems designers tended to cram as much information into these charts as possible. Such charts were filled with many kinds of special symbols to represent card readers, CRTs, printers, disk files, tapes, and so on. As a result these diagrams got so complicated that clients were unable to understand them. Clients could not determine whether the system being designed for them really would do what they wanted. Thus, they did not know what they were getting until they got it, by which time it was very expensive or impossible to change.

Because graphics are supposed to be so easy to understand, clients were reluctant to admit they did not understand these charts. Many customers did not ask the necessary questions for fear of admitting their ignorance. They tended to just say "yes, yes" and assumed that at least the people who drew the flowcharts knew what they were doing. Thus system flowcharts were often used to intimidate and hide the fact that the people who drew them did not understand what they were doing either. System flowcharts thus reversed the traditional saying, making it, "One lousy picture can obscure a thousand words."

DeMarco's approach was something like the following: Let's make diagrams that clients can understand so they can tell us as early as possible whether what we are designing is what they want. The earlier they can tell us, the less effort will be wasted building something they do not want, which will then have to be changed. The earlier we catch false assumptions, the less effort will be wasted pursuing them.

To make the diagramming techniques simple, DeMarco reasoned, we'll use only three generic types of symbols. Instead of making distinctions between different types of equipment by using many different types of symbols, we could assign labels to these generic symbols so that the viewer can look up the details elsewhere. The data flow diagram, therefore, shows only the connections between processes, data flows, and files and acts as a table of contents for more detailed information.

DeMarco showed how to use a hierarchy of simple diagrams to represent even complicated systems. We will discuss this hierarchy method later in this chapter. He also pointed out that one should be able to make a logical diagram of the system that shows only the information to be handled without having to commit to the type of physical device that will do the processing. As we noted in Chapter 2, this picture is called a logical description. The physical devices that process this data can be considered later in the physical description.

DeMarco's Data Flow Diagram uses a bubble to represent a process. Lines with arrows, called **arcs,** point from one process to another to show data flows. A straight line is used to represent a data store. In the example in Figure 3.2 capital letters *A* through *H* are processes, lowercase letters *a* through *m* are data flows, and *I* is a data store. The input to process *D* is data flow *d;* the outputs are data flows *e* and *f,* which go to other processes, and data flow *l,* which goes to a data store.

DeMarco says his diagrams should be kept simple by including only data flow. He absolutely refuses to have any conditions or control information put into his diagram. For instance, DeMarco's diagram would not show whether process *D* might produce data flow *e* under one condition and data flow *f* under another condition. He carries this to the point of insisting that one should not be able to add such information to his diagrams later.

But this restriction against added information has its price. When we deal with a more experienced client or have additional information available, we are forced to change to another method of representation to be able to show it. Development has thus become a patchwork quilt with different types of diagrams for different types of information. In this patchwork quilt approach we must learn several methods of representation and how to transform between them. And we must spend time making transitions between methods. With each transformation we are in danger of losing information and continuity. Using several different methods also makes it more difficult to use automated tools throughout the whole process because we have to reinput the information for each of the various types of representation.

Those who advocate extending DeMarco's method agree with DeMarco that initially basic diagrams should be kept simple but they believe that these basic diagrams can be used as a basis upon which additional information can be added later. By varying the amount of complexity and detail, the same basic diagram can be presented to a variety of audiences. This provides a continuity of information as one moves into the system design phase and more detail is required. Information is less likely to be lost in transforming from one method to another.

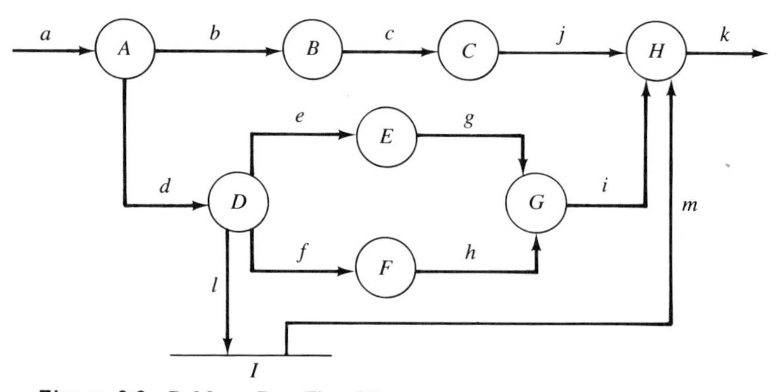

Figure 3.2 DeMarco Data Flow Diagram.

Much of the discipline of data flow diagramming remains the same whether we are using DeMarco Data Flow Diagrams or an approach that goes beyond his original creation. We have chosen to discuss the diagramming discipline in the context of the extended method.

Two-Entity Data Flow Diagrams

In this book we will emphasize Two-Entity Data Flow Diagrams because: 1. they can be used with the same simplicity as the DeMarco version, but 2. they can also be extended as further detail is needed later in the project. This additional detail may also be of interest to more sophisticated clients.

We recognize that in some work environments we will still be asked to deal with the DeMarco Data Flow Diagram. Therefore we have introduced the basic data flow diagramming concepts in the context of the DeMarco version and describe its unique features. Now we will describe the Two-Entity Data Flow Diagram and use it to introduce more advanced diagramming techniques. However, we should point out that many of these techniques can be used with both types of diagramming.

Basic Diagram

The **Two Entity Data Flow Diagram,** or **TE-DFD,** is an extension of the concepts introduced in the DeMarco Data Flow Diagram. Two-Entity Data Flow Diagrams can be used for the simplest representations used to communicate with clients who may be unsophisticated about information systems. But the same diagrams can also be extended to more elaborate diagrams representing subtle aspects of the system for clients who know a great deal about information systems and want to see the details. The producer will certainly want to see that detail in the later stages of the systems development.

The premise behind the Two-Entity Data Flow Diagram is that systems can be represented by showing connections between two basic types of entities: processes and data flows. A rectangle is used to represent a process, and an oval stands for a data flow. Further information about the type of process or data flow is shown, not by the invention of many special symbols, but by writing a label in the rectangle or oval. Thus the data flow diagram describes not the details of these processes and data flows but only how they are connected. The data flow diagram and the labels used for the processes and ovals represent a "table of contents" to direct the viewer to more detailed information about the process or data flow to be found elsewhere.

As you can see in Figures 3.3 and 3.4, the symbol used for a file is a variation on the process symbol, a rectangle open on the right end.

Note that in a TE-DFD we are able to write certain conditions on the arcs leaving a process. These conditions, which occur within the process, determine

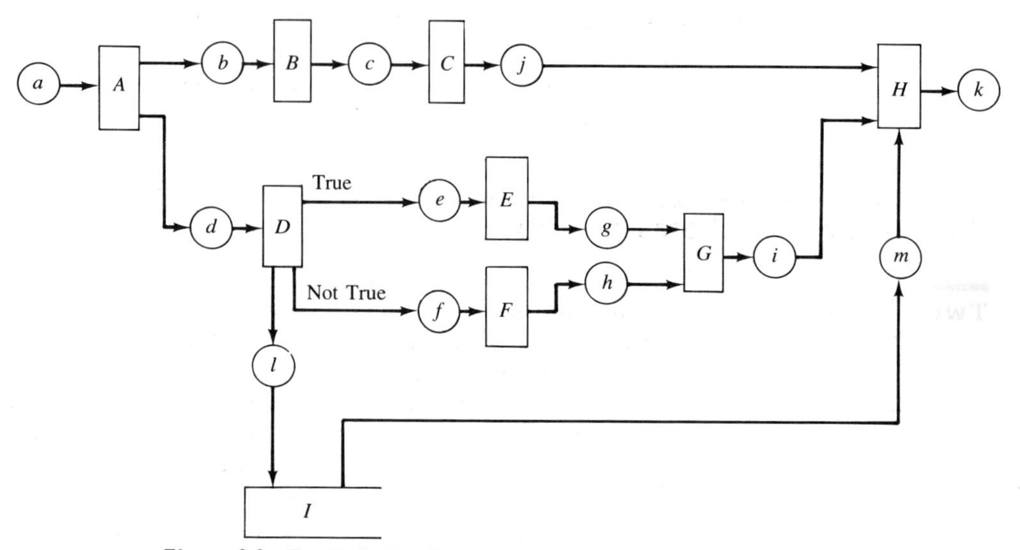

Figure 3.3 Two-Entity Data Flow Diagram. This small example corresponds to the DeMarco Data Flow Diagram in Figure 3.2.

when the data flow occurs. Conditions can be written on the arcs because the data flows are no longer written there but are contained within ovals.

Figure 3.4 is a simple two-entity data flow diagram for an order entry system. "Process Order" and "Purchasing" are processes; "Invoice" and "Reorder List" are data flows; and "Open Order" is a file.

Processes and data flows follow the **sandwich rules:** Between two processes there is a data flow, and between two data flows there is a process. Data can flow between a process and a process, or between a file and a process. A file is represented as an open-ended rectangle to remind us that files and processes play the same role in the sandwich rule. This treats a file as a process that consumes or produces data, which is consistent with some modern thinking about data base machines. The diagram can begin or end with either a process or a data flow. The two entities in Two-Entity Data Flow Diagrams come from the two positions in the sandwich rule: process and file are one type of entity, data flow the other.

Someone trained in mathematics will note that processes are like **operators;** data flows are like **operands.** (Operators are like verbs; they describe what's done. Operands are like nouns; they describe what the operator does it to and what is produced as a consequence.) An operator takes operands as inputs and produces other operands as outputs. Similarly, a process takes data flows as input and produces data flows as output.

Extensions to the Basic Diagram

When you show clients a data flow diagram that contains a lot of information, they may say, "That's too confusing" and refuse to look at it any further. On the

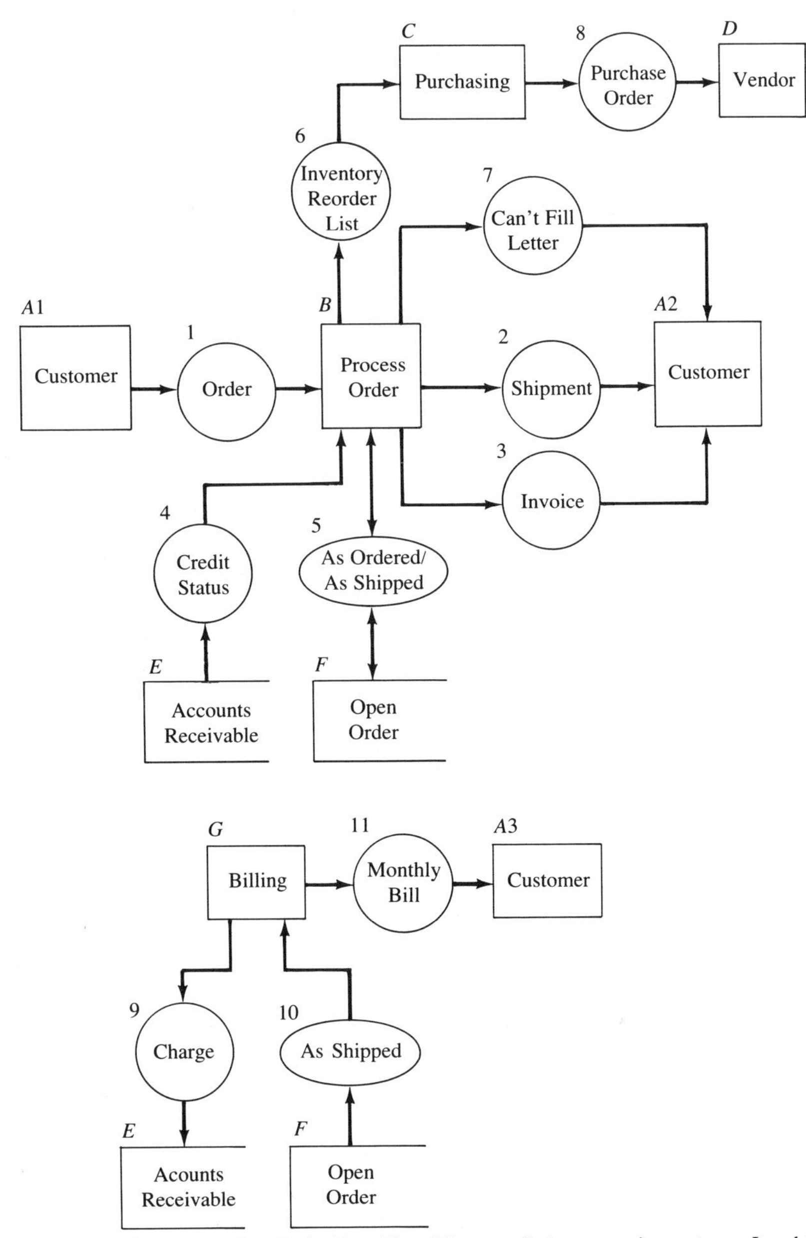

Figure 3.4 Basic Two-Entity Data Flow Diagram. Order processing system—Level 1.

other hand, if you show them a simple diagram, they are likely to complain, "Your diagram doesn't show when things are done. For example, there is nothing to show that you write the letter only if the order can't be filled. Your diagram makes it look like all of those outputs occur every time."

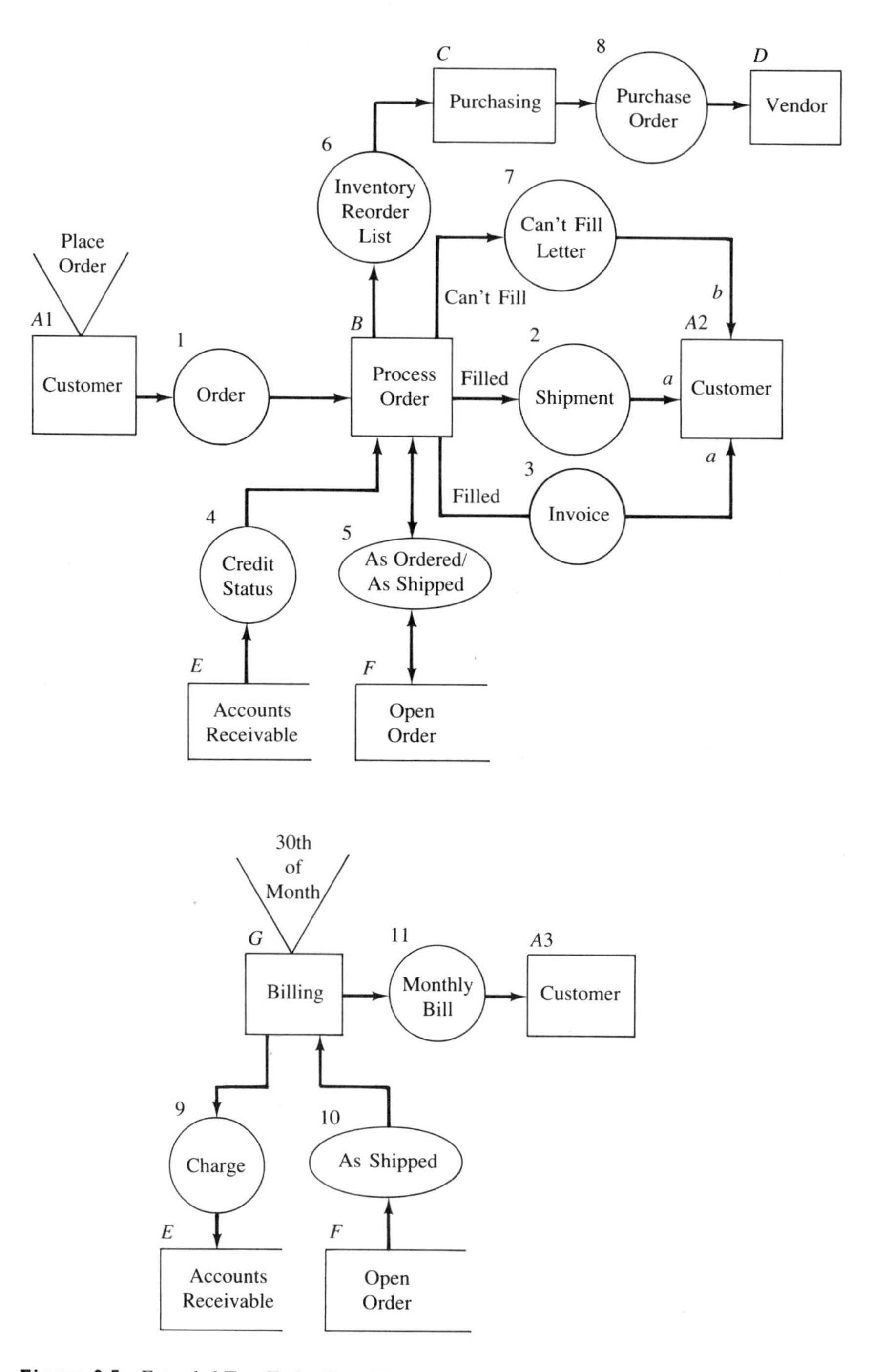

Figure 3.5 Extended Two-Entity Data Flow Diagram. Order processing system—Level 1.

Thus we would like to be able to create a simple version of the data flow diagram that contains only the data flow, not logic and control. But we would also like the capability to add logic and control when that is appropriate without having to redo the original diagram. To illustrate both versions Figure 3.4 shows a simple Two-Entity Data Flow Diagram, and Figure 3.5 shows this same diagram with more elaborate information added to it. Let us discuss this added information and extend our interpretation of the meaning of data flows and processes.

Agents and Processes

The label in a rectangle can describe either a process or name an **agent** that performs a process. The agent can be a person or a machine. For example, in Figure 3.4, "Customer" is an agent, a person who performs the process of generating an order. The same agent may appear more than once in the diagram, representing different processes performed by that agent. For instance, in Figure 3.4, the agent "Customer" appears once to represent the process of the customer generating an order and again to represent the process of the customer receiving the shipment.

When the same agent performs more than one process in our system, this can be shown in either of two ways. The agent can be shown as separate rectangles representing his or her performance of different processes. In this case a common piece of the identification number or letter in the upper left corner of the rectangle shows the reader that there is another rectangle somewhere else representing the same agent (see Figure 3.6).

In another method, we could show one rectangle for the agent who performs both actions. Numbers are put on the arcs entering or exiting the rectangle to distinguish the separate processes performed by that agent and to show the sequence in which those processes occur (see Figure 3.7).

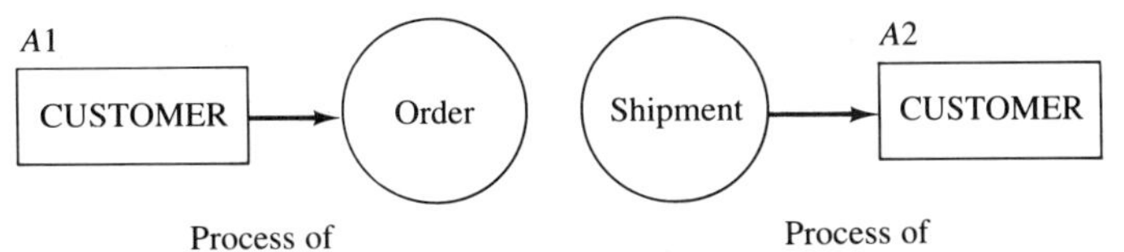

Process of
Customer Placing Order

Process of
Customer Receiving Merchandise

Figure 3.6 Same agent as more than one process. Agent repeated for each process.

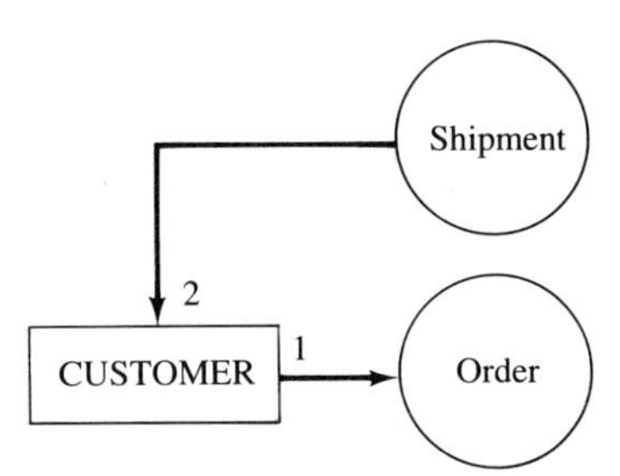

Figure 3.7 Same agent as more than one process. Agent shown once with sequence numbers.

Figure 3.7 shows that the customer performs the process of generating an order and at some later time performs the process of receiving the shipment. When people look at data flow diagrams, they know that there is some sequence to the processes and often feel uncomfortable if the diagram does not acknowledge that sequence.

Other Types of Operands and Operators

An oval can be used to represent other types of operands as well as data flows. An operand could be any one of the following:

- Flow of data or information
- Object
- Person
- Material
- Location, state, or condition of someone or something

The rectangle would then show the operation of making some change in the data flow, object, person, material, or in the position or condition. In this way we can represent a large variety of systems that may involve mixtures of data flows and other types of operands. Let's look at some examples.

When describing the operation of a blood bank, we may wish to describe not only the processing of information about the donor but also what the donors are doing. Figure 3.8 shows that the donor changes position from standing in one line to standing in another, a change in location.

Similarly, Figure 3.9 represents a change in the condition or state of the donor. Figure 3.9 illustrates another point—processes should be conservative. What flows in should in some shape or form be accounted for by what comes out. Here we are missing the accountability for one pint of blood. The diagram should be replaced by Figure 3.10.

This conservation and accountability of what goes in and comes out of each process can help us in our analysis of a system to see that nothing is lost or shows up mysteriously without explanation.

(These concepts of how operators and operands can be interpreted apply equally well to both Two-Entity and DeMarco DFDs.)

Control Information

Conditions may be written on the outputs of a process, showing the circumstances under which that output is generated by the process. For instance, in Figure 3.5, three arcs come out of the process entitled "Process Order." These arcs go to data flows and then to other processes. Each arc shows the conditions under which that output would occur. If the order can be filled, the outputs are "shipment" and "invoice." If the order cannot be filled, the output is a "can't fill letter." Thus we can write on the arcs exiting a process the conditions that occur within the process that cause this output to be produced.

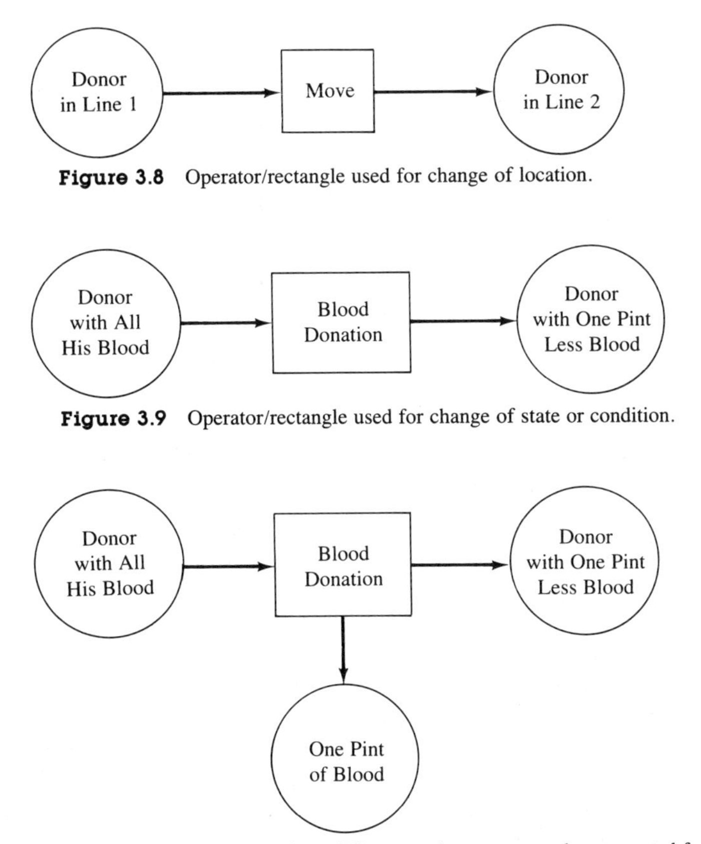

Figure 3.8 Operator/rectangle used for change of location.

Figure 3.9 Operator/rectangle used for change of state or condition.

Figure 3.10 Conservation—What goes in or out must be accounted for.

A **trigger** is what causes something to happen, such as the trigger on a gun causes it to fire. We say a process is **fired** when the necessary circumstances occur that cause the process to start. Information can be put on the diagram to show the triggers that cause processes to be fired. The three types of triggers are external triggers, data flow triggers, and control triggers. An external trigger takes place when an external event causes the process to fire. A data flow trigger occurs when a process is fired by the availability of the data needed by the process. While data flows can trigger processes, files cannot. Files are data waiting to be requested by a process. A control trigger occurs when one process sends a trigger signal to another process without passing any data. The data flow is labeled C for control.

An external trigger is shown with a label in a V above the process rectangle. In Figure 3.5 "place order" is the trigger that initiates the process represented by the agent "Customer."

We may also want to show how a process is triggered by certain combinations of the data flows. We can do this by putting letters on the arcs entering the

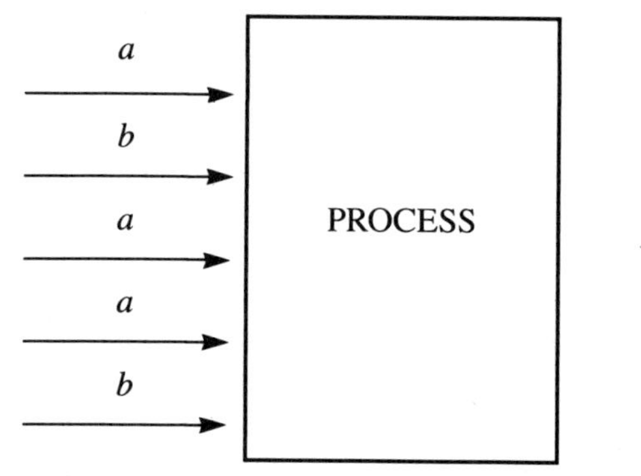

Input logic of triggers
(*a* and *a* and *a*) or (*b* and *b*)

Figure 3.11 Logic of input to trigger process.

process. When the process receives all the data or control flows with any one letter, the process fires.

For example, in Figure 3.11, this process would fire if it got all three inputs labeled "a," no matter how many "b" 's it had, or it would fire if it got both inputs labeled "b," no matter how many "a" 's it had.[1]

In Figure 3.12 we make a digression from our order entry example to show the simplest possible diagram needed to illustrate the use of conditions and triggers. A **transaction** is the processing that occurs for a particular set of input. A transaction going through this system first goes through process *A*, then splits into two parts that go through subsystems that can run in parallel. These two parts finally come back together to go through process *H*. Process *H* is triggered only when it gets results from both parallel subsystems. This is shown by the letters *a* on both inputs to *H*.

Within the subsystem *DEFG* the transaction goes through process *D* and then branches, either going through process *E* or through process *F* before going through process *G*. Since a transaction either goes through *E* for the true condition or through *F* for the not-true condition, it does not go through both. Process *G* is triggered, therefore, when it gets results from either *E* or *F*. This is shown with the *a* and *b* on the two inputs to *G*.

If we show how long it would take to execute each process, and assume that all the outputs produced occur at the same time when the process is finished, then using these triggers and the critical path scheduling concepts we will discuss in Chapter 7, we could compute how long it would take to do all of the processes for each combination of conditions.

[1]Someone familiar with logic would recognize that this use of letters to show the combinations of inputs to fire a process represents a disjunctive canonical form. With a negation, which we might show with a bar over a letter, all logical possibilities could be represented. This technique also allows us to use a Two-Entity Data Flow Diagram to represent much of the information that could otherwise be represented in a precedence diagram (Sholl and Booth, 1975) or by a Petri network (Peterson, 1977), which are sometimes used to represent real-time systems. Those familiar with Petri networks will recognize that ovals in the TE-DFD correspond to places in a Petri network and the rectangles correspond to triggers. There are no markers used in the TE-DFD, as are used in Petri networks.

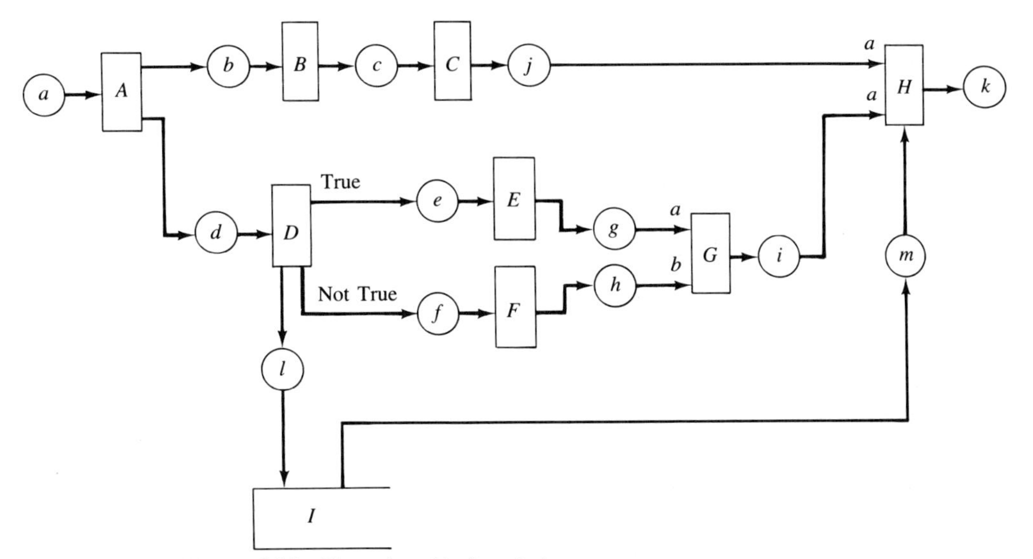

Figure 3.12 Illustration of logic and triggers.

Hierarchy of Data Flow Diagrams

Even using only the basic TE-DFD (or D-DFD) symbols and techniques, we can still make diagrams that have so many processes, data flows, and files that they get too complicated to understand easily. We exceed our Miller Principle limits in trying to figure out what is going on. Too many symbols crowd onto a page, or the diagram continues from one page to another. What we need is some hierarchical way of representing data flow diagrams. Although a data flow diagram does not have the natural hierarchical form of a Warnier-Orr Diagram or Tree (Chapter 4), we can still develop a set of data flow diagrams hierarchically. (These hierarchy methods apply to both TE-DFDs and D-DFDs.)

We get hierarchical structure in data flow diagrams by making simple diagrams that fit onto one sheet of paper using fair-sized symbols and without being crowded. The processes shown in that diagram may have more detailed processes going on inside, but these more detailed processes are shown on separate data flow diagrams on other pages. Thus no single data flow diagram goes beyond the limits of one piece of paper. Once we understand what is on one sheet, then by using the labels on the processes, we can find other sheets that describe the process in more detail.

Figures 3.13, 3.14, and 3.15 show the hierarchical representation of the data flow diagram we showed in Figure 3.12. Figure 3.13, which is called a **context diagram,** shows the whole system as one process, along with the immediate interfaces that connect it to other systems—that is, its environment. It is said to be a level 0 of the hierarchy. The first expansion of it is called level 1. Expansions of processes at level 1 are called level 2, and so on.

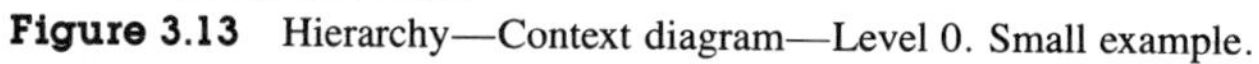
Figure 3.13 Hierarchy—Context diagram—Level 0. Small example.

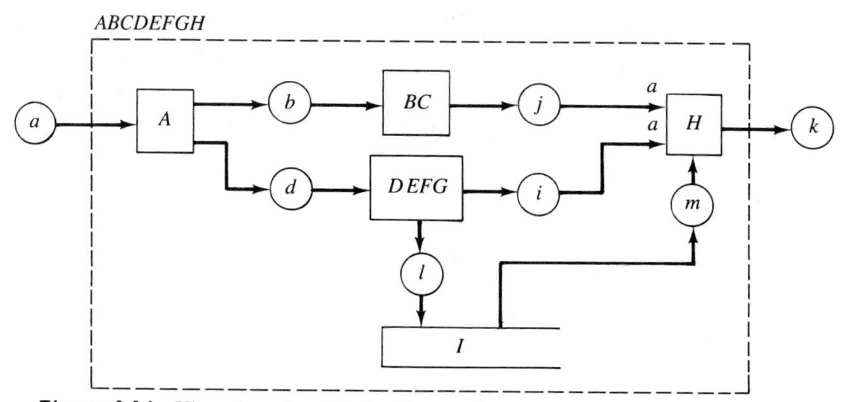

Figure 3.14 Hierarchy—Level 1. Small example.

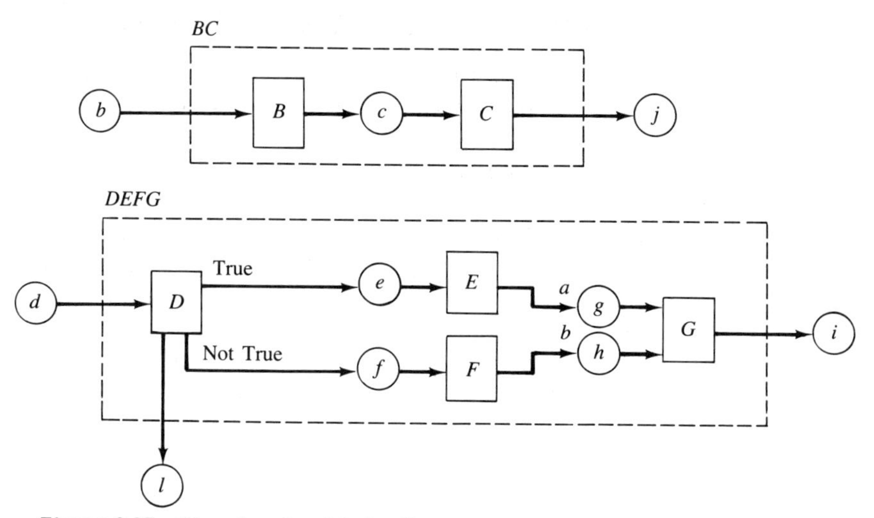

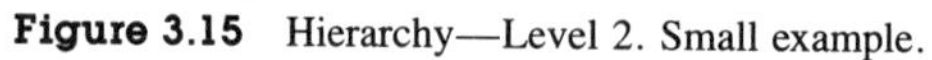
Figure 3.15 Hierarchy—Level 2. Small example.

Thus DFDs for complex systems can be developed as hierarchies of diagrams such that no single diagram is larger than one page. (Obviously, in this small example, each diagram is much less than a page.) We may stop this detailing when we get to processes for which we already have an available module or when we choose to convert the processes into Trees, where we will develop additional detail. (We will discuss Trees in Chapter 4.)

If we so choose, we can translate the whole DFD into a Tree. Translating from DFDs to Trees is essentially a mechanical process, but still an inconvenience that could be avoided if we were to develop the whole system from the very beginning as a Tree. But some clients feel more comfortable with data flow diagrams than they do with Trees, in which case we may translate the DFDs into Trees as soon as we get to a level of detail beyond which the client is interested.[2]

Figures 3.13, 3.14, and 3.15 illustrate the rules for expanding the data flow diagram into more detailed diagrams. Each diagram should be simple enough that it can be understood easily without concern for disturbing details. This usually means that it can fit on one 8½ by 11 page. Each process rectangle in each diagram is designated by a letter or sequence of letters and can be expanded into another diagram. The expanded diagram has the same designation as the process rectangle it represents. As we'll see in Figures 3.16 and 3.17, each process in the expansion can be designated by its parent's designation with a letter appended to distinguish it from the other processes in the expansion.

A diagram that is the expansion of a process has a boundary drawn corresponding to the boundary of that process rectangle in the higher-level diagram. Outside this boundary are just the immediate inputs and outputs of this process as shown in the higher diagram. Within this boundary is the expansion of the process. At each level of expansion we must account for all the inputs and outputs that appear at the higher level. A file should be shown outside a process if it is

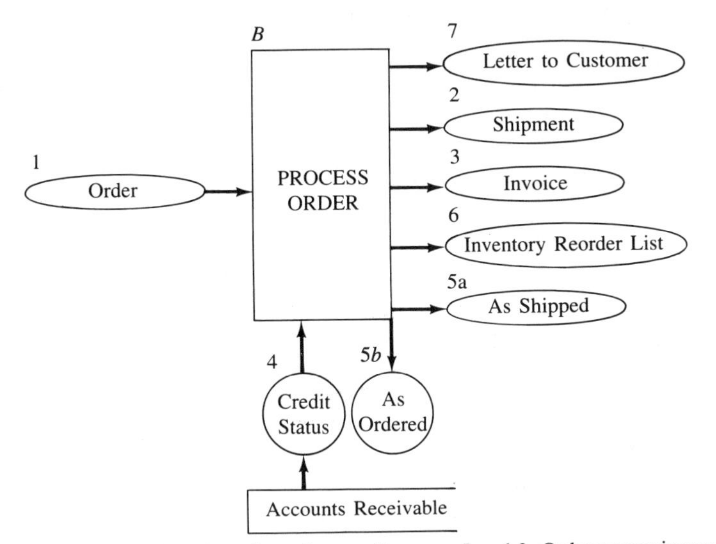

Figure 3.16 Hierarchy—Context diagram—Level 0. Order processing system.

[2]Note that hierarchy is shown in the data flow diagrams by a set of diagrams. In the Tree, the hierarchy is shown more naturally as the branching of the Tree.

to be used by another process. Otherwise it can be shown either inside or outside.[3]

Now let us return to our order entry example. Figure 3.16 shows the context diagram for Figure 3.5, and Figure 3.17 shows the expansion of the process "Process Order" shown in Figure 3.5. It is as if we were to open up the box "Process Order" in Figure 3.5 to look inside to see the details.

Logical DFD before Physical DFD

The purpose of an information processing system is to produce certain output information. Producing that output information requires that certain input information be transformed in well-defined ways. It is often useful to consider only the information and its transformations and defer until later all considerations of what media the information is recorded on, and who or what does the transformation.

A DFD may be a **logical DFD** or a **physical DFD.** In a logical DFD we focus only on the information and the changes that are made to it, not on how the information is recorded or what makes those changes. A physical DFD is created just by adding the description of the physical medium holding the information or the description of what or who does the processes.

If we use the convention of drawing the rectangles, as in Figure 3.18, we can show the designation in the top section, its description in the middle, and the physical device on the bottom (Gane and Sarson, 1979).

When we design a new system, we usually begin by doing a logical design, which considers just the information. The physical decisions are made only later. So in design the order is logical before physical.

But when we study an existing system, we do it in the opposite order—physical then logical. When we look at an existing system, the physical characteristics (what machines or who is doing what processing) can be easily recognized. It is easier to talk to clients in terms of the physical parts of the system with which they are familiar. Once we have agreed with the clients about what the system looks like physically, then we abstract just the information aspects of the system—that is, we remove from consideration all of the physical considerations.

Annotation

The processes, data flows, and files in the DFD can be described in more detail by supplementary **annotation** forms. Figure 3.19 shows such a form for process annotation filled in for the process "Order Entry" *(Ba)*. Figure 3.20 shows a form that can be used for either data flows or files; it has been filled in for the data flow "Pick-slip" (*B*2). Note that each annotation form is keyed to the DFD.

[3] In the literature on DeMarco Data Flow Diagrams (DeMarco, 1978), this hierarchical development is referred to as "partitioning." This usage is, however, completely contrary to the way the word *partitioning* is used in the mathematical literature—that is, to draw boundaries to separate parts into distinct sets. In fact the term "partitioning" would better describe precisely the reverse process, starting with a large diagram and partitioning its parts to develop a higher-level diagram. Thus we prefer the term "detailing" rather than "partitioning" for expanding the processes into more detailed processes.

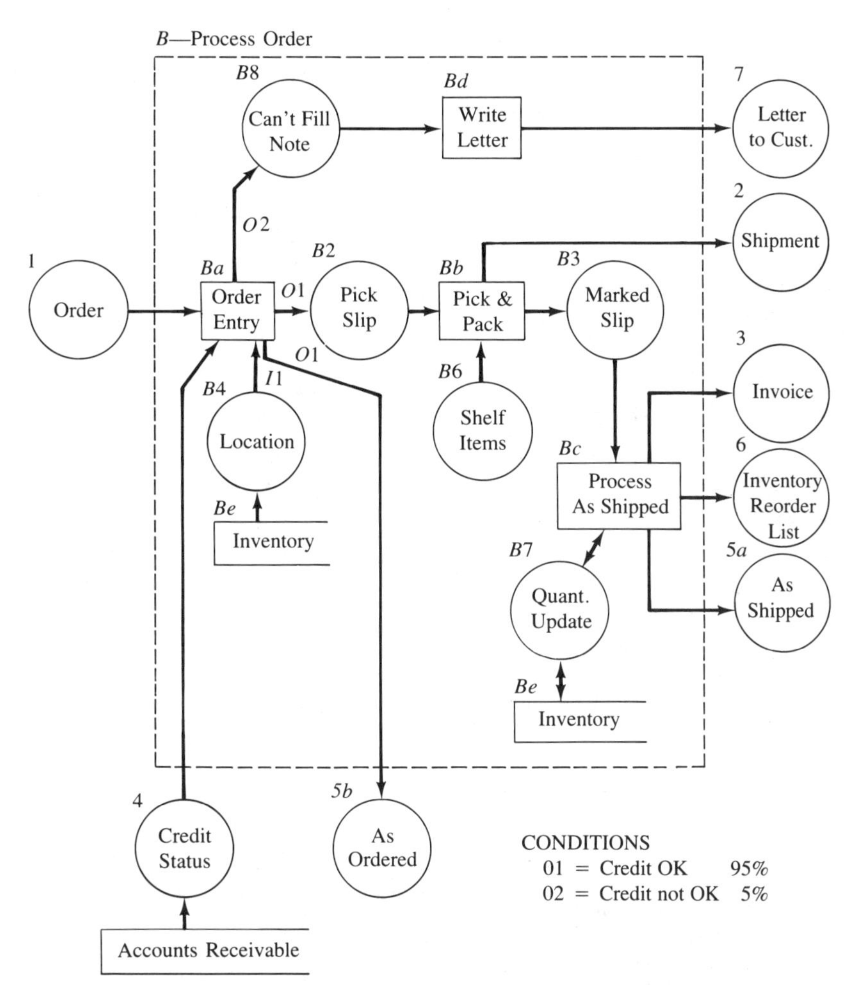

Figure 3.17 Hierarchy—Level 2. Order processing system.

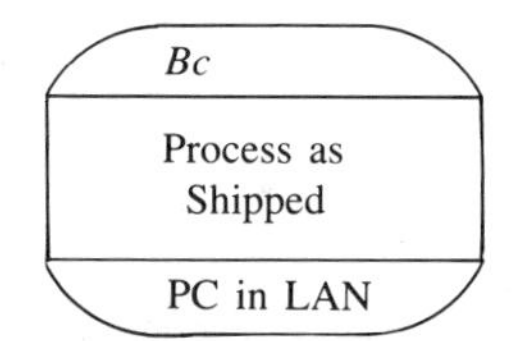

Figure 3.18 Alternative representation of "process rectangle."

The DFD acts as a table of contents to the annotations and shows the relations between the parts being annotated. These figures show just one example of what such forms might look like. There are many other possible ways of formatting this information. The forms may be maintained on a computer.

PROCESS ANNOTATION
Indexed by the Data Flow Diagram

WRITTEN BY: D. V. Steward DATE: June 16, 1987
DFD IDENT: Ba
TITLE: Order Entry
FUNCTION: To accept order into the order processing system

THIS APPEARS IN HIGHER LEVEL: B
PARTS OF THIS EXPANDED TO LOWER LEVEL: (none)
EXTERNAL TRIGGER: (none) FREQ. (na)

NARRATIVE/PLAY SCRIPT	ACTOR	CONDITION	FREQ.
Enter order on terminal	Clerk		
Ck A/R file for active customer	Prog		
Ck Inven file for available item	Prog		
Issue ''Can't fill note''	Clerk	Cust not in file	1%
		or items not avail	4%
Enter order in Open Order file	Prog		
Issue pick-slip	Prog		

EXCEPTIONS Computer down--Clerk performs actions manually

PERFORMANCE:
BY: Order Entry Clerk & Order Entry program
FREQUENCY/VOLUME: 350 orders/day/clerk
DURATION/RESPONSE: 60 sec/order
RESOURCES: Order entry terminal
FACILITIES: Order entry program

WHO HAS AUTHORITY TO MAKE CHANGES?: J. W. Smith

Figure 3.19 Process annotation for order entry.

Process Annotation

The process annotation (Figure 3.19) has a DFD identification (Ba) that ties it to a corresponding process in the DFD. It has a title ("Order Entry") and a brief statement of its function ("To accept order into the order processing system"). If there were an external trigger, it would appear here and in the data flow diagram. The process annotation would also show how often it fires. In this example Ba is triggered by input, which is shown on the DFD. The input and output data flows and their conditions and triggers are shown on the DFD diagram.

DATA FLOW/FILE ANNOTATION
Indexed by the Data Flow Diagram

WRITTEN BY: DV Steward DATE: June 16, 1987
Check DATA FLOW X or FILE ____
DFD IDENT: B2
TITLE: Pick-slip
FUNCTION: To tell pickers what items to pick for order

DATA ELEMENT DICTIONARY

NAME	CONTAINS	CONT.	APPEARS	FORMAT	RANGE/CODE	UNITS	MNEMONIC	ALIAS
Order Number		-	once	I6				
Cust. Number		-	once	I8				
Cust. Name		+	once	A25				
Cust. Name	Last_Name	-						
Cust. Name	First_Name	-						
Cust. Name	Middle_Init	-						
Last Name		-	once	A10				
First Name		-	once	A14				
Middle Init		-	once	A1				
Cust. Address		+	once	A40				
Cust. Address	Street Addr	-						
Cust. Address	City Addr	-						
Street Addr		-	once	A20				
City Addr		-	once	A20				
Item #		-	no. items	I6				
No. Ordered		-	no. items	I6				
Location		-	no. items	A3				
		-						

MEDIA: NCR 3-part paper ENCODING: alphanumeric
NO. RECORDS: 1000/day LOCATION: not stored
ORGANIZATION/SORT/KEY: not sorted or retrieved
RETENTION PERIOD: until used to ship order
BACKUP: Open Order File
ACCESS/SECURITY: (none)

WHO HAS AUTHORITY TO MAKE CHANGES? J. W. Smith

Figure 3.20 Data flow or file annotation.

A description of the process is written in narrative form. The narrative can be written like the script for a play, showing who or what does it (actor), the condition under which it is done, and how often that occurs (frequency). Any exceptions are indicated, such as in this case what happens if the computer is down. Also noted is physical information, such as who or what performs the process (this appears only if this is a physical description), the volume of orders the systems does or must be able to handle, or the response time (in this case the time to enter an order). The annotation should list necessary resources or facilities, such as the use of a computer terminal and a specific computer program, special forms, special office facilities, and so on. It is also important to know who has the authority to approve of changes to the process.

Data Flow or File Annotation

Both data flows and files are made up of records. These records are described by a **data element dictionary,** which shows what data elements make up the record, their structure and format. The structure describes how often a data element appears, and how it might be made up of other data elements contained within it. This data element dictionary appears as part of a data flow or file annotation. This same annotation form can be used for both data flows and files by checking the appropriate block. As in the process annotation, a DFD identification ties the annotation to the corresponding data flow or file on the DFD. The title and function give it a label and briefly describe its purpose.

A line of the data element dictionary table (Figure 3.20) either shows another data element contained within it, or it indicates its format and how often that data element appears in its parent. In Figure 3.20 the first line of "Cust. Name" shows that it appears once in the record. The + under "CONT" indicates it contains smaller data elements. A line appears below for each contained data element. A − under "CONT" indicates it contains no smaller data elements. "Item #" APPEARS "no. items" times. The parent of a data element is another data element that contains the first element; thus "Cust. Name" is the parent of "Last Name." If a data element is not contained in another, its parent is the whole record.

We can see that "Cust. Name" is made up of "Last Name," "First Name," and "Middle Init." From the "APPEARS" column we can see that "Cust. Name" appears once in the record and "Last Name" appears once in its parent "Cust. Name." "Item #," "No. Ordered," and "Location" appear for each item that occurs on the order. The format here is shown as *I* for integer, or *A* for alphanumeric, followed by the number of character positions (for example, six alphabetic characters would be shown as *A*6, or two character byte positions representing an integer would be shown as *I*2). In Chapter 4 we will see how such a data structure can also be represented by a data tree.

Other items of information may occur on such forms. If a variable is continuous, such as a temperature or cost, it may have a range of valid values, as 0 to 100. If the variable is discrete, such as a code for a credit status, the codes themselves would be used to distinguish the different values, such as 0 for bad credit, 1 for poor credit, and 2 for good credit. This could be shown in a field called "RANGE/CODE." The variable may have units, such as degrees Celsius or dollars, which are particularly important in scientific and engineering programs. These designations would be indicated in a column entitled "UNITS." There could be a field to show the "MNEMONICS" used for this data element in the programs and, if the data element has other names in other places, these alternative names can be shown in an "ALIAS" field.

"MEDIA" refers to the material the data are recorded on (for example, 5¼-inch diskette, NCR three-part paper, and so on) and "ENCODING" refers to how it is represented (ASCII, EBCDIC, alphanumeric, and so on). The NO. RECORDS and LOCATION are shown. The file may be organized by a SORT on a KEY, such as ascending by employee number. There is a place to record the RETENTION PERIOD (for example, one month, until end of tax year, and so on), means of BACKUP (backed up on tape each evening) and ACCESS/SECURITY (access to payroll clerks only/removed from vault only for monthly processing). It may also be shown who has the AUTHORITY to change the data flow or file design.

During the early design phase we may work with only a logical description of what is done to what data. We then do not show the physical aspects of whether it is done manually, which computer is being used, or whether the data are recorded on paper or diskette. Thus some of this information on the annotation forms may be left blank during the logical design of the system and filled in later once decisions have been made about how it will be implemented physically.

Matrices and Data Bases

Data flow diagrams are useful tools for tracing the movement of information through a system. Often, though, we may want to see the same information in the form of a matrix. As we'll see in this section, matrices relate data flows with processes or data elements with data flows, processes, and files. These relations, plus other combinations, can be made part of a computer data base, giving designer and programmers computer access to this information. For example, if we wish to change the format of a data element, we may wish to retrieve all the processes and files where this particular data element is used.

Data Flow Matrix

The Two-Entity Data Flow Diagram that we've discussed thus far can also be represented as a **matrix,** known as a **data flow matrix,** or **DFM.** A data flow matrix can be shown on paper or stored in a computer. (Incidentally, this matrix is just an array of rows and columns. It does not have any mathematical properties defined for it, such as sums or products.) In the matrix the rows represent processes

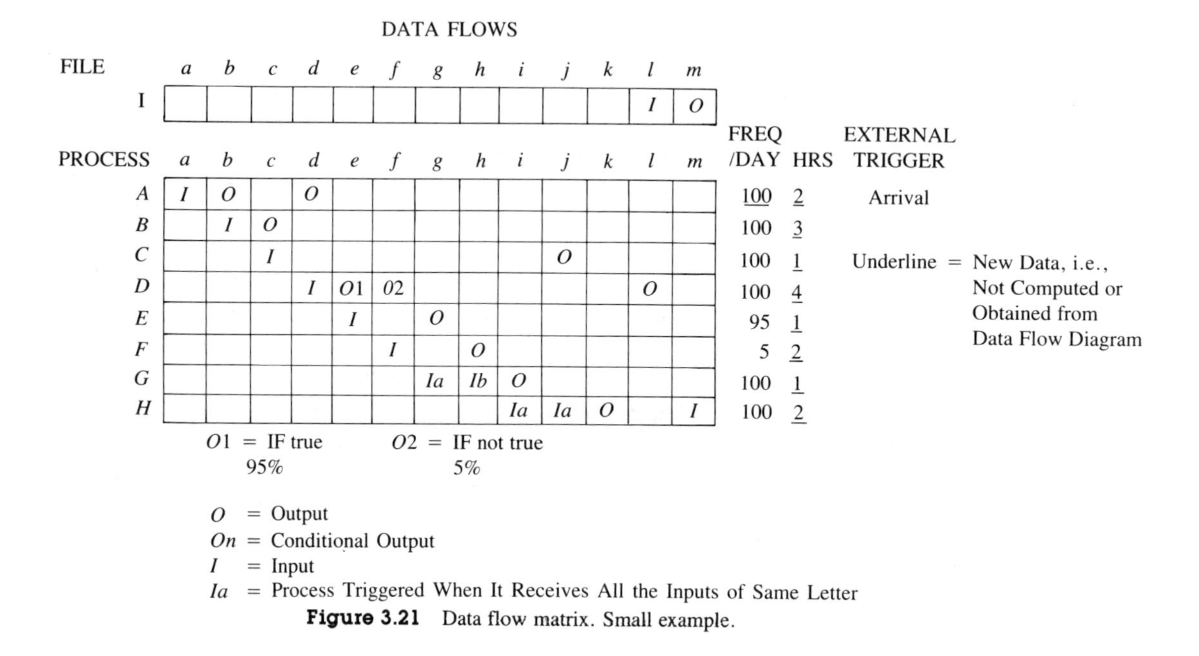

DATA FLOWS

FILE	a	b	c	d	e	f	g	h	i	j	k	l	m
I												*I*	*O*

PROCESS	a	b	c	d	e	f	g	h	i	j	k	l	m	FREQ /DAY	HRS	EXTERNAL TRIGGER
A	*I*	*O*		*O*										100	2	Arrival
B		*I*	*O*											100	3	
C			*I*							*O*				100	1	
D				*I*	*O*1	*0*2						*O*		100	4	
E					*I*		*O*							95	1	
F						*I*		*O*						5	2	
G							*Ia*	*Ib*	*O*					100	1	
H									*Ia*	*Ia*	*O*		*I*	100	2	

Underline = New Data, i.e., Not Computed or Obtained from Data Flow Diagram

*O*1 = IF true 95% *0*2 = IF not true 5%

O = Output
On = Conditional Output
I = Input
Ia = Process Triggered When It Receives All the Inputs of Same Letter

Figure 3.21 Data flow matrix. Small example.

or files and the columns represent data flows. The matrix is divided into two sections, the process rows and the file rows. *O*'s and *I*'s are used to show where a data flow is the output or input of a process or file. Figure 3.21 shows the data flow matrix corresponding to the TE-DFD for the small example shown in Figure 3.12. The *I* in row *A* column *a* indicates that data flow *a* is an input to process *A*. Similarly the *O*'s in the same row in columns *b* and *d* say that data flows *b* and *d* are outputs of this process. (The data flow matrix is similar to a technique developed by Borge Langefors in Sweden [Langefors and Sundgren, 1976].)

A number after an *O* in a matrix cell means that there are conditions that occur within the process that determine whether this data flow is produced. Since there is not room in the cell of the matrix to write the conditions, one looks outside the matrix to see what the condition is. For example, the output of process *D* of data flow *e* shows an *O*1. Below the matrix we see that this output occurs under the condition "IF true." Similarly data flow *f* is produced by this process under the condition "IF not true." Numbers on the *O*'s thus indicate conditions.

Letters after the *I*'s indicate how the data flows trigger the process. For the process *G*, the *I*'s in columns *g* and *h* are followed by different letters. This means that process *G* is triggered when it gets either of the data flows *g* or *h*. But for process *H* both the *I*'s in columns *i* and *j* are followed by the same letter. Thus *H* is triggered only after it gets both data flows *i* and *j*. This corresponds in the DFD to how letters are used on the arcs entering processes to show the data flow triggers. Thus, if there are letters after the *I*'s, then the process is triggered when all the inputs of any one letter are available. If all the inputs are *I*'s without a letter, then the process is triggered when all the inputs are available. External triggers are noted next to the row for the process triggered. For example, process *A* shows "Arrival" as its external trigger. Remember that processes are not triggered by files, since files only hold information until it is called for by a process. These conventions in the data flow matrix (DFM) correspond to the conventions we discussed before for the TE-DFDs.

From the data flow matrix we can quickly see what data flows are used by what processes and vice versa.

Figure 3.22 shows the data element matrix for our order processing example. Here we not only have *I*'s and *O*'s with numbers and letters, but *U*'s and little *a*'s and *s*'s. A *U* indicates Update, which is the same as both an *I* and an *O*. The *s* with the *I* or *O* indicates that each time the process is triggered, a single record is read or written. An *a* would indicate that the process reads or writes all the records in the file (as in a sort), and thus the number of records processed would be the full number of records in the file.

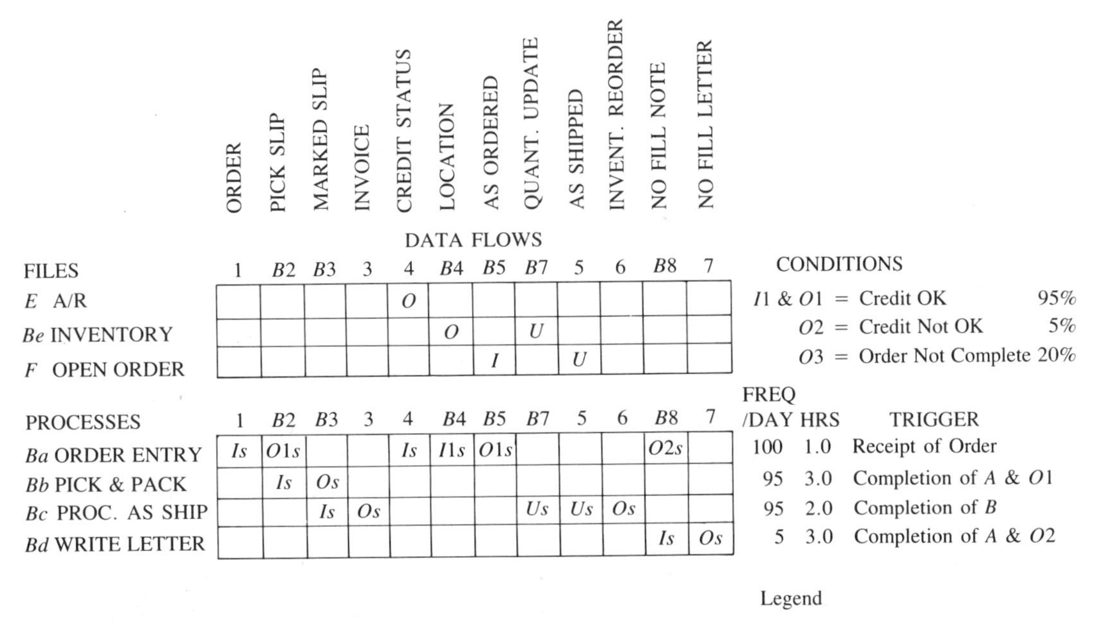

DATA FLOWS

FILES	ORDER 1	PICK SLIP *B2*	MARKED SLIP *B3*	INVOICE 3	CREDIT STATUS 4	LOCATION *B4*	AS ORDERED *B5*	QUANT. UPDATE *B7*	AS SHIPPED 5	INVENT. REORDER 6	NO FILL NOTE *B8*	NO FILL LETTER 7
E A/R					*O*							
Be INVENTORY						*O*		*U*				
F OPEN ORDER							*I*		*U*			

PROCESSES	1	*B2*	*B3*	3	4	*B4*	*B5*	*B7*	5	6	*B8*	7	FREQ /DAY	HRS	TRIGGER
Ba ORDER ENTRY	*Is*	*O1s*			*Is*	*I1s*	*O1s*				*O2s*		100	1.0	Receipt of Order
Bb PICK & PACK		*Is*	*Os*										95	3.0	Completion of *A* & *O*1
Bc PROC. AS SHIP			*Is*	*Os*				*Us*	*Us*	*Os*			95	2.0	Completion of *B*
Bd WRITE LETTER											*Is*	*Os*	5	3.0	Completion of *A* & *O*2

CONDITIONS

*I*1 & *O*1 =	Credit OK	95%
*O*2 =	Credit Not OK	5%
*O*3 =	Order Not Complete	20%

Legend

s = Single Record Processed
a = All Records Processed
U = Update, i.e., Both *I* & *O*

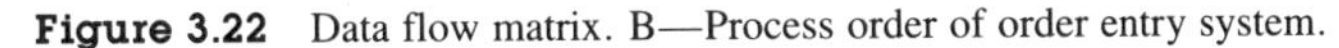

Figure 3.22 Data flow matrix. B—Process order of order entry system.

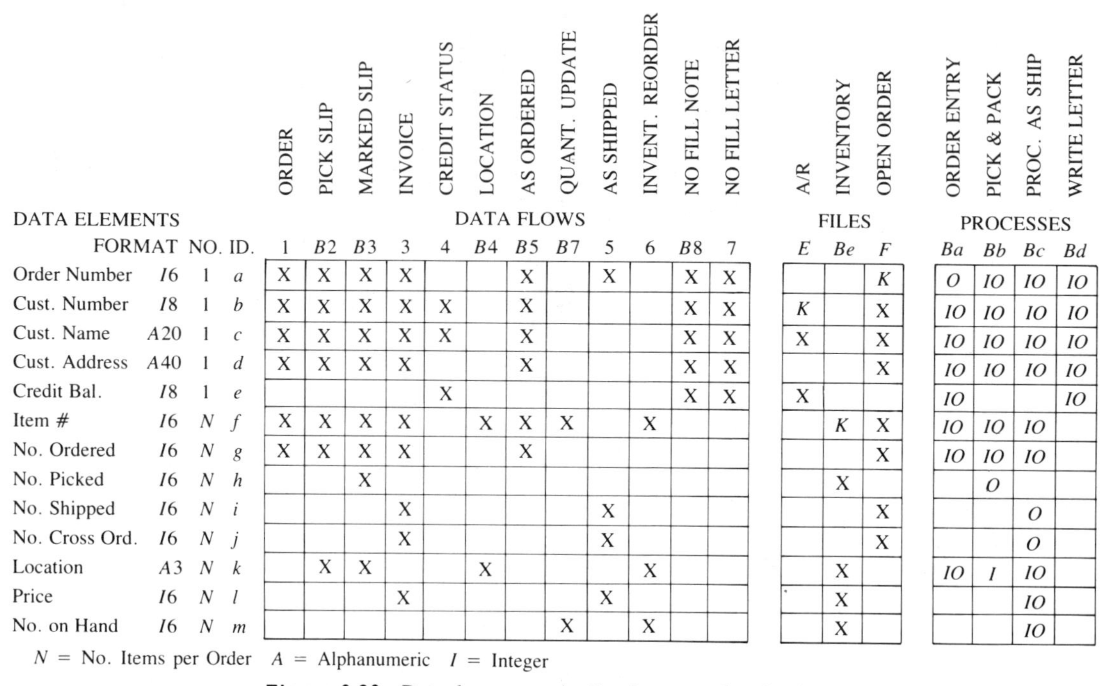

DATA ELEMENTS	FORMAT	NO.	ID.	ORDER 1	PICK SLIP B2	MARKED SLIP B3	INVOICE 3	CREDIT STATUS 4	LOCATION B4	AS ORDERED B5	QUANT. UPDATE B7	AS SHIPPED 5	INVENT. REORDER 6	NO FILL NOTE B8	NO FILL LETTER 7	A/R E	INVENTORY Be	OPEN ORDER F	ORDER ENTRY Ba	PICK & PACK Bb	PROC. AS SHIP Bc	WRITE LETTER Bd
				DATA FLOWS												FILES			PROCESSES			
Order Number	*I*6	1	*a*	X	X	X	X			X		X		X	X			*K*	*O*	*IO*	*IO*	*IO*
Cust. Number	*I*8	1	*b*	X	X	X	X	X		X				X	X	*K*		X	*IO*	*IO*	*IO*	*IO*
Cust. Name	*A*20	1	*c*	X	X	X	X	X		X				X	X	X		X	*IO*	*IO*	*IO*	*IO*
Cust. Address	*A*40	1	*d*	X	X	X	X			X				X	X			X	*IO*	*IO*	*IO*	*IO*
Credit Bal.	*I*8	1	*e*					X						X	X	X			*IO*			*IO*
Item #	*I*6	*N*	*f*	X	X	X	X		X	X	X		X				*K*	X	*IO*	*IO*	*IO*	
No. Ordered	*I*6	*N*	*g*	X	X	X	X			X								X	*IO*	*IO*	*IO*	
No. Picked	*I*6	*N*	*h*			X											X			*O*		
No. Shipped	*I*6	*N*	*i*				X					X						X			*O*	
No. Cross Ord.	*I*6	*N*	*j*				X					X						X			*O*	
Location	*A*3	*N*	*k*		X	X			X				X				X		*IO*	*I*	*IO*	
Price	*I*6	*N*	*l*				X					X					X				*IO*	
No. on Hand	*I*6	*N*	*m*								X		X				X				*IO*	

N = No. Items per Order *A* = Alphanumeric *I* = Integer

Figure 3.23 Data element matrix. B—Process order of order entry system.

Data Element Matrix

We can also use a matrix to show what data elements occur in each data flow and each file. This matrix is called a **data element matrix, DEM.** The data element matrix shows part of the information we would expect to see in a data element dictionary.

Let the columns represent data flows and files, and the rows represent data elements. A mark, for example X, then shows where a data flow or file contains a specific data element. For each data element, the matrix shows the element's format (*I* for integer, *A* for alphanumeric), its size in characters (bytes), the number of times it occurs in the record, and its identification mnemonic. Figure 3.23 is the data element matrix for the order entry example. Notice that for the files we can also show which data element is used as a key for retrieving or sorting the records by using a *K* for that element instead of an *X*.

Another set of columns shows the data elements that are used in each process. We might think that we can tell what data elements are used as input and output of each process by noting which data flows into and out of the process and the data elements in those data flows. However, it is possible that a data flow is used by more than one process and thus a process might not use all the data elements in the data flow.

DATA ELEMENTS	FORMAT	NO.	ID.	1 ORDER	B2 PICK SLIP	B3 MARKED SLIP	3 INVOICE	4 CREDIT STATUS	B4 LOCATION	B5 AS ORDERED	B7 QUANT. UPDATE	5 AS SHIPPED	6 INVENT. REORDER	B8 CAN'T FILL NOTE	7 CAN'T FILL LETTER	E A/R	Be INVENTORY	F OPEN ORDER	Ba ORDER ENTRY	Bb PICK & PACK	Bc PROC. AS SHIP	Bd WRITE LETTER
				DATA FLOWS												FILES			PROCESSES			
Order Number	I6	1	a	X	X	X	X			X		X		X	X			K	O	IO	IO	IO
Cust. Number	I8	1	b	X	X	X	X	X		X				X	X	K		X	IO	IO	IO	IO
Cust. Name	A20	1	c	X	X	X	X	X		X				X	X	X		X	IO	IO	IO	IO
Cust. Address	A40	1	d	X	X	X	X			X				X	X			X	IO	IO	IO	IO
Credit Bal.	I8	1	e					X						X	X	X			IO			IO
Item #	I6	N	f	X	X	X	X		X	X	X		X				K	X	IO	IO	IO	
No. Ordered	I6	N	g	X	X	X	X			X								X	IO	IO	IO	
No. Picked	I6	N	h			X											X			O		
No. Shipped	I6	N	i				X					X						X			O	
No. Cross Ord.	I6	N	j				X					X						X			O	
Location	A3	N	k		X	X			X				X				X		IO	I	IO	
Price	I6	N	l				X					X					X				IO	
No. on Hand	I6	N	m								X		X				X				IO	

No. means number of times it appears in record

N = No. Items per Order A = Alphanumeric I = Integer

DATA ELEMENT DICTIONARY/MATRIX

DATA FLOW MATRIX

FILES	1	B2	B3	3	4	B4	B5	B7	5	6	B8	7
E A/R					O							
Be INVENTORY						O		U				
F OPEN ORDER							I		U			

CONDITIONS

$I1$ & $O1$ = Credit OK 95%

$O2$ = Credit Not OK 5%

$O3$ = Order Not Complete 20%

PROCESSES	1	B2	B3	3	4	B4	B5	B7	5	6	B8	7	FREQ /DAY	HRS	EXTERNAL TRIGGER
Ba ORDER ENTRY	Is	O1s			Is	I1s	O1s				O2s		100	1.0	Receipt of Order
Bb PICK & PACK		Is	Os										95	3.0	
Bc PROC. AS SHIP			Is	Os				Us	Us	Os			95	2.0	
Bd WRITE LETTER											Is	Os	5	3.0	

Legend

Underline = Given Data Not Computed

s = Single Record Processed

a = All Records Processed

U = Update, i.e., Both I & O

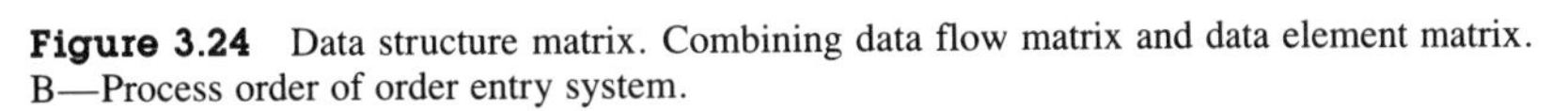

Figure 3.24 Data structure matrix. Combining data flow matrix and data element matrix. B—Process order of order entry system.

Data Structure Matrix

The two matrices discussed thus far—the data flow matrix (Figure 3.22) and the data element matrix (Figure 3.23)—can be put together (Figure 3.24) by making the data flows in the columns of the two matrices correspond. The result is called a **data structure matrix.**

The data structure matrix pulls together a great deal of information about the system so that it can be analyzed for consistency and completeness. The matrix can also be used to trace the effect of changes and who should be notified. For example, if a data element is changed, the matrix can be used to show all the process modules and file structures that might have to be changed. If we add to the data base the people who are responsible for these modules and files, we can tell who should be consulted or notified about the change.

The matrix can also be used to make estimates of required storage volumes and processing volumes and times. We will discuss these calculations later in this chapter.

Putting the Matrix in a Computer Data Base

The data structure matrix can be stored in a computer data base so that it can be kept up to date in one place. The people working on the project can thus refer to the matrix through their terminals or personal computers. The data structure matrix can be implemented in any data base management system that has the following properties:

1. A name that appears in a record in one file can be used as a key to retrieve a record in another file.
2. There can be more than one key per record.
3. Files can be indexed on more than one key and retrieved as though the records were contiguous in this key.

DBASE II or III or most any relational data base could be used to store these matrices.

Information associated with a row or column (for example, data element, process, data flow, or file) is represented in the data base by a record with one key. Each data flow, for example, would have a record in which the name of the data flow is the key. Data flows appear as the columns of both the data flow matrix and the data element matrix, but they would need to be described by only one file. The information that appears in a cell of the matrix is stored as a record with two keys—one for its row and the other for its column. One can retrieve all the occupied cells in any row using the index on the row key, or in any column using the index on the column key.

Each matrix shows a relation between two objects. For example, the data flow matrix shows the relation between processes or files and data flows. While the relations between these objects can be shown as a matrix, just the structure of the relations between objects can be described by an undirected graph with vertices and edges (undirected lines) between vertices. Each vertex is an object, such as a data flow, process, file, or data element.

For example, the structure of the set of data structure matrices we have discussed in this chapter can be portrayed in Figure 3.25. Each edge in this graph

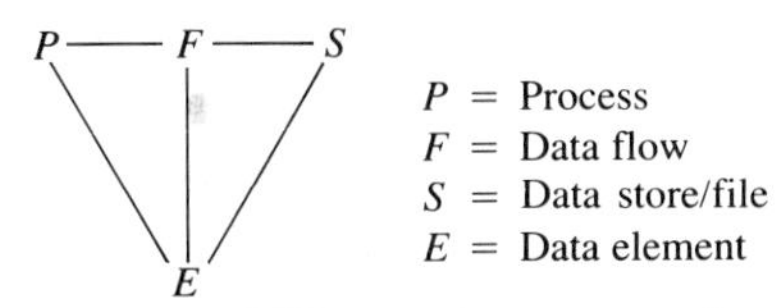

Figure 3.25 Data base structure for data structure matrix.

corresponds to a matrix, and the matrix is represented in the data base by a file of records with the two keys corresponding to the ends of the edge in the graph. For example, the edge between P and F is the part of the data flow matrix relating process to data flows. The edge between F and S is the part of this matrix relating data stores to data flows. And the edge between F and E is the matrix for the data element dictionary.

The data element matrix shown both in Figures 3.23 and 3.24 is rather simplistic. It assumes that all data elements occur at the same level with no hierarchical structure. Let us consider how the same data base management system can be used to represent a hierarchical data structure.

Consider a file that contains data on a doctor's patients. There is a record for each patient showing Name, Social Security Number, and Service. The Name is composed of First Name, Middle Initial, and Last Name. For a patient there can be any number of Service subrecords. Each Service subrecord is composed of a Diagnosis, Treatment, Recall, and Charge. The Recall is composed either of a Date when the patient is to be recalled or the word "COMPLETE." DeMarco has a method that would show this structure as in Figure 3.26.

```
Patient data = { Name + Soc Sec + {Service} }
Name = F_Name + M_Name + L_Name
Service = Diagnosis + Treatment + Recall + Charge
Recall = [ Date | ``COMPLETE'' ]
   where:
      + = concatenate
      | = option
      {} = repeated
```

Figure 3.26 DeMarco representation of data structure.

This example can be represented in data base form as Figure 3.27. The interpretation of these fields is the same as for Figure 3.20. The same example will be used in Chapter 4 to show how data structures can be represented by trees.

Now we can extend our data base by adding these hierarchical data structures. A process, data flow, or file/store can refer either to a data structure containing other data structures, or directly to a primitive data element (see Figure 3.28).

Making Calculations on the Matrix

By adding some numbers to the data structure matrix we can compute a set of other numbers that may be useful in analyzing the system. Clearly we can add up the number of bytes for all the data elements in a data flow or file to determine

NAME	CONTAINS	CONT.	APPEARS	FORMAT
Patient data		+	repeated	
Patient data	Name	+	once	
Patient data	Soc Sec	–	once	9N
Patient data	Service	+	repeated	
Name		+	once	
Name	F Name	–	once	15A
Name	M Name	–	optional	2A
Name	L Name	–	once	15A
Service	Diagnosis	–	once	40A
Service	Treatment	–	once	40A
Service	Recall	+	once	
Recall	Date	+	optional	
Date	MM	–	once	2N
Date	literal	–	once	``/''
Date	DD	–	once	2N
Date	literal	–	once	``/''
Date	YY	–	once	2N
Recall	literal	–	optional	``COMPLETE''
Service	Charge	–	once	4.2N

Figure 3.27 Data structure for data hierarchy.

the size of the records involved. If we are also given the number of records to expect in a file, we can estimate the amount of storage required. As we will see in a moment, if we are given how often per unit of time certain processes are triggered, we can compute how often the other processes and the data flows occur.

In data processing often how long it takes to run a process is determined by how long it takes to do the input and output. Thus if we know how many records are processed in a given time and how large the records are, we can sometimes make good estimates for the times to run the processes. If we know the times to run each process and thus how long a transaction must wait in the queue for each process, we can use critical path methods (see Chapter 7) to compute the time a transaction takes to get through the system. For example, in our order entry system we could estimate the time for an order to be processed.

Figure 3.29 is a worksheet for the data structure matrix in Figure 3.24. It is used to compute the sizes of records and files and the frequency of use of each

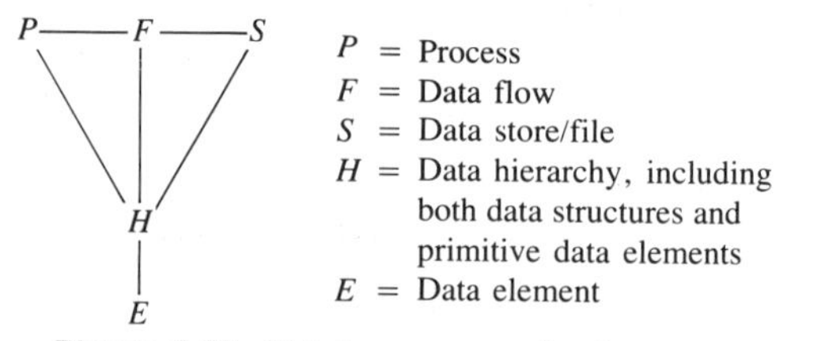

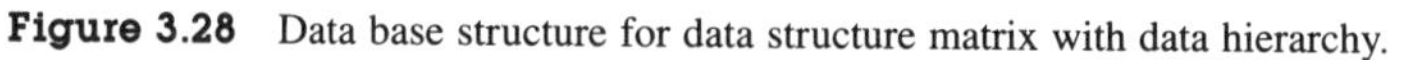
Figure 3.28 Data base structure for data structure matrix with data hierarchy.

	DATA FLOWS												FILES			Obtained from
	1	*B*2	*B*3	3	4	*B*4	*B*5	*B*7	5	6	*B*8	7	*E*	*Be*	*F*	
CHAR/RECORD for $N = 20$	314	374	494	674	36	180	314	240	366	300	82	82	36	420	554	*a* DEM
No. of Thousands of Records in File													2	5	.5	*b* Given
No. of Thousands of Characters in File													72	2100	277	$c = a \times b$

	DATA FLOWS												
	1	*B*2	*B*3	3	4	*B*4	*B*5	*B*7	5	6	*B*8	7	
OCCURS/DAY	100	95	95	95	100	95	95	95	95	95	5	5	*d* DFM
RECORDS/OCCURRENCE	1	1	1	1	1	20	1	20	1	0.1	1	1	*e* Given
RECORDS/DAY	100	95	95	95	100	1.9	95	1.9	95	9.5	5	5	$f = d \times e$
						K		*K*					

Figure 3.29 Worksheet for Figure 3.24.

data flow and each process. The CHAR/RECORD row shows the number of characters in each record of a data flow or file. This is computed from the data structure matrix by adding up the number of bytes in each data element that appears in that record multiplied by the number of times it appears. The number of bytes is shown as the number after the letter in the FORMAT. For example, Order Number has 6 bytes. The number of times it occurs appears in the NO. column. Going down the column for data flow 1, called ORDER, we see that it contains data elements of sizes 6, 8, 20, and 40, which are multiplied by one, and data elements of sizes 6 and 6, which are multiplied by N. With N given as 20 (the average number of items ordered per order), there is a total of 314 bytes per average order record.

Additional information has been added to Figure 3.24 from which we can determine how frequently the processes and data flows are used. For each process that is triggered externally, we have shown what triggers the process and how many times a day (or hour or second, and so on) it is triggered. For example, Figure 3.24 shows that process Ba ORDER ENTRY is triggered by "Receipt of Order." Under the FREQ/DAY column we see that it occurs 100 times a day. The 100 is underlined here to show that this is information that is given rather than information that can be calculated from other data in the diagram. For each process the matrix shows how long it takes from when a transaction enters the queue for that process until it is completed. For example, the matrix shows that the process Ba takes one hour. These duration numbers are also given data and are thus underlined.

The number of times per day (or hour or second, and so on) that a process runs or a data flow occurs can be computed from the external triggers, or the data flows that trigger it, and how many times those data flows occur. To illustrate these calculations we go back to Figure 3.21, which has been especially set up to show several situations that do not arise in the example in Figure 3.24. If we know that a process, such as process D, runs 100 times a day and that a particular data flow *(e)* is produced on a condition that occurs 95% of the time, we can conclude that the data flow *(e)* occurs 100 × .95 = 95 times a day. If that data flow is the only trigger of another process *(E)*, we can conclude that each time the data flow occurs, the process is run (for example, process *E* runs 95 times a day.) If we look at process *G*, we note that it runs when it gets either input *g* or input *h*. It gets *g* 95 times and *h* 5 times, so *g* runs 95 + 5 = 100 times. Looking at process *H*, we note that for it to be triggered it must get both inputs *i* and *j*. It gets *i* 100 times and *j* 100 times. Thus it gets both *i* and *j* 100 times, so *H* will run 100 times per day. We put the number of times a day the process runs in the FREQ/DAY column in the data structure matrix. We put the number of times a day a data flow occurs in the worksheet in the OCCURS/DAY row. When the same data flow is written and read, we count it as occurring only once.

In Figure 3.29 we have also shown a row for RECORDS/OCCURRENCE. This must be given because it cannot be computed from other numbers in the figure. But by reason we can see what this number might be. For every order received by process Ba ORDER ENTRY, the process will retrieve a record from the inventory file for each item ordered. Thus, each time this data flow occurs, on the average twenty records will be passed. Multiplying the number of OCCURS/DAY by the RECORDS/OCCURRENCE, we get the RECORDS/DAY. A *K* below the entry indicates the value is in thousands. The "Obtain from" column shows how to get the numbers in each row—whether the number is derived from the DEM (data element matrix) or the DFM (data flow matrix), and whether the number is given or computed from other rows.

The timing of transactions flowing through the system can be determined by critical path calculations using the duration in the HRS column. Critical path will be discussed in Chapter 7.

If the matrix is maintained in a data base, these numbers can be computed by the computer.

The Lano Matrix

The information that we have shown in a data flow matrix can also be shown in the form of a Lano N-squared matrix (Lano, 1977), which we may simply call a **Lano Matrix**. Using the following figure, we can show the relation between these two approaches. Figure 3.30 shows the data flow diagram, and Figure 3.31

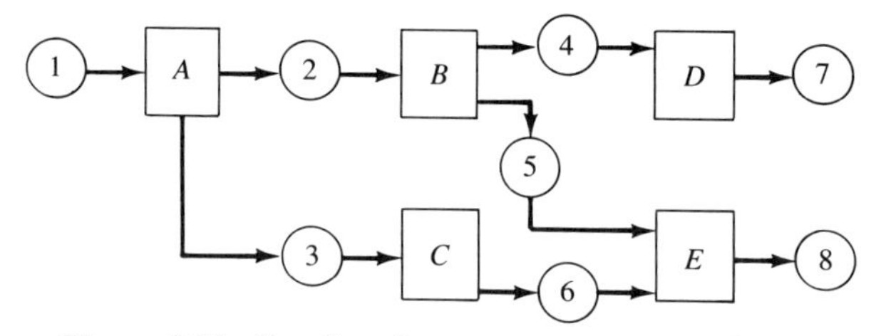

Figure 3.30 Data flow diagram.

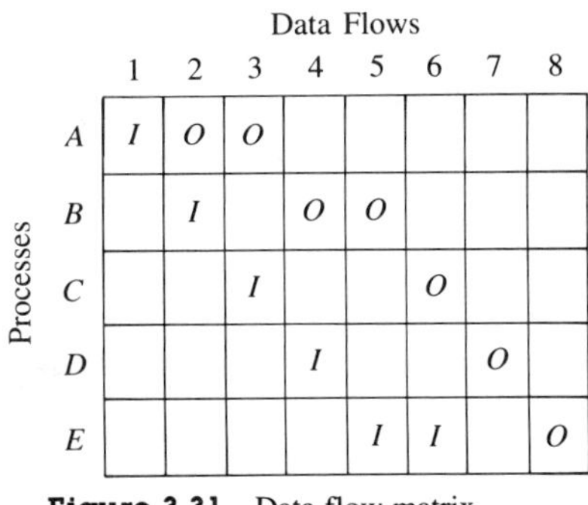

Figure 3.31 Data flow matrix.

shows its data flow matrix. Figure 3.32 is the Lano Matrix for the same data flow diagram.

While the rows and columns in the data flow matrix are processes and data flows, in the Lano Matrix both the rows and columns are processes. Lano puts the data flows that link the processes into the cells of the matrix. He then adds a row labeled IN to show flows coming into the system, and a column labeled OUT to show flows going out of the system.

Note that a data flow matrix is like a Lano Matrix turned inside out. While in the Lano Matrix the data flows are inside the cells, in the data flow matrix the data flows are exposed on the outside as columns. This allows us to write information about the data flow on the outside of the matrix rather than squeezing it into the cells. It also makes it possible with the data flow matrix to match the columns representing the data flows with the columns representing these same data flows in the data element matrix.

Either matrix can be equally well represented in a data base. A data base can tie information together and put information in the cells in ways that cannot easily be represented on paper.

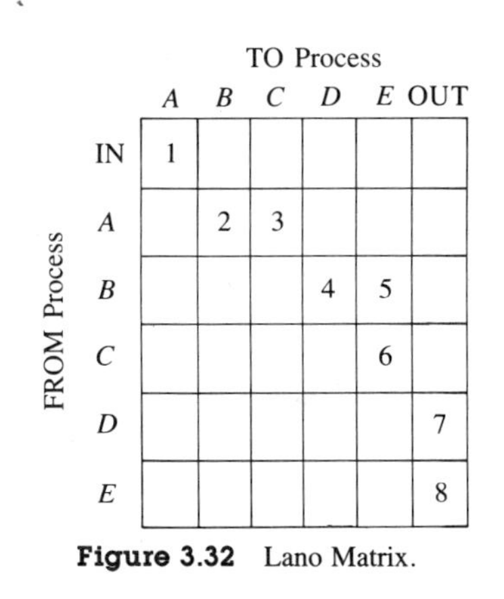

Figure 3.32 Lano Matrix.

It is easier to use the data flow matrix when we show the data flow matrix and the data element matrix together on the same two-dimensional sheet of paper. The Lano Matrix is sometimes used to develop an initial perception of the interactions between parts of the system without yet being concerned with the data flows that cause these interactions.

Summary

Clients and developers need an effective graphic method by which they can describe the system they are creating. This method must be one that a client who is not prepared to become an expert in information systems can learn quickly. Data flow diagrams are such a method.

Data flow diagrams show how data flows, the processes that transform them, and how files are connected. Symbols are used for data flows, files, and processes; labels in these symbols refer to descriptive information in attached documents called annotations.

The original and still most widely used form of the data flow diagram is the DeMarco Data Flow Diagram. The charm of a DeMarco Data Flow Diagram is its simplicity. A variation called the Two-Entity Data Flow Diagram can be used with the same simplicity, but can also be used as part of an integrated system, capturing additional details as the system becomes more completely defined.

This book focuses on the Two-Entity version because it lends itself to an integrated approach. It provides for control information and can be represented, along with a data element dictionary, as a matrix in a computer data base. When changes are made to the items described in this data base during the course of the system development, the data base can be used to identify other items that will be affected. The matrix is also handy in estimating file sizes and data handling volumes.

Exercises

1. For each of the following data flow diagrams (A and B):
   a. Draw the hierarchical Two-Entity Data Flow Diagram.
   b. Draw the DeMarco Data Flow Diagram.
   c. Develop the data flow matrix.
2. Draw a Two-Entity Data Flow Diagram for a blood bank. Show the donor waiting in a queue to register, then moving to another queue to make a donation. Show the collection of information about the donor's medical history and its entry into a file, and a check being made to determine whether he or she meets the qualifications of an acceptable donor.

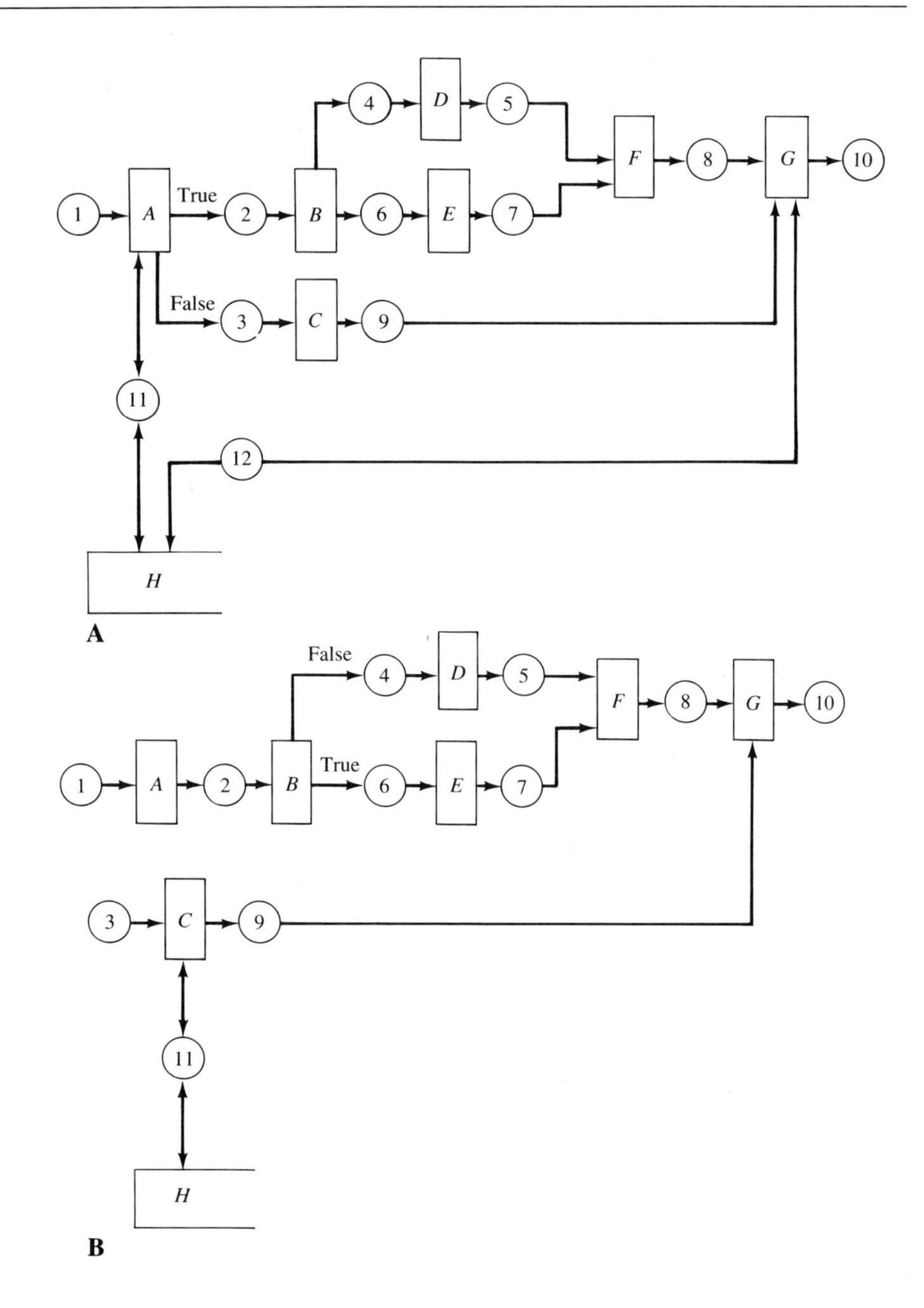

3. Develop a level 3 Two-Entity Data Flow Diagram detailing process Ba Order Entry. See the narrative/play script in Figure 3.19 to see what processes might be involved.

4. Make up a process annotation for the credit check process that appears in the detail of Ba Order Entry.
5. Draw a Two-Entity Data Flow Diagram for the current system in the Capert case study in Appendix A.
6. Working with the example given in the discussion on Lano Matrices, show how the Lano Matrix can be generated mechanically from a data flow matrix using a process very much like matrix multiplication. First make two binary matrices, one showing the outputs and one showing the inputs.

   The data flow matrix with binary 1s for the outputs and rows added for input and output with the outside of the system is as follows. We will call this matrix O.

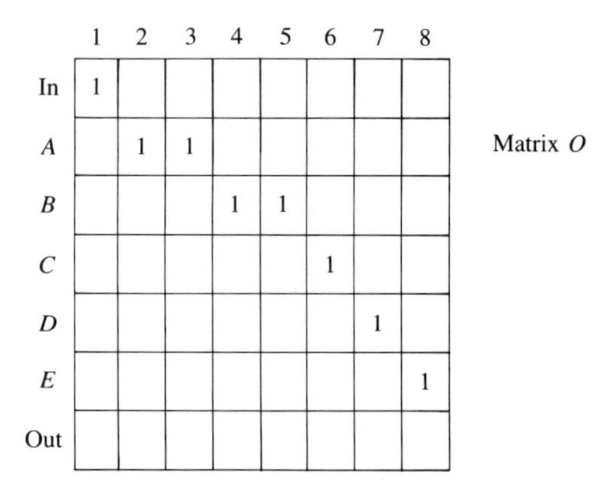

	1	2	3	4	5	6	7	8
In	1							
A		1	1					
B				1	1			
C						1		
D							1	
E								1
Out								

Matrix O

The matrix showing binary 1s for the inputs and the same added rows is as follows. We will call this matrix I.

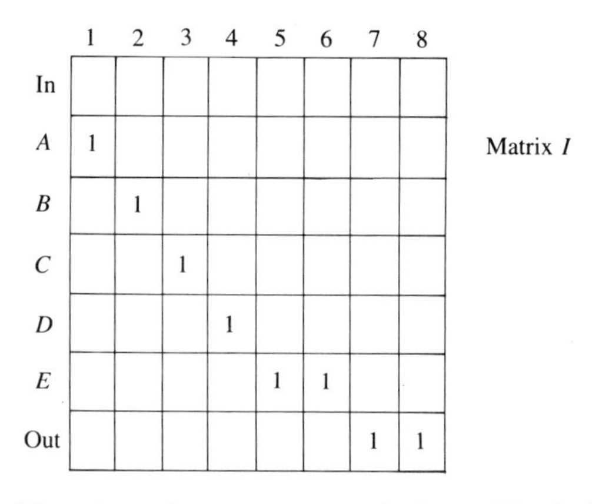

	1	2	3	4	5	6	7	8
In								
A	1							
B		1						
C			1					
D				1				
E					1	1		
Out							1	1

Matrix I

Now show that you can get the Lano Matrix by multiplying matrix O on the left by the transpose of matrix I on the right. Every time a 1 would be generated in the resulting matrix, instead of writing the 1, write the number of the data flow that would have produced the 1.

7. Construct a data base to represent the data structure matrix and data hierarchy using an available data base management system, such as DBASE II or III. Write programs to insert information into the data base, to make inquiries, and to show what data flows and processes are affected when a data element is changed.

References

DeMarco, T. *Structured Analysis and System Specification*. New York: Yourdon Press, 1978.

Dickinson, Brian. *Developing Structured Systems: A Methodology Using Structured Techniques*. New York: Yourdon Press, 1981.

Gane, Chris, and Trish Sarson. *Structured Systems Analysis: Tools and Techniques*. Englewood Cliffs, N.J.: Prentice-Hall, 1979.

Langefors, B., and B. Sundgren. *Information Systems Architecture*. New York: Petrocelli/Charter, 1976.

Lano, R. J. *The N Squared Chart,* a TRW internal report, Redondo Beach, Calif., 1977.

Martin, James, and Carma McClure. *Diagramming Techniques for Analysts and Programmers*. Englewood Cliffs, N.J.: Prentice-Hall, 1985.

Page-Jones, M. *The Practical Guide to Structured System Design*. New York: Yourdon Press, 1980.

Peterson, James L. "Petri Nets." *ACM Computing Surveys* 9, no. 4 (1977): 223–252.

Sholl, Howard. A., and Taylor L. Booth. "Software Performance Modeling Using Computation Structures," *IEEE Trans. on Software Engineering*, vol. SE-1, no. 4 (Dec. 1975): 414–420.

Steward, Donald V. "A Tale of Hope for Anyone . . . Lost in the Forest Primeval." *Computerworld,* March 19, 1984.

Tree Structures

- Three methods of representing systems using tree structures are structure charts, Warnier-Orr Diagrams, and Trees.
- Structure charts are shaped like a tree with the root at the top; large modules are broken into smaller modules as one progresses down the chart.
- Warnier-Orr Diagrams show the hierarchical structure of a program with braces representing successive program domains branching to the right.
- Trees portray programs with branches and limbs indicating statements and their sequence. Sufficient detail may be shown that programs can be compiled from these structures.
- Trees can illustrate the data flow on the limbs and branches. Data flow on the limbs toward the root should be limited, and all activities within modules should be closely related.
- Depending on the client's sophistication, the designer may begin with a Two-Entity Data Flow Diagram and transform it to a Tree, or begin directly with a Tree.
- Tree structures follow a top-down development, beginning with the general and proceeding to the specific. By carefully following guidelines regarding correct interfaces, designers and programmers can build programs with fewer chances for mistakes.
- Progress on a project can be measured and presented to management throughout the whole project, from requirements specification to code generation, by counting branches of the Tree.

Data flow diagrams focus on the data flows and their processing. As valuable as they are for representing systems, they are not suitable for portraying a system down to the level of detail such that programs can be written from them. A hierarchy is somewhat forced on the data flow diagram by expanding process blocks into other diagrams. Also, to write programs, we need to show the control. DeMarco Data Flow Diagrams do not indicate the control at all, and Two-Entity Data Flow Diagrams do not do so with sufficient detail. When designing a system, if we start with data flow diagrams, we will probably have to switch to another method before we begin to write code.

Another type of representation is based on a tree structure. Showing a hierarchy is more natural in a tree structure than in a data flow diagram. Also, tree structures focus more on control. This chapter discusses three methods of tree structures: structure charts, Warnier-Orr Diagrams, and what are simply called Trees.

As we'll see, one common approach to system design is to start with a DeMarco Data Flow Diagram during the requirements phase, then transform the D-DFD into a structure chart to do the design, and finally to use some pseudo code or high level language to do the implementation. We will show an integrated method in which all these steps can be done with just one Tree.

We will also discuss some issues concerned with the design of systems and programs because they are closely associated with the methods of representation we will be discussing. A further discussion of design occurs in Chapter 15.

Structure Charts

Structure charts form a tree with the root at the top (see Figure 4.1). The whole program—in this case, a payroll program—is broken into modules. These modules are further broken into smaller modules. Not only does a structure chart show the way programs and modules are broken into smaller modules, but it also portrays the way the program is controlled by having each higher level module call and pass parameters to and from the lower level modules under it.

The structure chart is used to analyze how information is passed by the module calls. An open circle shows the passing of a parameter such as "gross pay." A closed circle shows the passing of a condition flag such as a flag that shows that a match has not been found. A diamond shows when a call to a module depends on a condition. A half circle through the lines going to called modules indicates that these modules are iterated. In the discussion of Trees later in this chapter we will see how this information about data flow is used to evaluate how well we have arranged the program into modules by looking at the information that is passed between the calling and called modules.

Warnier-Orr Diagrams

In the mid 1970s a Frenchman named Jean-Dominique Warnier developed a method for constructing programs from the structure of the data with which the

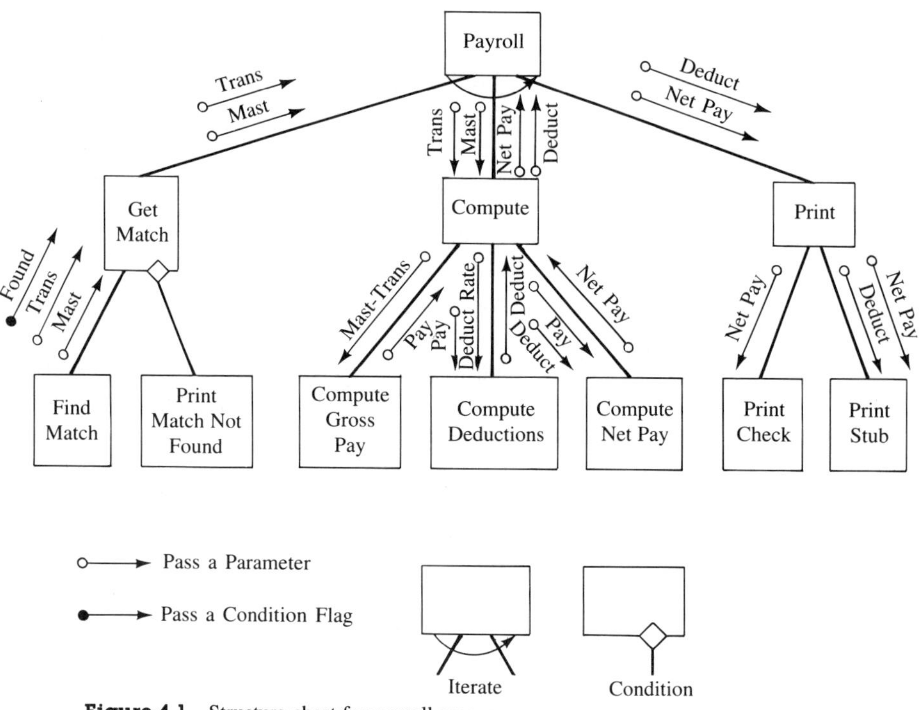

Figure 4.1 Structure chart for payroll program.

program works (Warnier, 1980). His method was widely adopted in Europe. Subsequently, an American named Kenneth Orr has developed variations of the Warnier technique (Higgins, 1983; Orr, 1981). These latter variations have led to structures known as **Warnier-Orr Diagrams.** (It should also be noted here that Michael Jackson in England has developed a related methodology [Jackson, 1975].)

Warnier-Orr Diagrams use braces to show how either data or programs are arranged hierarchically. Figures 4.2 to 4.5 show the Warnier-Orr Diagrams for the data used in a payroll, and Figure 4.6 shows the corresponding Warnier-Orr Diagram for the program that works with that data. The techniques for representing both data and program are similar. At the left of the page a large brace represents the domain of the whole data structure or program. As you move to the right, you notice that smaller braces are used to break the whole into smaller domains representing more detail. This represents a hierarchy in both the data structure and program.

Numbers within the parentheses show estimates of how often a structure is to be iterated or whether it is an option. One number shows how often the structure is iterated. Two numbers show there is an option; the structure is repeated either the number of times given by the one number or by the other, depending on some condition. A simple two-way decision is thus shown as (0,1). (See Figure 4.6.)

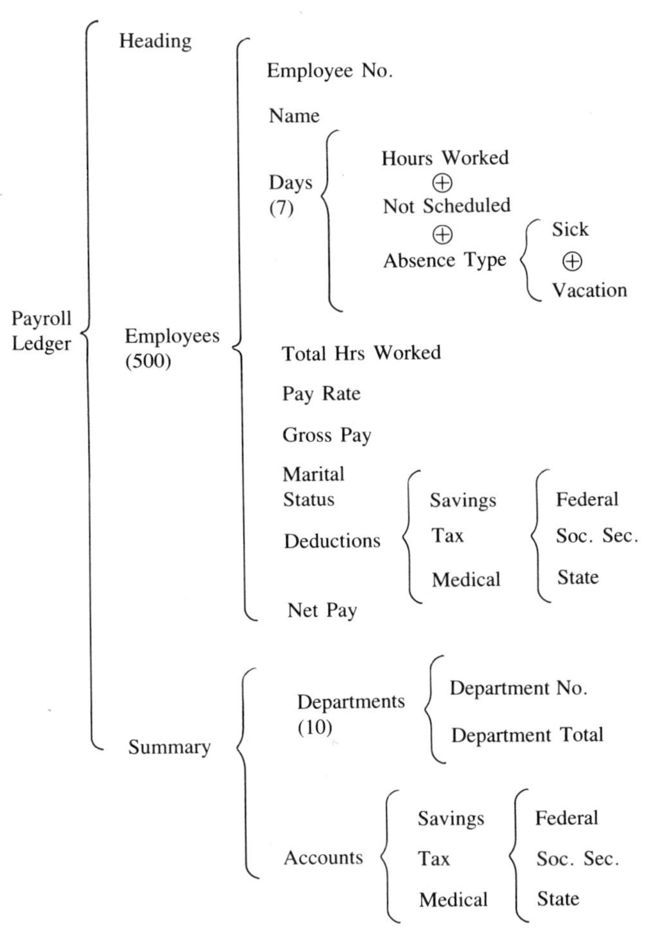

Figure 4.2 Payroll ledger. Warnier-Orr Diagram for data structure.

A statement before a brace can describe the circumstances under which that brace occurs in the data or is executed in the program. A plus between braces means OR, and a circled plus means Exclusive OR (that is, one or the other, but not both).

Note that in the program structure of a Warnier-Orr Diagram we do not fully describe the logic of iteration. For example, we do not use a statement like FOR $I = 1$ TO 10 to show advancing an index used to distinguish elements within the loop. Thus, the Warnier-Orr Diagram has no provision for some of the information that would be needed to generate a program automatically from the diagram.

The original intent of the designers of the Warnier-Orr Diagram was not that their diagram would represent a program in a form from which the program could be compiled. The Warnier-Orr Diagram was intended only to be used for communication between the designer and the client to determine whether the diagram

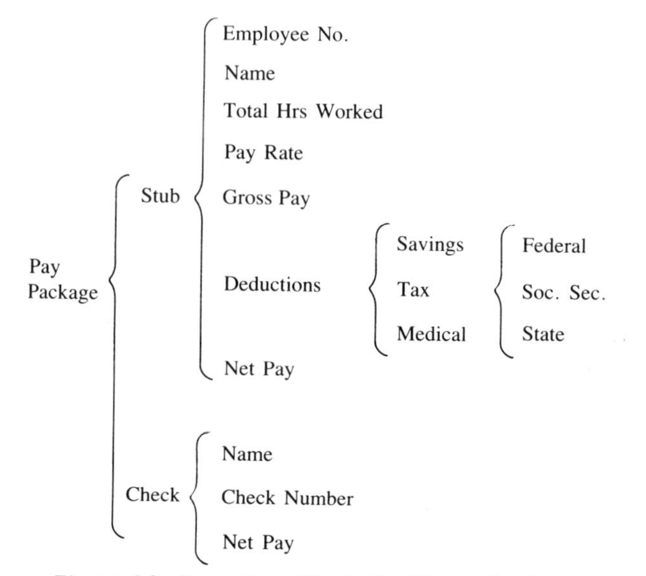

Figure 4.3 Pay package. Warnier-Orr Diagram for data structure.

represents what the client wants. Thus, as with data flow diagrams, there was an emphasis on making the representation easy for inexperienced clients to understand. Once the client approves the diagram, the developer can use some other method such as a structured programming language or pseudo code to design the program. (Pseudo code is a way of representing programs in a more rigorous way than English, yet not as rigorous and confusing as a programming language would be to the client who may review it.)

Despite its original concept, the Warnier-Orr Diagram does in fact come close to being able to represent a program with sufficient completeness that a program could be compiled from it. If you look at how programmers plan their programs using Warnier-Orr Diagrams, you will often see them cheat by inserting statements like FOR $I = 1$ TO 10. Sometimes this information is added by putting a reference on the diagram to a description of the iteration or decision process, which is written to the side. By this device we can extend the Warnier-Orr

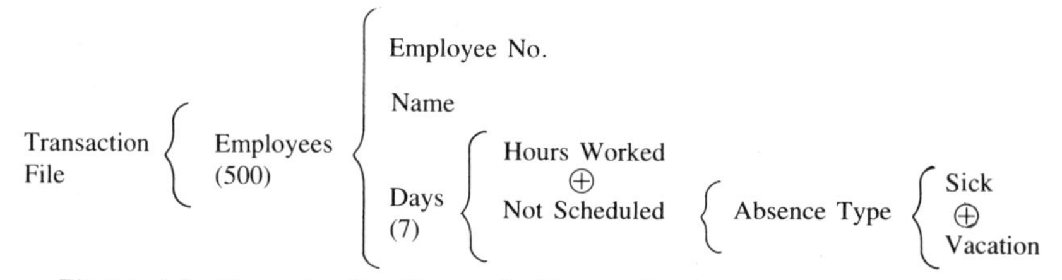

Figure 4.4 Transaction file. Warnier-Orr Diagram for data structure.

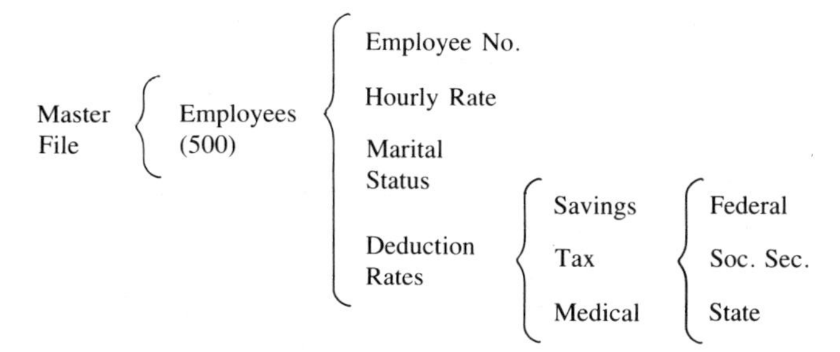

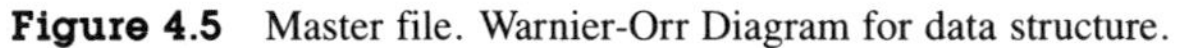

Figure 4.5 Master file. Warnier-Orr Diagram for data structure.

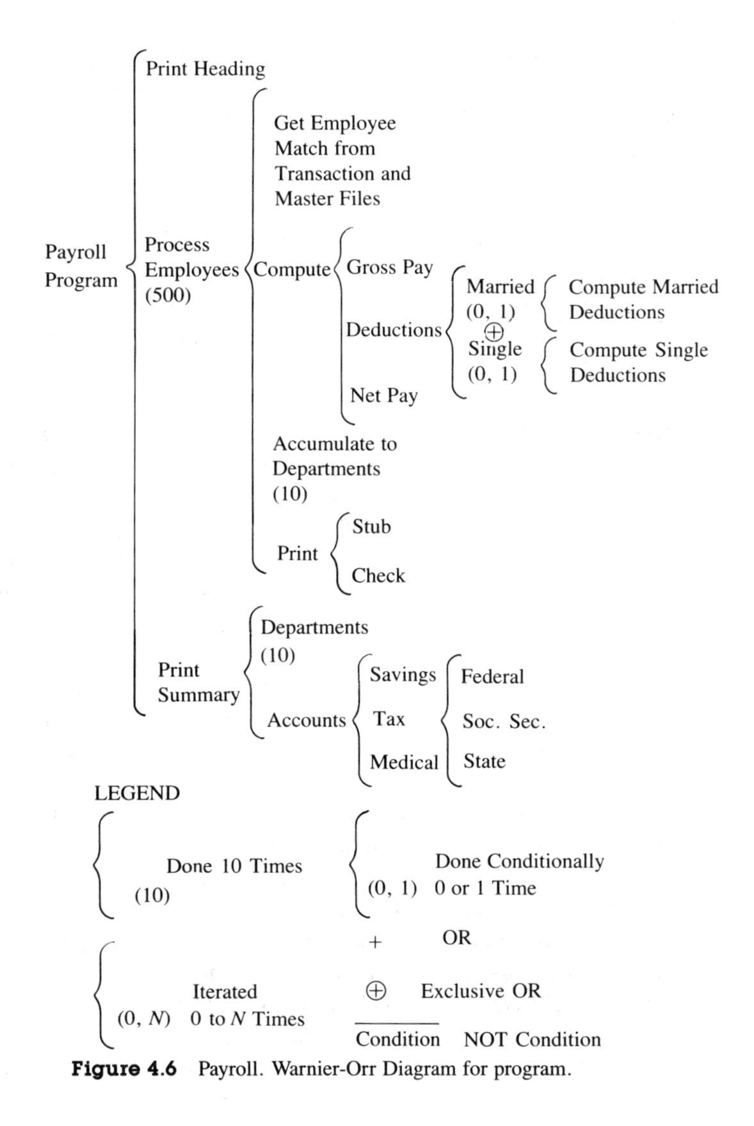

Figure 4.6 Payroll. Warnier-Orr Diagram for program.

diagramming technique beyond just the requirements phase. The Tree, which we discuss next, expands on this idea and handles these extensions more systematically.

Trees

The third type of treelike design structure is called simply a Tree (with a capital T). Trees depict program statements and sequence on a series of branches and limbs extending from left to right. They are far more flexible than structure charts or Warnier-Orr Diagrams and have the capacity for holding all the details of a full program. Since Trees can indicate data flow, they can be used compatibly with data flow diagrams. The way Trees are built, using a top-down principle, helps to encourage sound construction from the general to the specific.

Figure 4.7 shows a simple program represented by a Tree. The structure begins on the left with a broad definition of what the program does. The parts of the program are broken down into greater detail as the branches of the tree proceed to the right. Ultimately the structure produces a complete description of the body of the program. The program can be compiled and run from this graphic representation and the data description. Even without reading the formal rules for the Tree, you can get a good understanding of how this particular program works just by looking at Figure 4.7.

To explain how tree structures are created, we'll begin by presenting some basic definitions and conventions. As we will see, these guidelines are probably not quite what we would expect from our knowledge of nature's trees. For instance, our Tree is drawn from left to right on the page to make it easier to write the various descriptions horizontally. Thus the **root** appears on the left side of the page and is considered the top of the Tree (see Figure 4.8). On the root there is a brief description of what the program does. The Tree is made up of horizontal lines called **branches** and vertical lines called **limbs.** A Tree or subtree begins with a branch and is said to be entered or exited through that branch. **Subtrees** are subprograms within the program itself. A branch has on its right end either a **leaf** or a limb. A leaf is the end of a branch where no further branching occurs. Limbs have branches off them to the right. Up and down on a limb refers to up and down as seen on the page.

The execution of the Tree begins and ends at the root. Executing a branch means executing the leaf or subtree entered through that branch. After execution, control returns back through the same branch. Three types of limbs are used to build a Tree: single, wiggly, and double (see Figure 4.9). A single limb is used to depict sequential execution. The branches off a single limb are executed in sequence from top to bottom. A wiggly limb shows either sequential or parallel execution. Branches off a wiggly limb could be executed in parallel if the system were so capable. A double limb depicts mutually exclusive execution. Only one branch off a double limb may be executed. Which branch is executed depends

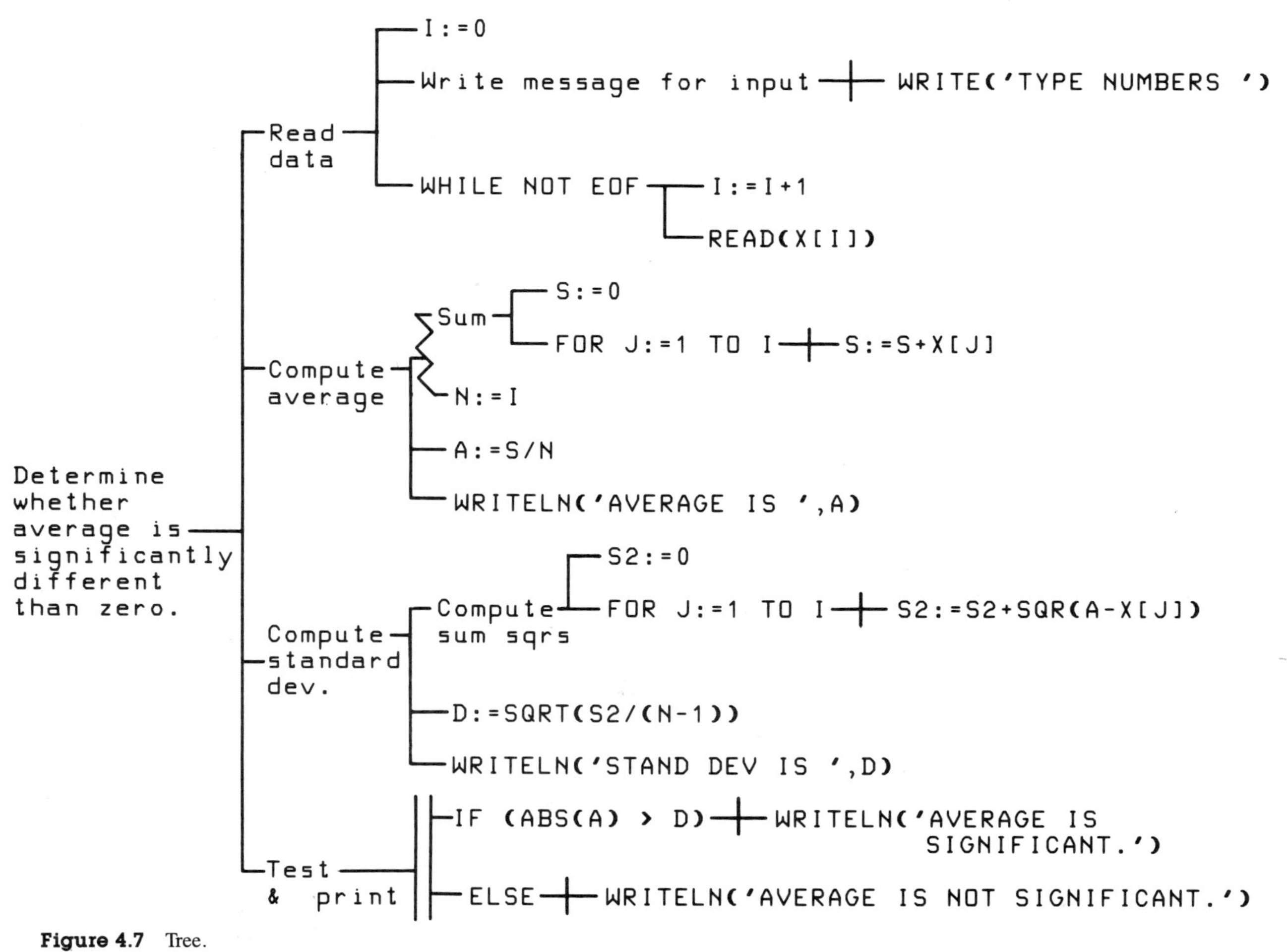

Figure 4.7 Tree.

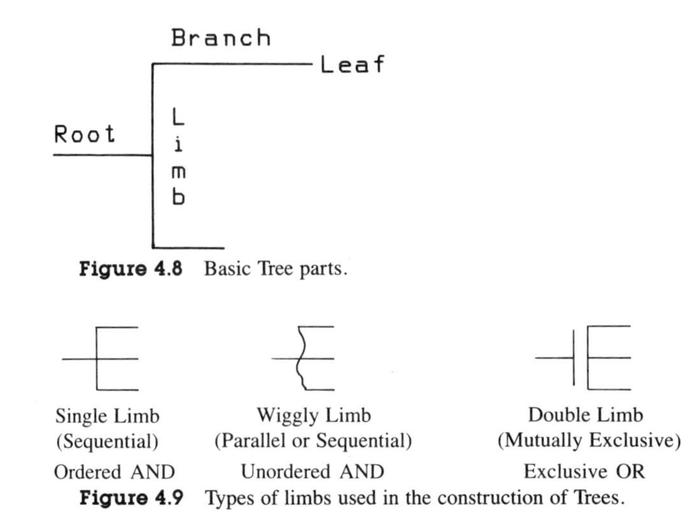

Figure 4.8 Basic Tree parts.

Figure 4.9 Types of limbs used in the construction of Trees.

upon conditions written on the branches. Each of these types of limbs may have any number of branches on the right.

A branch can carry either a remark describing what is to be done by the subtree entered through that branch, or a control statement, or both. A control statement describes whether or how often its subtree is to be executed. The possible control statements are:

IF condition
WHILE condition
FOR variable = expression-1 TO expression-2 STEP expression-3
UNTIL event

The IF, WHILE, and FOR statements are interpreted just as they are in any structured language. We'll describe the "UNTIL event" statement in a moment. The leaves carry executable statements such as READ, WRITE, and variable = expression statements, or a procedure call. Remarks are written in lower case letters beginning with an initial capital letter. Control statements and the executable statements on the leaves are written in all capitals.

IF-THEN-ELSE and CASE

The "IF conditional THEN statement-1 ELSE statement-2" is represented in a Tree as in Figure 4.10.

A case statement can be represented in a tree as shown in Figure 4.11.

UNTIL Event

"UNTIL event" is a less familiar but very powerful control statement that is useful in a Tree. The subtree entered through the branch with the "UNTIL event"

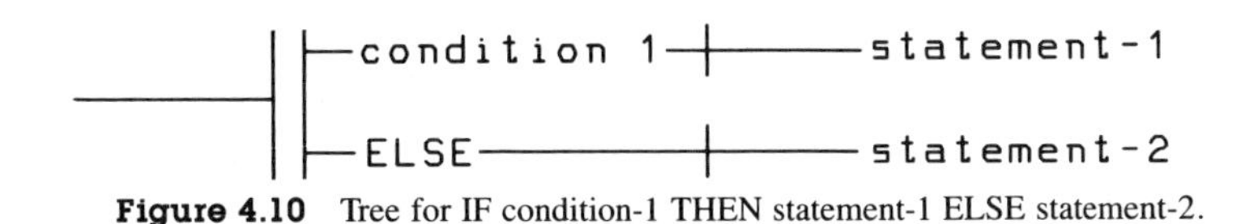

Figure 4.10 Tree for IF condition-1 THEN statement-1 ELSE statement-2.

will be executed until an event statement of that same name is executed somewhere in its subtree. Control then exits immediately through the UNTIL branch. The event statement is shown as the event name preceded by # (for example, #end_of_file). It will usually be executed as a consequence of an IF condition on an end_of_file or error that requires special handling at a higher level in the program.

Do not confuse the "UNTIL event" with the "REPEAT . . . UNTIL" statement available in several languages. "REPEAT . . . UNTIL" only allows an escape from the end of a program block. (A program block corresponds here to a subtree.) The "UNTIL event" allows an escape from anywhere that a special condition occurs, which requires handling higher in the Tree. But the "UNTIL event" also should not be confused with the "EXIT" statement available in several languages. The conventional "EXIT" statement provides only for an exit to the

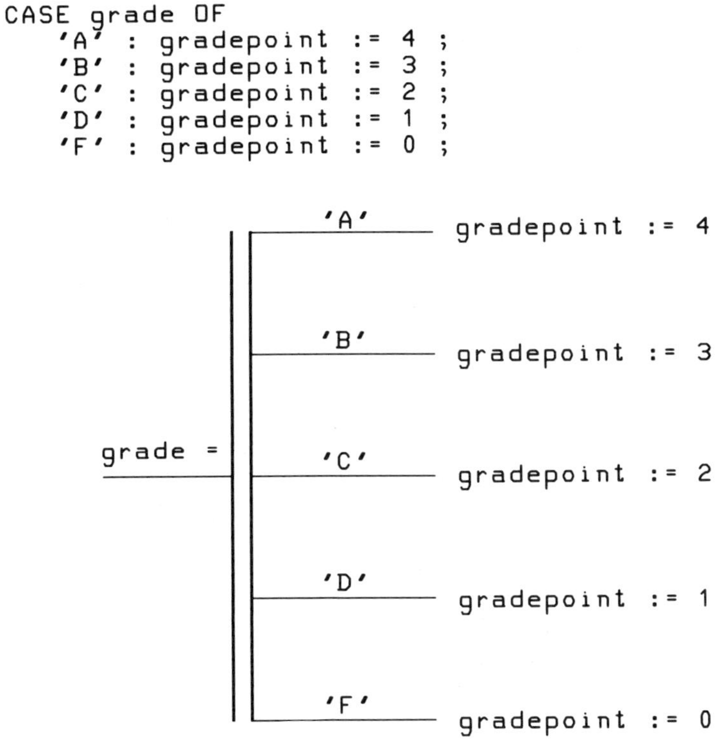

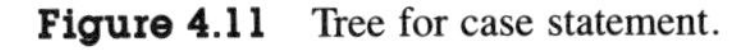

Figure 4.11 Tree for case statement.

```
   BOOLEAN: eofevent, errorevent ;
   LABEL: 1 ;

  'UNTIL EOF or Error '
   eofevent := false ;
   errorevent := false ;
1: WHILE not (eofevent or errorevent) DO
      BEGIN

     'Cause eofevent to occur'
      IF "EOF occurs" THEN BEGIN eofevent := true ; GOTO 1 END ;

     'Cause errorevent to occur'
      IF "Error occurs" THEN BEGIN errorevent := true ; GOTO 1 END ;

      END
```

Figure 4.12 UNTIL event simulated in Pascal.

next higher level. The "UNTIL event" shows what level the escape is to—where the consequences of the event can be handled. The "UNTIL event" will handle as special cases the functions of either the "REPEAT . . . UNTIL" or the "EXIT." (See Figure 4.28 for an example of how an "UNTIL event" may be used in a program.)

The "UNTIL event" is not yet a standard structured language construct, so it must be simulated in these languages using a GOTO. The simulation of an "UNTIL event" should be the only purpose for which a GOTO is ever needed. If the "UNTIL event" construct were part of the language, then no GOTO would be needed nor should it be necessary to write clumsy code to avoid using a GOTO.

An example of how an "UNTIL event" exiting from within a block on either an end_of_file event or an error event could be simulated in Pascal is shown in Figure 4.12.

Note that the "UNTIL event" acts like a "Come From," showing where the control transfers to when the event occurs (Clark, 1973). This feature is what allows an exit to any higher level where one is prepared to handle the consequences of the exit.

State Diagrams and Trees

A state transition diagram can often be useful for describing the behavior of a program, particularly an interactive program. We will explain the basic concepts of these diagrams and show how their state transitions can be represented in a Tree with the help of the "UNTIL event".

We may think of a program as being in any one of a number of possible states. The state the program is in determines what the program will do when it receives certain input and what state it will go to next.

Figure 4.13 is an example of a state transition diagram for a simple editor with three states: an edit state, a menu state, and an operating system state. Circles are used to show the possible states. Directed lines between the states show the transitions from one state to another. A label on the top of the line shows the input that causes the transition, and a label below the line shows the action taken when the transition occurs.

We can see from this state transition diagram that if the program is in the edit state, depressing the I key will insert an I into the text. But if the program is in the menu state, depressing the I key will cause the program to execute the command corresponding to the I in the menu. Pressing the Ins key while in the edit state will cause the program to display the menu and go to the menu state. We will assume that while in the menu state the I will indent a paragraph, a U will undent the paragraph, and an X will cause an exit to the operating system. After performing an I or U, the program returns to the edit state.

This behavior can also be shown as a Tree (see Figure 4.13). The first double limb sends control to the correct state, depending on the value of the variable called state. The UNTIL's on the branches into and out of the double limb return control to this double limb after a transition. When there is both an IF and an UNTIL on the same branch, by convention we show the IF above and the UNTIL below the branch. For each state a double limb selects the specific transition,

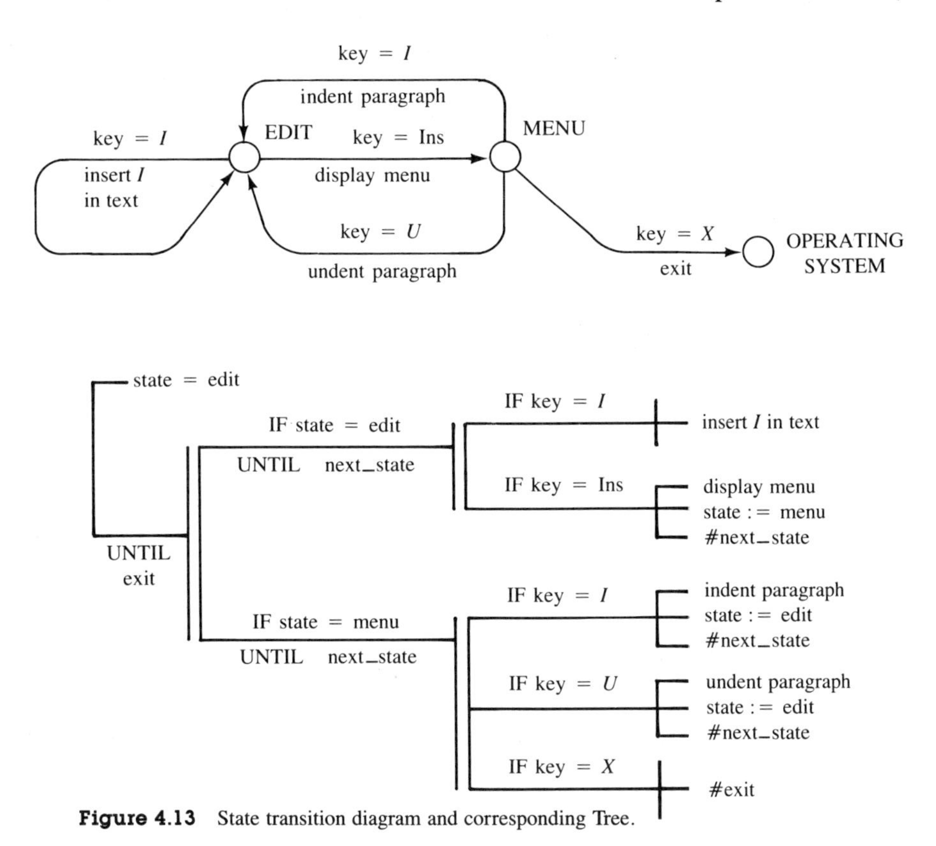

Figure 4.13 State transition diagram and corresponding Tree.

depending on the input. Then a subtree describes what happens when the transition occurs: The action to be taken, if any, the setting of the state variable to the next state, and the #next_state event which returns control to the UNTIL next_state.

Functions and Procedures

At some stage in the development of a program we may wish to show in the Tree what subtrees will be packaged as modules—that is, what parts of the program are implemented as procedures or functions. This information can be shown with a box on the branch entering the subtree that is to be programmed as a module (see Figures 4.14 and 4.15).

Once this module is given a name, it would be shown as in Figure 4.15.

A call to a module is shown by putting the module name in a box or circle on a leaf (see Figure 4.16). The distinction between the box and circle calls will be discussed later in this chapter when we talk about asynchronous processing. A Tree for a module shows a box with the module name on the root (see Figure 4.17). A module that calls itself is recursive.

Although our example is a trivially simple program used to illustrate the concept, the top-down principles apply just as well to large systems of hundreds of thousands of lines of code. In fact, the larger the system, the more important it is to use a top-down approach.

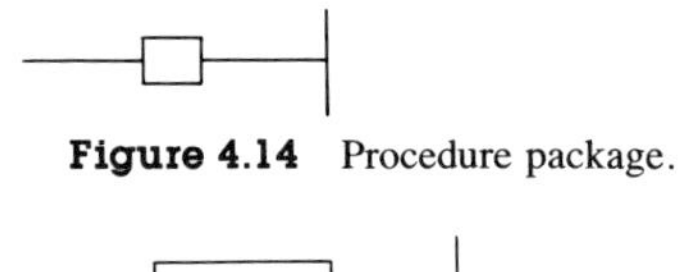

Figure 4.14 Procedure package.

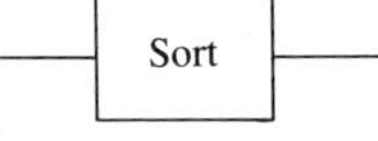

Figure 4.15 Procedure package with label within larger Tree.

Figure 4.16 Call to sort procedure.

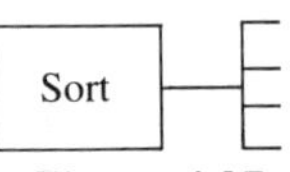

Figure 4.17 Tree for sort procedure.

Data Flow and Structured Design in the Tree

One of the advantages of the Tree as a design structure is that the data flow can be shown. This allows us to relate the Tree to a Two-Entity Data Flow Diagram and to analyze the data flow to produce well-structured programs. To understand how this feature works, let's first look at how the data flow is shown in a Tree.

Data in a Tree flows into a branch going to the right. This data flow may be written along the top of the branch. After the subtree is executed, the flow is exited out the branch flowing to the left (which can be shown along the bottom of the branch). Between branches data flows down the limbs (shown on the right of the branch). Data flows are written within parentheses. We will show the data flow on the Tree only when it is useful to do so. (See Figure 4.18.) Next to the limb segments between branches we can show in parentheses the data elements that are produced above and used below that limb segment. Figure 4.19 shows the same Tree as Figure 4.7, but with data flows on the limbs.

Putting the data flow in the Tree is useful for two reasons:

1. The depiction provides a compatibility with the information in Two-Entity Data Flow Diagrams, which makes it possible to transform from one to the other.
2. The depiction can be used to analyze the data flows for various possible hierarchical designs to get one such that the modules are reasonably independent and easy to maintain.

This latter reason leads us to the subject of structured design.

Structured design focuses on developing programs that have low coupling (that is, low data flow between modules) and high cohesion (that is, parts working intimately together to satisfy a common goal within modules) (Page-Jones, 1980; Yourdon and Constantine, 1979). The modules are our subtrees. Low coupling and high cohesion have been found to contribute to making programs easier to develop and maintain. Changes made in one module can affect other modules only through the data flows between them. Thus, keeping the data flows between modules to a minimum makes it easier to keep track of the consequences of a change on other modules.

The data flows shown on the limbs of the Tree make the coupling clear. A well-structured program will have a small volume of data flow on the high limbs

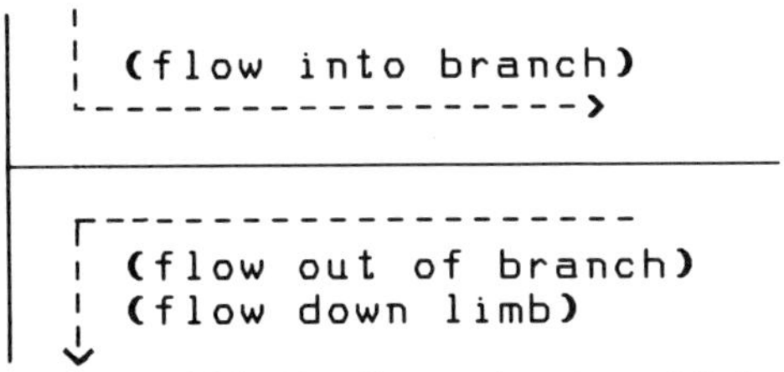

```
|  |
|  | (flow into branch)
|  '------------------>
|
|-----------------------------
|
|  r-------------------
|  | (flow out of branch)
|  | (flow down limb)
|  v
```

Figure 4.18 Data flows on branches and limbs.

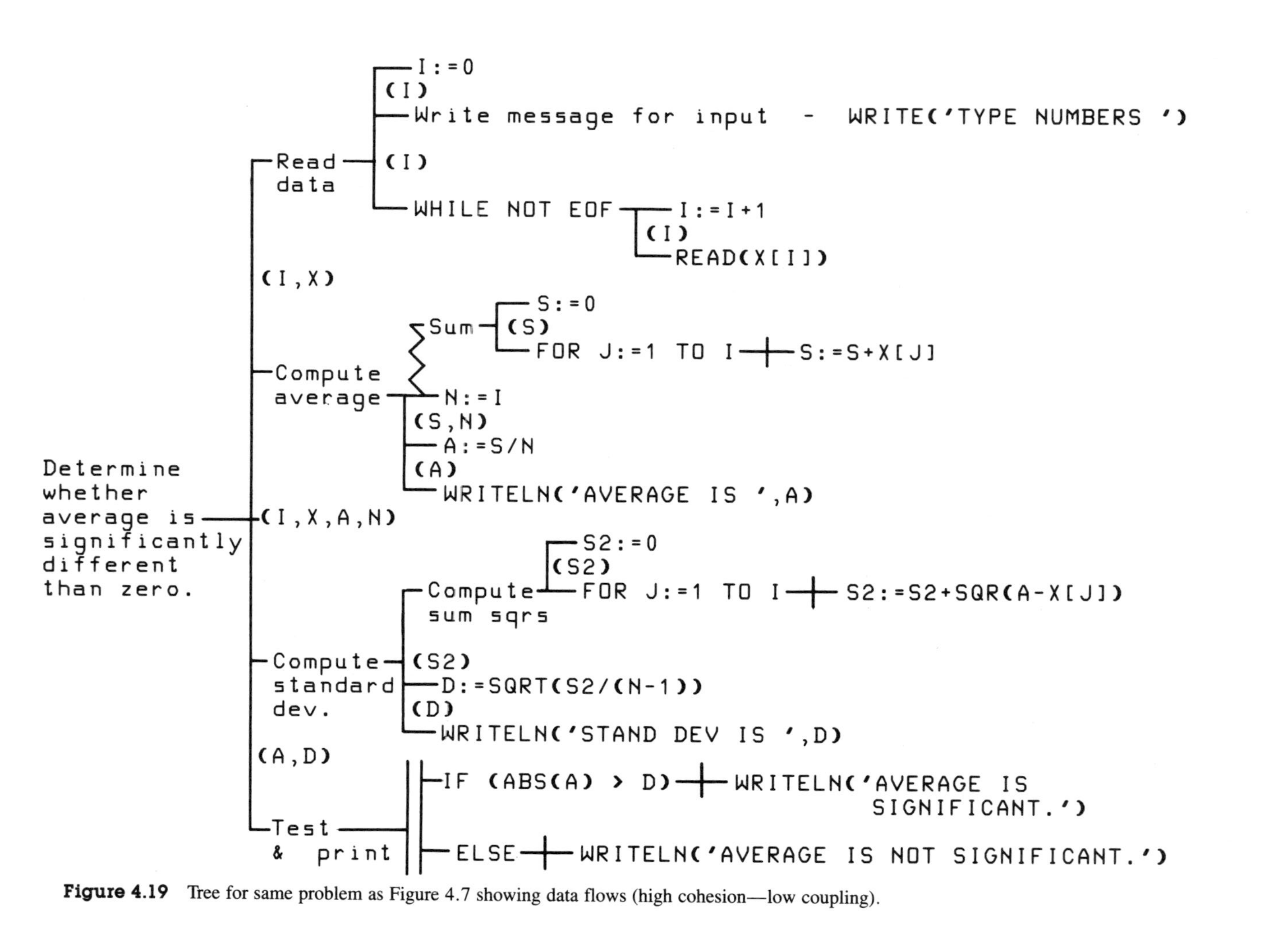

Figure 4.19 Tree for same problem as Figure 4.7 showing data flows (high cohesion—low coupling).

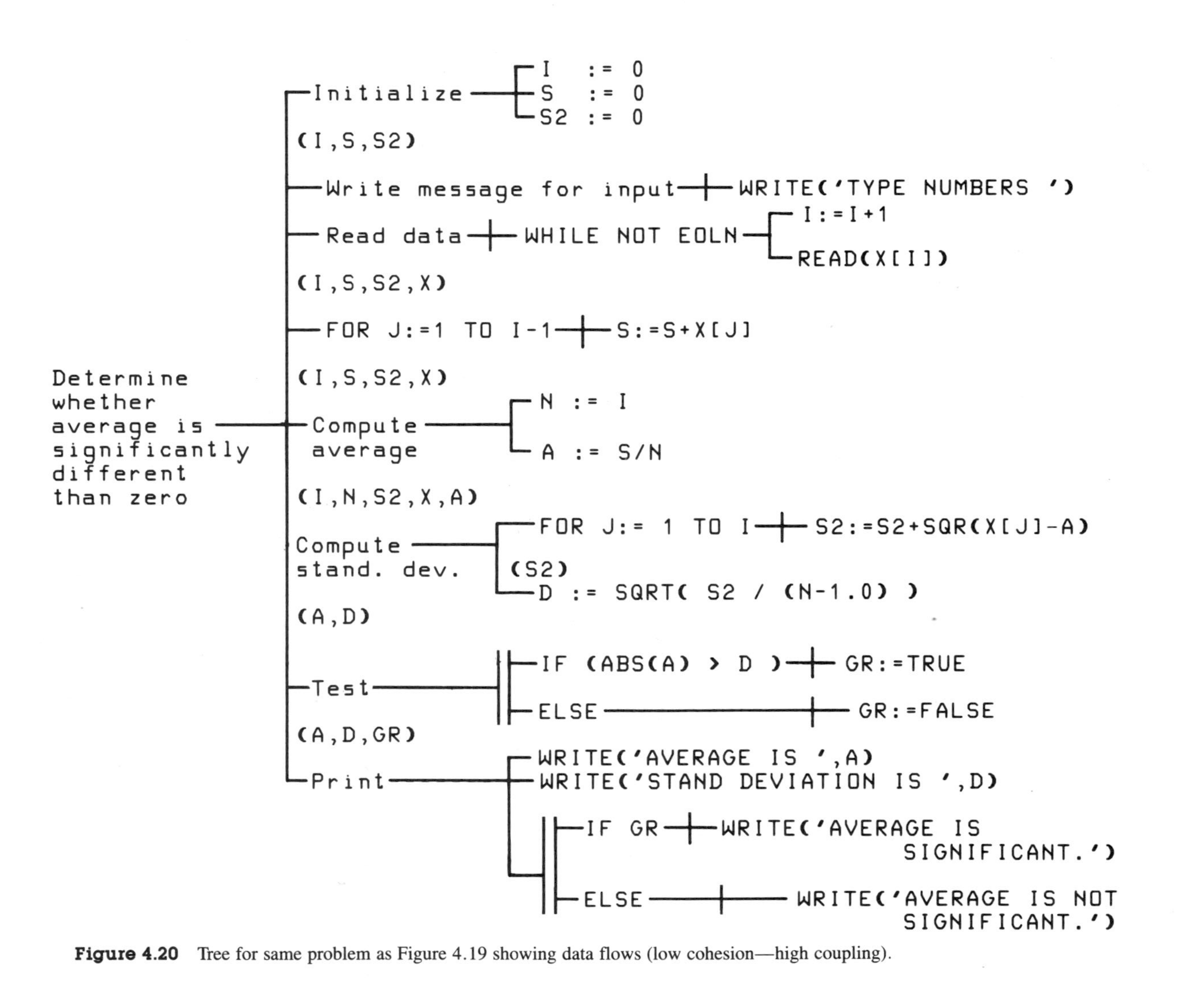

Figure 4.20 Tree for same problem as Figure 4.19 showing data flows (low cohesion—high coupling).

(to the left), resulting in a low coupling between modules. High cohesion requires that everything done within the module is closely related in its purpose. This is indicated by whether what is done within the module can be given a short description on the branch leading to it. It does not have high cohesion if it has to be described by a list of different types of activities. Lower coupling between modules tends to result in higher cohesion within the modules (Page-Jones, 1980).

Comparing the two versions of programs to perform the same function shown in Figures 4.19 and 4.20, we can easily see that the program design in Figure 4.19 is a better design, having less data flow along the higher limbs, thus demonstrating less coupling and implying more cohesion.

Using the following rules with Trees will generally produce good structured program designs:

1. Minimize the data flows on the high limbs.
2. Let the major decisions be made toward the top.
3. Keep the data and operations at the top logically clean and simple. Keep the physical considerations about such matters as the format of the data structures on external media at the lower end of the Tree. Transform these messy data structures so they are clean data structures when they reach the top of the Tree.
4. Introduce extra levels of limbs and branches as needed to keep from having too many branches exiting from any one limb. Five or seven branches off one limb is a reasonable maximum (remember Miller's Principle?). An exception might be a double limb representing a case statement.
5. Initialize a variable and finish using it within as close a proximity as possible.
6. Make sure that the actions collected within one subtree tend to have an integrity of purpose so they can be described easily in a simple statement without having to make a list of all the different kinds of things it does.

Transforming Data Flow Diagrams into Trees

Two-Entity Data Flow Diagrams can be rewritten as Trees. One area that requires special consideration is the representation of asynchronous processes, which we will discuss in this section. Even before starting the requirements, though, the developer should consider whether to begin with a data flow diagram or go directly to a Tree.

Figures 4.21 and 4.22 show two basic patterns that appear in TE-DFDs and their representation as Trees. Recognizing these patterns can be useful in transforming TE-DFDs into Trees.

Figure 4.23 is a data flow diagram with circles drawn to show how parts of the data flow diagram correspond to parts of the Tree. A circle is drawn around the set of parallel paths from just before the process where they split to just after the process where they come together. If the split involves conditions, it is represented in the Tree by a double line. If not, it is represented by a wiggly line.

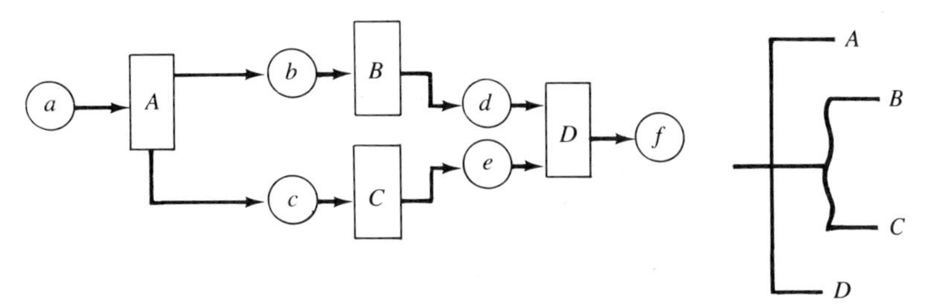

Figure 4.21 Mapping between TE-DFD and Tree—parallel or sequential.

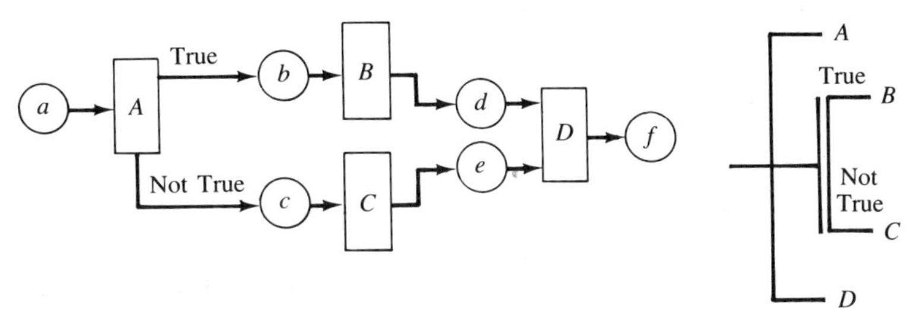

Figure 4.22 Mapping between TE-DFD and Tree—condition.

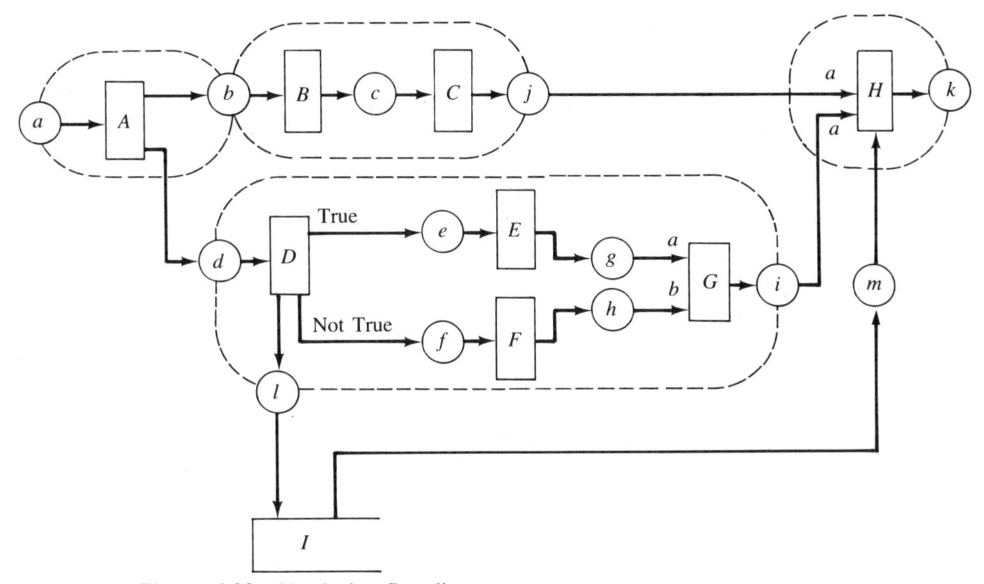

Figure 4.23 Simple data flow diagram.

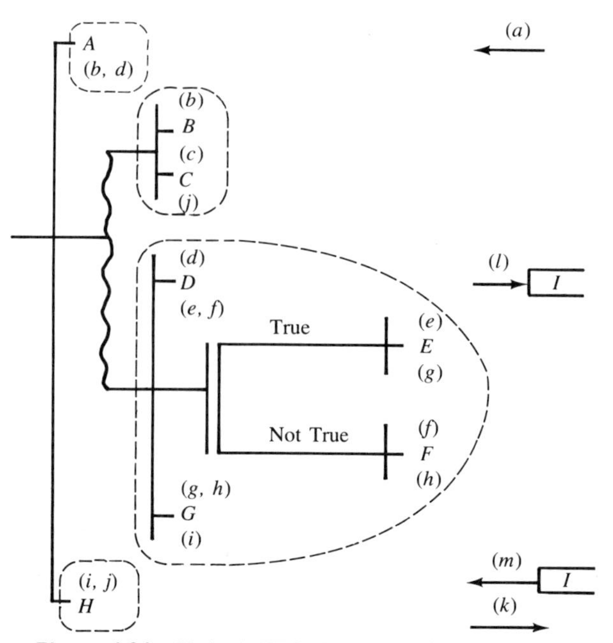

Figure 4.24 Circles in TE-DFD map transformed into circles in Tree.

The relation between the Two-Entity Data Flow Diagram in Figure 4.23 and the Tree in Figure 4.24 can now be readily seen. The processes are represented in the data flow diagram as rectangles, and in the Tree as subtrees. The data flows appear in the data flow diagram as ovals, and in the Tree as data flows on the limbs. Parallel processes are shown in the data flow diagram with multiple arcs exiting from the same process, and in the Tree with wiggly limbs. Sequences of processes or circles drawn around sets of processes correspond to branches off a single limb. The circled processes and the processes themselves can be related to subtrees. Two-Entity Data Flow Diagrams that can be converted to Trees by this circling process are called hierarchical TE-DFDs. Study Figures 4.23 and 4.24 until you are comfortable with the relations between data flow diagrams and Trees.

Note that in a Two-Entity Data Flow Diagram conditions can be shown on the arcs leaving the process. This must be shown in the Tree in two steps: as the process followed by the decision, as is shown for process *D* in Figures 14.23 and 14.24. We must resist the temptation to write this as *D*———|| because this would imply not that *D* is followed by the decision, as we intend, but that *D*, when described in greater detail, is the decision, which is not what we intend.

While data flows within the system are shown on the limbs, data flows into and out of the system are shown along horizontal arrows entering and exiting at the leaves. This is where the READs and WRITEs occur. Files are shown as open-ended boxes.

Wherever a double or wiggly limb obscures the data flow in or out of a process, a short limb called a **stub** is added to carry the data flow. For example, the data flowing in and out of E (e and g, respectively) are shown on the stub above and below E in Figure 4.24.

We put the data flow on the Tree only when needed to analyze the coupling, or to use the correct constructs for hierarchical development, described later in this chapter. If the Tree is developed using a Tree Editor (also described later in this chapter), the data flow can be developed by the computer from the Tree.

Asynchronous Processes

The data flow diagram may imply that more than one process can go on at one time. In Figure 4.23, B and C can occur simultaneously with D, E, F, and G. **Asynchronous processes** are those that can run in parallel (that is, simultaneously) or can be done one at a time in an arbitrary order. When programming with just one CPU, the programmer can make an arbitrary choice of which process to run on the processor before the other. As microprocessors become cheaper, we can afford to have many of them working asynchronously in the same system. And as we learn how to control them, we will be seeing more multiprocessing asynchronous systems. Asynchronous processing can be represented in both the Tree and in the data flow diagram.

Not all data flow diagrams are hierarchical TE-DFDs. Figure 4.25 shows a diagram for which this circling process will not work. To convert these data flow diagrams into Trees will require either that we sacrifice some information about what can be run in parallel, or that we introduce a new trick, the precedence procedure. If we only have one processor available and thus cannot use the asynchronous capability of the system, the former process is quite adequate. If we have many processors and wish to take advantage of all the asynchronous capability possible, we must use the precedence procedure.[1]

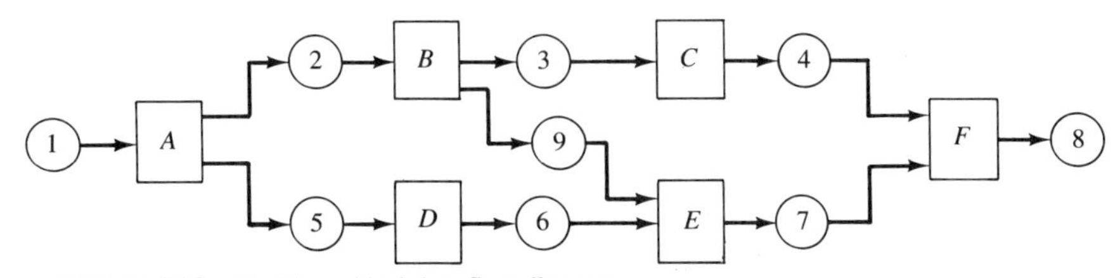

Figure 4.25 Nonhierarchical data flow diagram.

[1]Techniques have been developed for transforming data flow diagrams into structure charts. Structure charts are hierarchical and in some ways like Trees. These methods center on identifying two standard patterns in the data flow diagram, the transform and the transaction. What they accomplish is similar to what we have presented here, but they do not preserve the asynchronous capabilities of the data flow diagram. See exercise 6. (Page-Jones, 1980)

A **precedence procedure** is a procedure that is called from several different places in a Tree and is executed only when all the calls have been made. These are called **precedence calls.** There are two types of precedence calls: a call shown within a circle records the completion of a predecessor. A call shown within a box records the completion of a predecessor but also shows where the control returns to after all the predecessors have occurred and the precedence procedure has run. Figure 4.26 shows the Tree with its calls and the predecessor procedure that transforms the data flow diagram in Figure 4.25 into a Tree.

The *B* and *D* shown with the wiggly line above the precedence procedure is a way of stating to the procedure what predecessor calls must occur before this procedure is run.

If we have only one processor, we can choose arbitrarily an order in which to run one at a time those processes that could be otherwise asynchronous. We can represent this arrangement in a Tree without the precedence process, as in Figure 4.27.

In the extreme, every process can be represented by a separate precedence procedure. The precedences describe the sequence in which they will be executed. There are data flow machines that asynchronously execute processes described this way.

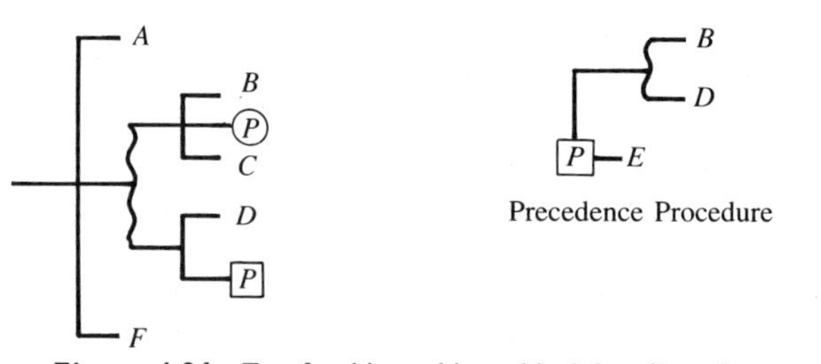

Figure 4.26 Tree for this nonhierarchical data flow diagram.

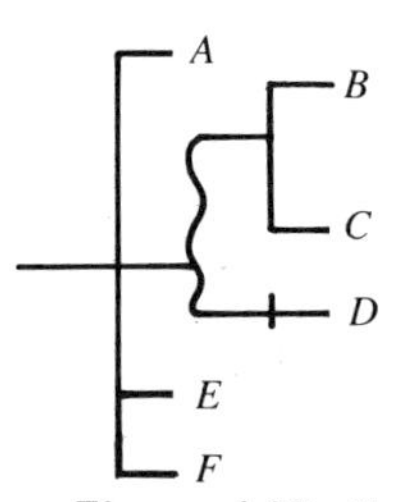

Figure 4.27 Tree for nonhierarchical asynchronous processes that could be used with a single processor.

Relation between Trees and Data Flow Diagrams

Data flow diagrams emphasize data flow, but the flow of data also has implications on the flow of control. Trees emphasize the flow of control, but the data flows can be shown on the limbs. By putting information about flow of control in the data flow diagram, and information about flow of data in the Tree, we draw the two closer together, making it possible to translate from the one to the other without loss of information.

Note that it is not necessary to show the control in the data flow diagram, or the data flow in the Tree. These descriptions can be left out for a high level view if it is felt the added detail would be confusing. But, when appropriate, this information can be added without using a different type of diagram.

We may want to start with a data flow diagram because it is understandable to clients with the least explanation. It is often the better tool for representing an overview of a whole system of many modules. Then we may want to transform the data flow diagram into a Tree and continue the development as a Tree. But if our client already understands Trees or is willing to spend the few moments needed to learn them, we may find it more convenient to work with Trees from the very beginning, starting with the discussions with the client to establish the requirements specification.

We can choose from among several approaches:

1. We can start with a DeMarco Data Flow Diagram for the requirements and system design, then transform it to a structure chart for the modular design. Finally, we would transform the structure chart to pseudo code or a high level language for implementation, thus making two transitions from requirements to code.
2. We can start with a Warnier-Orr Diagram, which usually requires that we transform into pseudo code or high level language for the implementation. This requires one transition.
3. We can start with a Two-Entity Data Flow Diagram and transform it into a Tree. This approach is made practical by the existence of a systematic way of transforming from the TE-DFD to the Tree such that information is not lost in the transformation.
4. We could begin by describing the requirements with a Tree and carry that same Tree all the way from requirements through design to code generation, thereby using one method of representation and no transitions. This last is an integrated approach, suitable for use with computer aids.

We recommend either approach 3 or 4.

Trees and Top-Down Development

Trees follow the top-down principle, progressing from general concepts to increasing levels of detail. This is not only a sound, logical way to develop systems and write programs but also a careful method with its own built-in checks. In this section we'll see how the use of prototypes, modules, and several building constructs used in conjunction with Trees helps to strengthen the integrity of program designs.

Remarks are used in top-down development to describe initially *what* is to be accomplished. Later the details showing *how* it is to be done are supplied as the Tree is expanded to the right.

Thus, as the tree develops from left to right, details get resolved. Clients can initially express what they want done by writing remarks on the branches. The analyst then expands this Tree to further levels of detail, and the programmer expands it further yet, until finally it can be translated mechanically into the source code of a structured language. All this work uses just one form of representation.

The Tree can be understood by the client, analyst, and programmer alike, so that each has some assurance that what the others are doing is consistent with what he or she is doing. Each can see his or her own contribution, and how it affects and is affected by everyone else's. If a change in the intent of what is to be done occurs at the top of the Tree, corresponding changes are forced on the implementation below. If there is a change in the implementation or it is finally determined that the requirements cannot be implemented, the effect of a different implementation must ripple up to the high end of the Tree. There any effects that change the requirements can be reviewed with the client. The whole process from conception to program is represented with growing detail, maintaining an integrity that comes from working within only one expanding diagram.

Figure 4.28 shows the Tree for the beginning of the development of a payroll program, and Figures 4.29 to 4.32 show the development from the top level down to more detailed levels. In this example the root tells us what the program is to do—"Generate payroll." Then the branches are filled in to show how to do this. Since these details may not be known at first, remarks are inserted to show what must be accomplished, allowing the details of how it is to be done to be filled in later. The order in which these details are developed is at the convenience of the developers, and need not follow the sequence in which the program executes them.

In the payroll program the basic structure comes from processing matched records from two files until one of the files runs out of records. This is shown by the UNTIL branch.

Moving away from the root in the Tree shows *how* to do it, and moving toward the root shows *why* it is done. Moving from top to bottom of the limbs shows *when* it is to be done. Basic algorithms occur toward the root. Adaptations to specific environments occur toward the leaves. Modifications to accommodate the program to new clients will usually affect only the leaf end. Since modifications toward the leaf end tend to have less effect on the rest of the Tree, they are easier to make.

In Figure 4.28, we can see that branches can sometimes be interpreted as states—that is, the set of circumstances that exist at that point in the program. "Get Match" is a state in which a match exists between a transaction record and

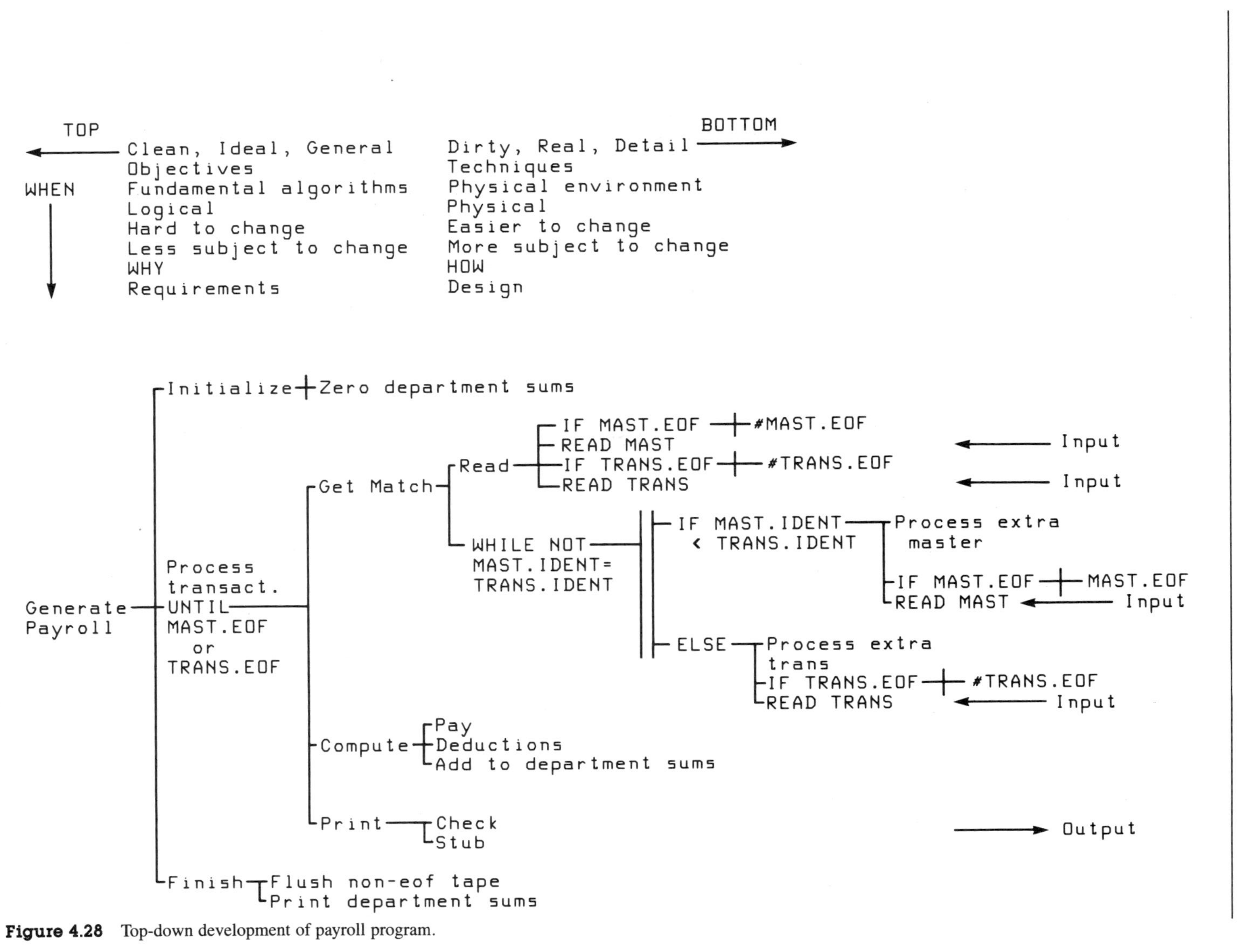

Figure 4.28 Top-down development of payroll program.

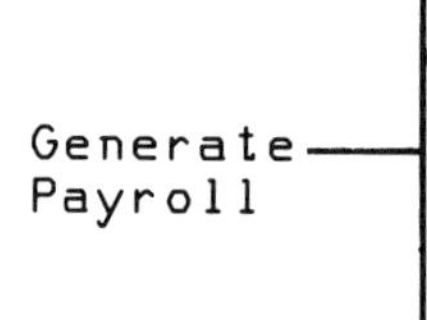

Figure 4.29 Step 1. Top-down development of payroll program.

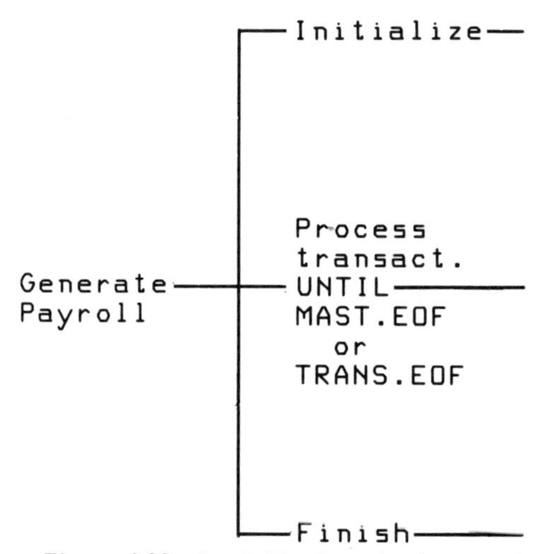

Figure 4.30 Step 2. Top-down development of payroll program.

a master record. When the subtree is exited through that branch, a match exists. Other parts of the program need to know that this match state exists, but they need not know how it was produced. The details of how the match is made occur in the subtree beyond the branch.

The Tree is like a management structure. At the top (root) everything is simple. At the bottom are the nitty-gritties. In between, the simple is matched to the nitty-gritty.

The major interfaces are defined toward the root. They must be defined before the modules they tie together are designed. Any change in these interfaces will tend to affect everything beyond. Changes at the bottom (that is, toward the leaves) are easier to make without affecting other parts.

We have used the word *logical* to describe just the information and how it is handled. The description of what medium the information is recorded on and

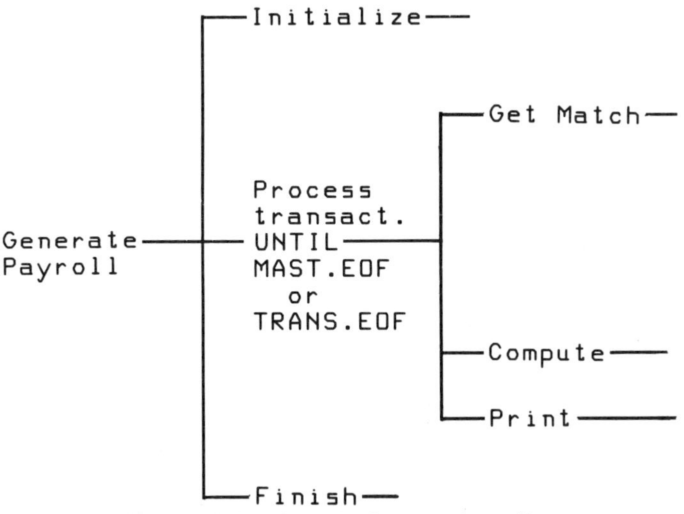

Figure 4.31 Step 3. Top-down development of payroll program.

who or what handles that information is called the *physical* description. The root end of the Tree tends to be a logical description—for example, the principles of an accounts receivable system, without concern for the details of who or what does it. The leaf end represents the details of how the principles are applied to a particular situation—say the accounts receivables for the J. J. Jones Co.—and tends to be a more physical description—describing the interfaces to the specific types of terminals used by J. J. Jones. It is also more detailed, such as describing the heading on the J. J. Jones Co. document.

Prototyping and Tailoring

An effective strategy for developing programs is first to develop a prototype based on the top (root end) of the Tree and selected lower subtrees. A **prototype** is a simpler program that has some of the properties of the final program being developed. Dummy modules, which may not do anything except print the parameters they were called with and return, can be used to substitute for the parts not yet implemented. Such a prototype will incorporate and test the interfaces. The user can see how certain parts of the system work and propose changes. If these changes affect the basic structure at the top, that structure gets resolved before the rest of the Tree is developed.

The prototype based on the top of the Tree may be very simple, because it works only with clean data. This is usually the easiest and quickest part of the program to write. By working only with carefully prepared, clean input data and hand-generated files, a designer can often very quickly write a prototype to demonstrate the principal features of the program. The details of formatting and

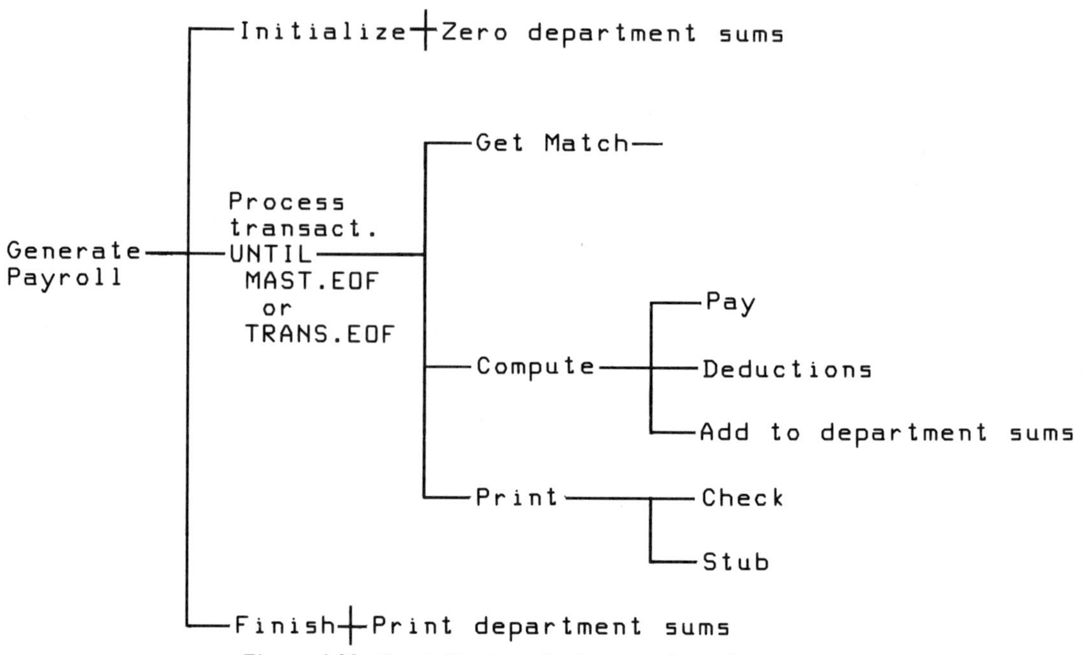

Figure 4.32 Step 4. Top-down development of payroll program.

fixing or rejecting bad input often represent a major portion of the program. They occur at the bottom of the Tree and need not be dealt with in the prototype. Prototyping will be discussed further in Chapter 9.

This strategy also works for developing general application programs that must be tailored to the specific needs and environments of each client. The general algorithms appear at the top of the Tree and remain the same from client to client. Only the bottom of the Tree is changed to accommodate each unique application. The bottom, which is subject to the most change, is also easier to change without affecting large parts of the program.

Modular Programming

A common programming technique is to program the modules one at a time and then put them together to form the whole program. But if the modules are developed before the interfaces are well defined, putting the modules together can be a disaster. The modules may have to be completely rewritten before they can talk to each other. Developing the system top-down as a Tree avoids this problem because the major interfaces must be defined before the modules are written. In Chapter 16 we will discuss developing the interfaces before the modules, which we refer to as the Mortar First—Bricks Later approach.

In a chief programmer team the chief programmer develops the top of the tree, then the programmers can develop the lower branches (Baker, 1972). (We will discuss this type of organization in Chapter 5.) Thus the chief programmer, who is concerned with the interfaces, coordinates the top, while the programmers work out the details at the bottom. The lower parts of the Tree have fewer interactions, allowing the programmers to work more independently.

Correct Hierarchical Development

We cannot easily rid complex programs of all errors by testing and correcting them. It is usually much less costly to spend the care to avoid the error than to find and fix it.

The best way to generate correct programs is by following a discipline of construction. This discipline says that a correct program is produced by putting together correct subprograms using correct constructs. To follow this construction we use the concept of a statement and the building constructs Join, Include, and Or. The Join, Include, and Or correspond, respectively, to the single, wiggly, and double limbs in a Tree. Hamilton and Zeldin (1976) have developed a method called Higher Order Software (HOS) based upon provably correct constructs for building programs. These constructs may be too difficult to be practical for the majority of program development, but they could be invaluable where high reliability is essential. Martin (1985) has brought these rigorous techniques within the reach of a broader group of programmers. The HOS constructs fit very well in the Tree structure and can be used as a useful guide in developing any software.

A statement is a function with input and output (see Figure 4.33). A statement can be constructed from other statements using the three building constructs: Join, Include, and Or. Care must be taken to see that the interfaces (the inputs and outputs) of each constructed statement are properly defined with reference to the statements from which it is constructed, and that the functions accept the defined input and produce the correct output.

These constructs can be used top-down to begin with a desired function and decompose it into smaller functions until reaching correctly implemented existing functions or language statements. The constructs can also be interpreted bottom-up as the way programs are built from existing functions or language statements.

According to convention the inputs and outputs of function*X* are shown as input*X* and output*X*. A list separated with commas means some or all of the items listed. We say that a higher function is being built from lower functions.

The Join builds a higher function by executing lower functions in sequence (see Figure 4.34). Each lower function can use the input provided to the higher function or previous lower functions in the sequence. The output of the higher function must come from the outputs of the lower functions.

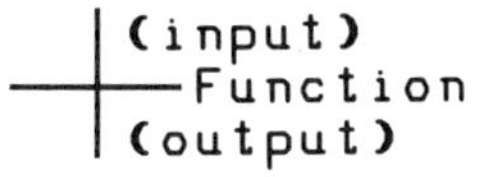

```
 |(input)
-+-Function
 |(output)
```

Figure 4.33 Statement.

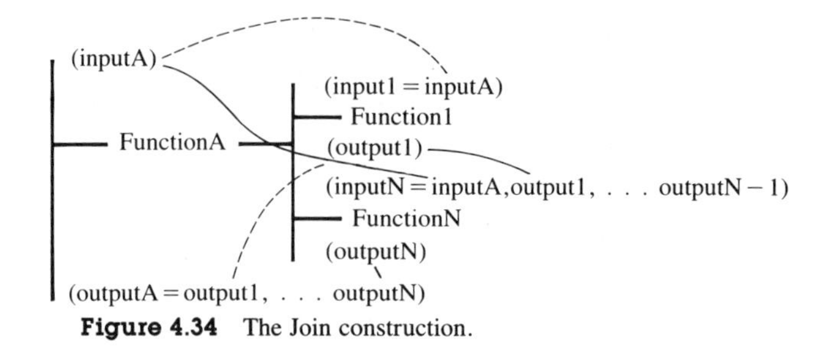

Figure 4.34 The Join construction.

The Include builds a higher function by executing lower functions asynchronously (see Figure 4.35). Each lower function takes the inputs or a subset thereof from the higher function. The output of the higher function is the set of outputs from the lower functions.

The Or builds a higher function from one of several lower functions chosen by a set of conditions (see Figure 4.36). A set of conditions determines which lower function is run. The inputs and outputs of each of the lower functions are the same as the higher function.

Figure 4.37 presents an example, created by James Martin (Martin, 1985), which shows how the statement, Join, Include, and Or constructs can be used to build a stool using the Tree representation. The dotted lines show the relations between the inputs and outputs of the lower and higher functions. We can see that whether this procedure will build a correct stool depends on careful definition and consistency of the interfaces. For example, (top, legs) are the interface between "Make parts" and "Assemble parts." The top must be correctly drilled and the legs correctly sized as the output of "Make parts," so they can be used as the input of "Assemble parts." Building programs using these constructs focuses our attention on those matters that must be ensured if the program is to be correct.

How effectively we will be able to build correct programs using these techniques will depend on how careful we are in ensuring the correctness of the:

1. *Interfaces*—Ensuring that the interfaces (inputs and outputs) are correct and consistent according to the constructions
2. *Functions*—Ensuring that the functions do work correctly with the stated inputs to define correctly the stated outputs

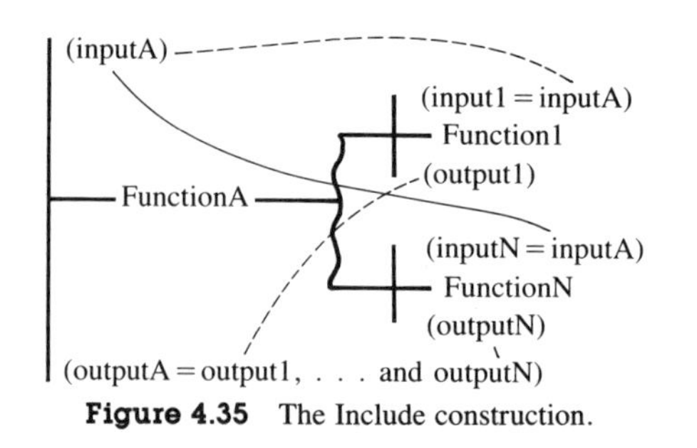

Figure 4.35 The Include construction.

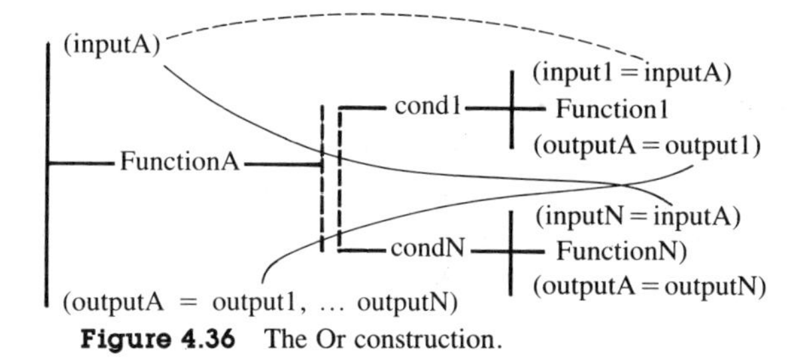

Figure 4.36 The Or construction.

To think through the correctness and consistency of interfaces, we consider:

1. *Structure and Format*—The data are recorded in the correct form (for example, as a single variable, array, or record; integer, string, or real; ASCII or EBCDIC).
2. *Interpretation*—The units and the range (that is, highest and lowest values) are correct.
3. *Value*—Given the structure, format, and interpretation, the correct values are used and produced.
4. *Timing*—In real-time systems, the data are provided at the time they are needed.

To think through the correctness of the functions, we consider:

1. *Inputs and Outputs*—Inputs and outputs are correctly defined with respect to the items we considered earlier for interfaces.
2. *Transformation*—The outputs are correctly produced as transformations on the inputs.

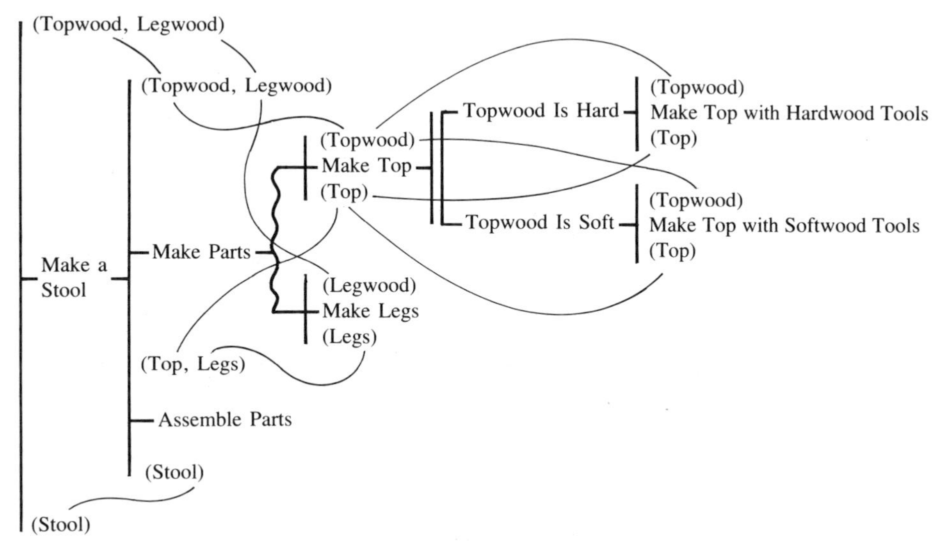

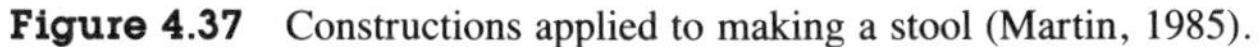

Figure 4.37 Constructions applied to making a stool (Martin, 1985).

3. *Use of Resources*—The function does not use more space for program—or memory and disk space for data—than allotted. Resources are allocated to the functions and interfaces. This may be done starting at the top and guessing at the resources needed by each function and interface. Then as the details are worked out, we can adjust these allocations as needed. Making these adjustments by subtracting resources from one place and adding them to another is referred to as making trade-offs.
4. *Timing*—In real-time systems, establish that if the inputs are provided when needed, the outputs will be provided when needed.

Using Trees for Data Structures

Figures 4.38 and 4.39 show how a Tree can also be used to represent data structures. Here *O* on a branch means "optional," * means "any number of times," and quotation marks enclose literals. Within parentheses is shown the number of numeric digits *(N)* or the number of alphabetic characters *(A)*. Only one branch is taken off each double vertical limb. In Chapter 15 we will discuss how the data structure tree can be used as a guide in developing the program Tree.

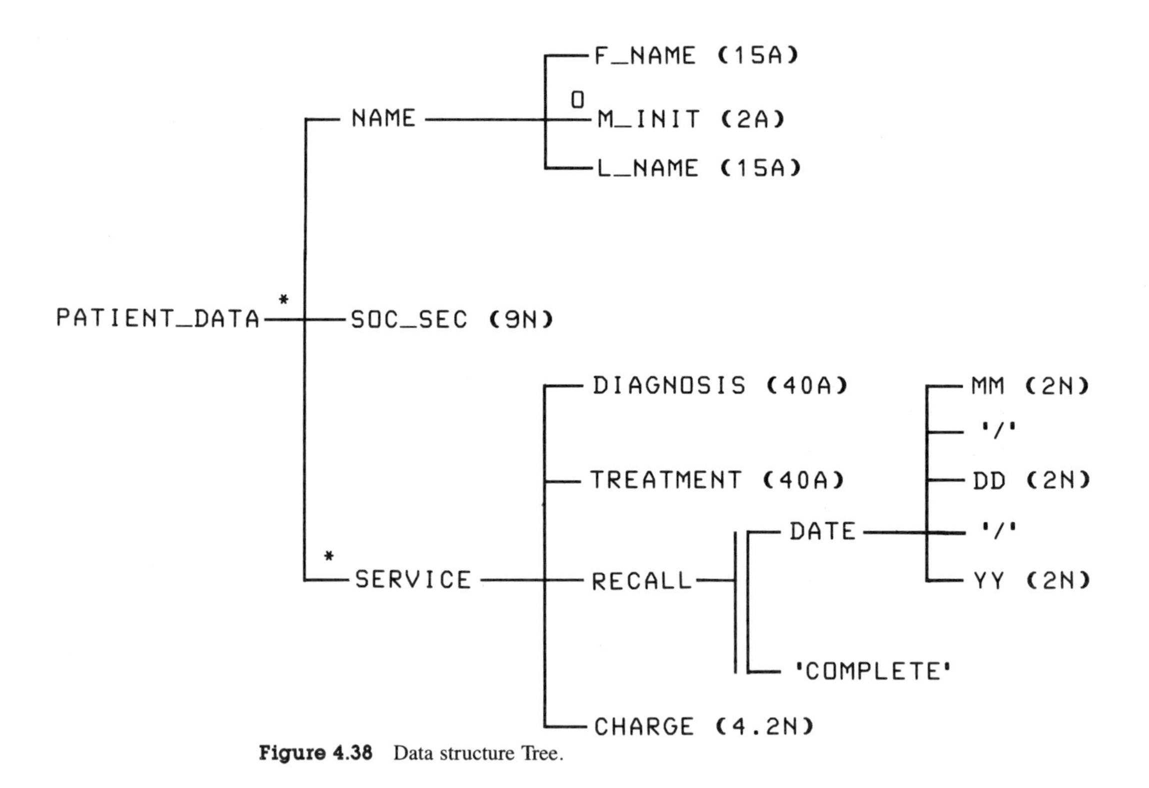

Figure 4.38 Data structure Tree.

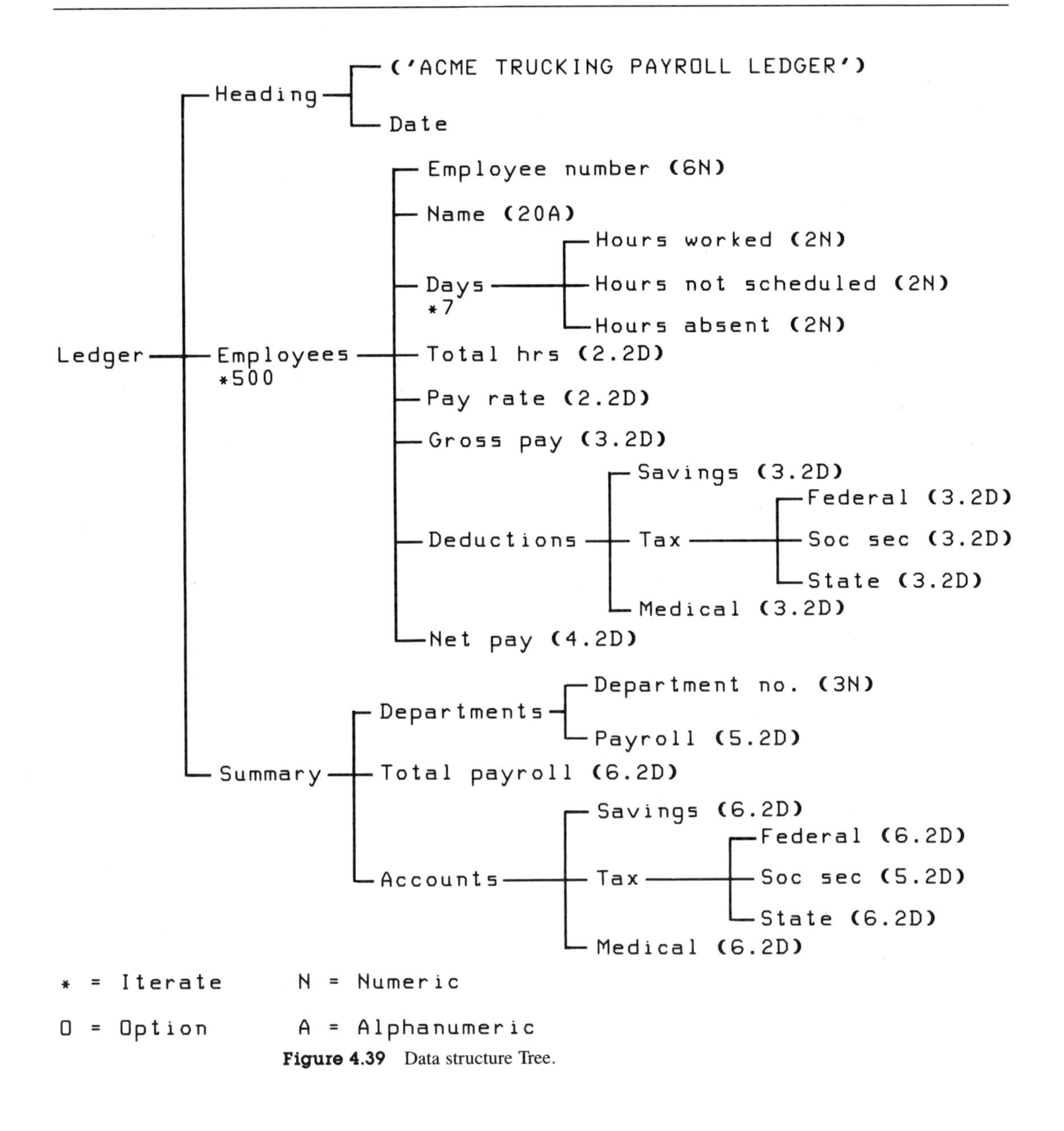

Figure 4.39 Data structure Tree.

Tree Metrics

Software engineering emphasizes the careful development of requirements and plans early in the project. But unless the progress can be measured during this planning, management is likely to get nervous. They will want to push the project too quickly into programming, where they can feel more comfortable about progress because they can count lines of delivered source code.

Thus to get management acceptance of software engineering, we need to be able to measure progress during the planning. In the early planning no lines of source code exist to be counted. But by capturing the results of the planning in a Tree, we can introduce metrics early in the project to show how much progress has been made.

Progress Metric

The Tree gives us a metric that can be applied beginning with the up-front requirements and be carried all the way to code generation. We just count the branches. This is a measure both of the size of the system and how much detail has been developed.

Complexity Metric

Two other measures can be applied to the Tree. McCabe (1976) has developed what he calls a cyclomatic number that is applied to something called the flow graph of a program to obtain an index of complexity. Some data show this number has been useful in predicting the difficulty of maintaining programs. A metric related to McCabe's index when applied to the Tree can be computed as follows:

> Count the number of branches with controls (for example, IF, WHILE, FOR). (Do not count ELSEs.) For each double limb include in the count only the number of branches off the limb minus one. This count is then a measure of the complexity of the program.

Coupling Measure

Another measure tells us how well we have designed the program to have low coupling. It is computed as follows:

> For each branch we count the number of data elements flowing into the branch, plus the number of data elements flowing out of the branch. We divide this count by the level number. Then we sum this result over all branches. Finally we divide this number by the same sum without dividing by the level number.

$$\frac{\text{SUM} \dfrac{\text{data elements in} + \text{data elements out}}{\text{level number}}}{\text{SUM} \quad \text{data elements in} + \text{data elements out}}$$

where SUMs are over all branches

A single variable or an array would be counted as one data element each. An array and an index would be counted as two data elements. The level number is the number of limbs passed to the right of the root.

This gives us a measure of data flow between modules that counts data flows on the higher (lower-level number) branches more heavily than the data flow on the lower branches.

If the Tree is developed on a computer, these metrics can be counted by computer programs.

Computer Editors for Working with Trees

A computer editor with built-in functions for handling outline or tree structures can be used to develop the Tree and record it in a computer file.[2] Using the editor, the initial requirements are entered and the design details are added until finally the program can be compiled from a file built by the editor. The computer, working on this file, can also count branches to measure progress and compute coupling and cohesion metrics. Since the same Tree is used throughout all the phases of the process, information doesn't have to be reentered when moving from one phase to another.

The editor displays the Tree on the screen as an outline (Figure 4.40), which is a collapsed form of the Tree. The branches leaving a limb in the Tree are indented in the outline under the branch entering that limb. As an option, the single, double, or wiggly limbs can be shown to the left of the indented lines as in Figure 4.40. Compare this figure with the Tree in Figure 4.7.

Figure 4.41 shows the top levels of the Tree with just the "Compute standard deviation" subtree expanded to show its details. The computer can be asked to display different levels of detail for different parts of the program so some parts can be seen while others are hidden. This makes the editor an interactive, living document; a great advantage over flipping through dead pieces of paper. If desired, the outline can be printed on the printer as a true Tree.

Summary

When we want more detail and control information than a data flow diagram shows, we can use a tree structure. Three types of tree structures were considered in this chapter: structure charts, Warnier-Orr Diagrams, and Trees (with a capital T).

There are three patterns that can be used in system development. When developers and clients first work together to draw up the requirements specifications,

[2]Brown Bag software of Campbell, CA offers a PC-Outline program for the IBM PC's and compatibles, which has been tailored for this application.

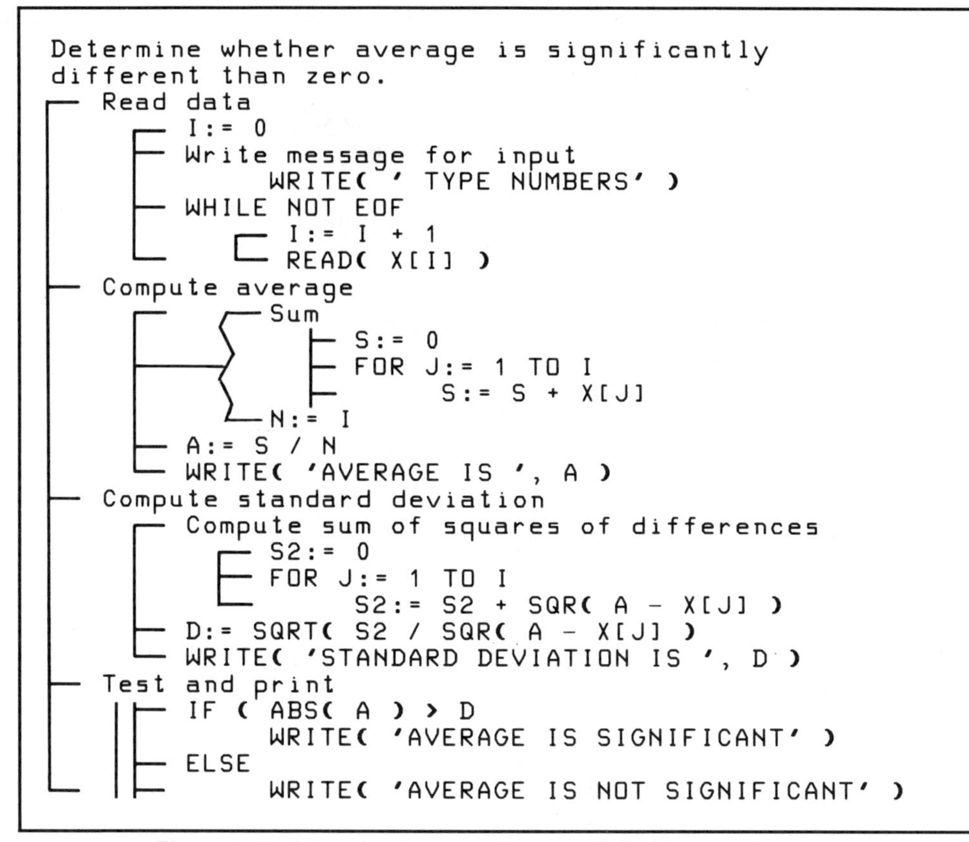

Figure 4.40 Indented outline form of Tree as edited with a tree editor.

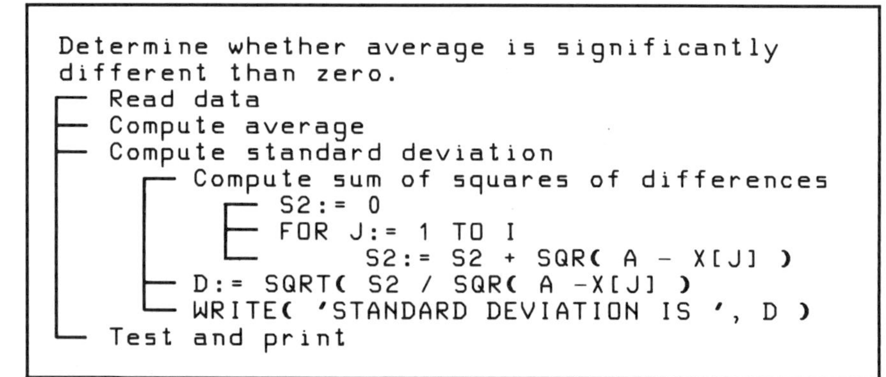

Figure 4.41 Indented outline form of Tree top level with "compute standard deviation" expanded.

they can use DeMarco Data Flow Diagrams. But then the developer must translate this information into structure charts to do the design, and translate again into a source language to compile the program. As an alternative, the developer and client can eliminate the data flow diagramming step by doing the requirements specifications with Warnier-Orr Diagrams. But these diagrams also must be translated into a source language before code can be generated.

A third and preferable way is to use Trees throughout the whole process. The developer and client state the requirements as a Tree. The developer designs the system to meet these requirements using the same Tree. And finally this Tree is used to generate code. No information or time is lost translating between representations. Progress can be measured by counting branches. If the Tree is developed using a computer editor, the computer documents the work, counts the branches, and helps in the analysis of perspective designs.

The Two-Entity Data Flow Diagram and the Tree were developed jointly so that if the client prefers to start with a data flow diagram, the transformation to a Tree is straightforward with no loss of information.

Tree structures are useful in designing programs so that the data flows between modules are small (low coupling) and the functions of the modules can be simply stated (high cohesion). This makes it easy to see what the program does when it is maintained, saving considerably on maintenance costs.

Top-down development requires that we go back and forth between requirements and design, a process called recursion. This becomes much easier when both requirements and design can be expressed in the same Tree. As the Tree grows, the general is turned into the specific, what is wanted is turned into how to do it, and logical descriptions of what happens to information are turned into the real constraints of input and output formats and physical devices.

Trees use branches to hold remarks or control statements, and limbs to show the order of execution. In the early steps of development, the leaves hold statements about what remains to be done. In the final steps, the leaves hold executable program statements. Data flows shown on the branches and limbs provide the compatibility between Trees and TE-DFDs, and make it possible to analyze for coupling and cohesion.

Trees can also be used to represent data structures and these structures can often be used to develop the Tree structure of the program itself.

Exercises

1. For each of the data flow diagrams given in the exercises at the end of Chapter 3, draw the equivalent Tree.
2. Show the data flows in the Trees you got in exercise 1.
3. Develop the Warnier-Orr Diagram for the standard deviation program shown as a Tree in Figure 4.7. What are the differences between what can be represented by the two techniques?

4. Develop a more complete Tree for a payroll program working from the example of the Warnier-Orr Diagram in Figure 4.6 and the partially developed Tree in Figure 4.28.
5. Can you develop examples of data flow diagrams of real systems that are not hierarchical?
6. Represent as a Tree the data flow diagram that Pressman (1979) uses to illustrate a program for a computer embedded in a car instrument panel. Compare the method he presents for converting a DeMarco Data Flow Diagram to a structure chart (related to our Tree) and consider:
   a. How systematic are the two methods?
   b. Do they give consistent results, or is there a great deal of freedom in how choices are made?
   c. Which method could you more easily automate?
   d. Which method preserves the asynchronous properties of the data flow diagram?
   e. Is a precedence procedure needed to preserve the asynchronous properties of the data flow diagram?

References

Alagic, Saud, and Michael A. Arbib. *The Design of Well-Structured and Correct Programs*. New York: Springer-Verlag, 1978.

Clark, R. Lawrence. "A Linguistic Contribution of GOTO-less Programming." *Datamation* 19, no. 12 (1973), 62–63.

Dijkstra, E. W. "GOTO Statement Considered Harmful." *Communications of the ACM* 11, no. 3 (1968), 147–148.

Halstead, Maurice H. *Elements of Software Science*. New York: Elsevier North Holland, 1977.

Hamilton, Margaret, and Saydean Zeldin. "Higher Order Software—A Methodology for Defining Software." *IEEE Transactions on Software Engineering* SE-2, no. 1 (1976), 9–32.

Hansen, Kirk. *Data Structured Program Design*. Englewood Cliffs, N.J.: Prentice-Hall, 1986.

Higgins, D. A. *Program Design and Construction*. Englewood Cliffs, N.J.: Prentice-Hall, 1983.

Jackson, M. A. *Principles of Program Design*. London: Academic Press, 1975.

Jackson, Michael A. *System Development*. Englewood Cliffs, N.J.: Prentice/Hall International, 1983.

Linger, R. C., H. D. Mills, and B. I. Witt. *Structured Programming: Theory and Practice*. Reading, Mass.: Addison-Wesley, 1979.

Martin, James. *System Design from Provably Correct Constructs*. Englewood Cliffs, N.J.: Prentice-Hall, 1985.

Martin, James, and Carma McClure. *Diagramming Techniques for Analysts and Programmers*. Englewood Cliffs, N.J.: Prentice-Hall, 1985.

Martin, James, and Carma McClure. *Action Diagrams: Clearly Structured Program Design*. Englewood Cliffs, N.J.: Prentice-Hall, 1985.

McCabe, Thomas J. "A Complexity Measure." *IEEE Transactions on Software Engineering* SE-2, no. 4 (1976), 308–320.

Orr, K. T. *Structured Requirements Specification*. Topeka, Kans.: Orr Associates, 1981.

Orr, Kenneth T. *Structured Systems Development*. New York: Yourdon Press, 1977.

Page-Jones, M. *The Practical Guide to Structured System Design*. New York: Yourdon Press, 1980.

Pressman, Roger S. *Software Engineering: A Practitioner's Approach*. New York: McGraw-Hill, 1979.

Steward, Donald V. "A Tale of Hope for Anyone . . . Lost in the Forest Primeval." *Computerworld,* March 19, 1984.

Warnier, J.-D. *Logical Construction of Systems*. 3d ed. New York: Van Nostrand Reinhold, 1980.

Wirth, N. "Program Development by Stepwise Refinement." Comm. ACM, vol. 14, no. 4 (April 1971), 221–227.

Yourdon, Edward, and L. L. Constantine. *Structured Design: Fundamentals of a Discipline of Computer Program and Systems Design*. New York: Yourdon Press, 1979.

Managing the Software Development Process

In Part 2, we've described how to represent the software product. In this part we'll discuss how to manage the process that creates the product.

Our focus, therefore, is project management. Within the wider area of management, project management is concerned with the use of resources to achieve specific objectives within a specified period of time. A project has a well-defined start and finish. Managing a project is like running a company that is formed at the beginning of the project and goes out of business when the project ends. The project lives within a company or larger organization that has a longer life and can provide services, such as personnel and accounting, to the project.

Chapter 5 begins with the people involved in the project: what are their expectations, how are those expectations negotiated, and how do they react to change? The principles of management discussed here are particularly pertinent to software engineering. Chapter 6 goes on to consider how to estimate the cost and time to do the work. Finally, Chapter 7 looks at how to plan, schedule, and control to get the work done.

5

Managing People and Expectations

- A project begins with the identification of failed expectations with the current system. The client and the project management develop descriptions and estimates that lead to new expectations.
- Managers must guide the project within constraints, which define what cannot be done, and capabilities, which define what can be done.
- The expectations of the client, who expects to receive a system, are linked to the expectations of project members, who expect to be rewarded for their contributions to the project. Managers have to delegate tasks and motivate subordinates so that their work combines to satisfy the client's goals.
- Often the patterns of communication within an organization inhibit new ideas or realistic appraisals from reaching management.
- Effective organizations have an atmosphere that is receptive to change and that accepts new ideas unless there is a compelling reason to reject them.
- In software development, management and subordinates may be arranged in any of several possible organizational structures, including hierarchical, matrix, or chief-programmer team.
- In addition to choosing a capable team, managers must motivate members of the project, wield power effectively, negotiate with clients and subordinates, and resolve conflicts.

In this chapter we will consider the people who will propose, accept, make, and live with the consequences of a change. We look at how their expectations are negotiated and linked together to accomplish the goals of the project.

Software engineering is concerned with projects of such scope that many people must be involved. Therefore we are concerned with how to coordinate the work of these people to accomplish the goals of the project. A project begins with expectations to be met and ends with an evaluation of whether they were met.

From Past Failures to New Expectations

As we said in Chapter 2, a proposal to make a change begins with the recognition of a failure of expectations. The client identifies a lack of match between a need to be satisfied and a means to satisfy it. The change has been completed when there is finally a satisfaction of these expectations, a recognition that these expectations cannot be satisfied, or a change in the expectations themselves. We'll begin this chapter by seeing how a project moves from the identification of a failure of expectations to a definition of new expectations.

A failed expectation can occur because:

1. The expectations for an earlier change had not been satisfied (for example, the new order entry system does not handle the volume of orders that was intended).
2. The system has changed its performance from what had been expected (for example, the order entry system no longer handles the volume of orders it once did because the data base has become so large that the response is now too slow).
3. Someone has changed his or her expectations of what the performance should be (for example, an analysis of other order entry systems shows that they perform better than ours, or a new boss has expectations that are different from his or her predecessor's).
4. A new opportunity or technology has changed the expectations of what can be done (for example, microcomputer costs have come down enough to warrant considering a new microcomputer-based system).
5. A change in the environment requires that we make a compensating change in what the system must do (for example, we are losing customers because our competition can provide faster order processing, or the government now requires that we keep records we had not kept before).
6. A change in mission requires the system to do something it had not been intended to do before (for example, the company now wants to handle foreign parts in its order entry system).

If we were selling a product that had already been made, like clothing off the rack, we could easily establish clear expectations of what our customer will receive and what the item will cost. The product is available to be seen and tested. We already know what it costs to make. But in systems projects we are selling a promise to make something that does not now exist. We may not know for sure whether we can make the product, and even if we can, we may not know what the product will cost. The entire process involves risks.

Before any work is started, clients usually want to know what benefits they can expect. Investors want to know the chances they will get their money back with a reasonable return. The people making the change, therefore, must estimate how long the project will take, what it will cost, how the system will perform, what it will cost to operate, and what the benefits will be once the change is made. They must also estimate the chances of their estimates being wrong so the client and investor can determine the risks they will be taking.

These estimates establish expectations. The success or failure of a project is determined by whether these expectations are met. Expectations are negotiated with clients before they give their consent to the project. Clients generally want to know what system they can expect to receive, what they can expect that system will do for them, how they must operate it, what benefits they can expect it will provide, and what they can expect it will cost them in time and money.

Once the change is made, it will be evaluated by how well it met those expectations. Thus, before a change is initiated, all parties must understand who has what expectations and must negotiate and agree on exactly what these expectations are. We will discuss in Chapter 12 how to define and negotiate these expectations, and in Chapter 14 how to evaluate them.

As the project proceeds, more of the work lies behind us, and less work lies ahead. Estimates for the cost and time for what remains to be done should improve. Of course, expectations can also change during the project, either because of some external influence on what the client expects, or because of what is learned as the project unfolds. These revised estimates and expectations are shared with the client in regular reviews. These reviews themselves may cause the client's expectations to change. Anytime the latest estimates of cost and benefits are not worth the remaining work, the project should be stopped, perhaps spending sufficient time to record the work done to date so that the lessons from this project can be invested in future projects. But when the project is finished, if it meets the client's expectations at that time, it is a success. If not, it is less than a success.

Working within Constraints and Capabilities

Managers must meet these new expectations while working within certain constraints that define what they cannot do, and with certain capabilities that define what they can do. These constraints and capabilities may be determined by policies established by the manager's own supervisor or by the client. Other constraints and capabilities are set by the available resources and by technical limitations.

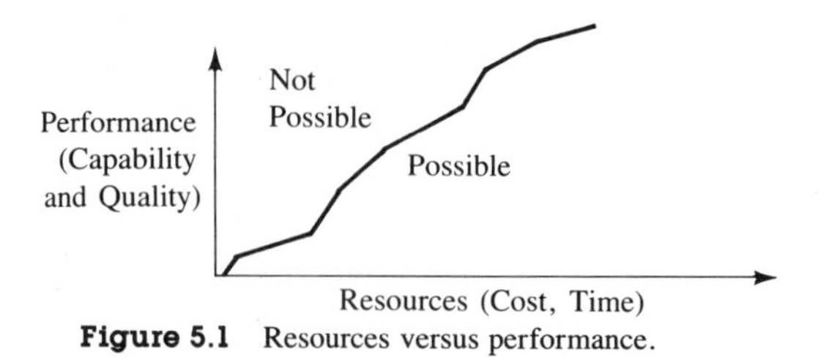

Figure 5.1 Resources versus performance.

Managers must allocate resources to where they are most needed, and monitor and control how these resources are used. They must work within the standard constraints of time, cost, capability, and quality. To diagram this concept so we can see it in two dimensions, we show cost and time merged as one dimension called "Resources." The other dimension, called "Performance," is a combination of capability (what the system will do) and quality (how well it will do it) (see Figure 5.1).

Here we can see the boundary between the area that we can work within, and the area beyond what we can reach because there is only so much we can do with certain resources. Obviously we want to get the greatest performance for the least expenditure of resources within the possible region.

We can stay within the time and cost constraints by not meeting the expectations for capability and quality, or we can meet the time constraint by increasing the costs. However, there are limits to how much we can reduce the time no matter what we are prepared to spend. It is extremely difficult to meet expectations on all the constraints simultaneously.

Before obtaining the resources, managers must estimate the amount of each resource required and when it will be needed. Managers negotiate to get these resources by giving the providers an expectation of what they will receive in return.

With the information available beforehand, we will often find it difficult to determine precisely what resources will be needed. Resources must be carefully monitored and controlled during the project to keep their expenditures within acceptable limits.

Managing Expectations

A common theme in the following sections will be to show the linkages of who has what expectations of whom. These linkages must be developed in order to make a change. A change is proposed because someone expects to benefit from it. But usually the person whom it benefits will not be the one whose effort is required to make the change.

Some changes to a software system are easy to make because no linkages are required. Only one person is affected; he or she is capable of making the change and does so. Other changes can be hard to make because there are so many linkages of expectations to be made and maintained. Without any one of these linkages the project can fail. The software engineer must be involved in making sure that these linkages are set up and maintained as long as needed to achieve the goals of the project.

This system of expectations must be established by communications and negotiations. Negotiations lead to a contract between people stating what expectations each has of the other. The contract can be formal or informal, written or verbal.

A fundamental premise of project management is the need to maintain expectations. As soon as you know you cannot maintain the original expectations, you must negotiate new ones. Thus, as the project progresses, a large part of the project manager's role is to manage expectations, those of the client as well as those of the project team. In this section we'll see how these expectations are connected together.

Linkage of Expectations

To create a product, the manager must bring together a number of people with different skills, and he or she must help them so that they work together effectively. Some of these people may have the technical knowledge to plan the system; others may have the skills to do the actual work, or the accounting background to monitor it, and so on.

The client is the one who gets the benefits of having the product. But if the product is to be built, the client must distribute some part of those benefits to the contributors who build it. Most of these contributors want their benefits as they contribute. But the client receives benefits only after the system is built and operational. So someone must finance the project by putting up the money now to pay the other contributors. In exchange, he or she expects to get back from the client the original money plus interest. Management has to make all these arrangements if the project is to succeed.

It is often said that in order for a project to be successful, the benefits of having the product must be higher than the costs of producing it. Actually a stronger condition must be met. Each person whose contributions are required to produce the product must perceive that the benefits bestowed upon him or her are worth the effort he or she contributes.

The benefits to the contributors may come in many forms. The most obvious benefit is money. Once the contributor has at least a certain minimum amount of money, other factors can become important: the opportunity to take pride in the product, the chance to obtain more and better experience, or the possibility of sharing in the rewards when the product is finished.

Thus before the change can be made, management must set up a system for distributing benefits from those who will receive the benefits in the future to those who will demand their share of the benefits now if they are to make the

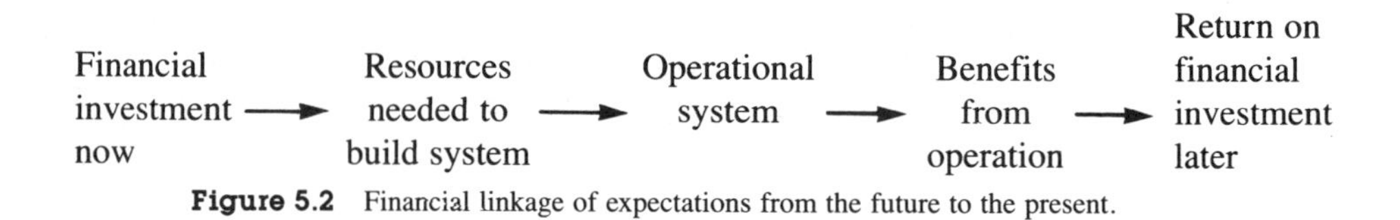

Figure 5.2 Financial linkage of expectations from the future to the present.

change. This distribution is handled by establishing links between people through a system of expectations.

These linkages have two dimensions: benefits moving from person to person, and benefits moving in time from the future to the present. Moving benefits between people is done with the exchange of money. Moving benefits in time is done with investment. An investment is money that someone puts up now with the hope of receiving some benefit in the future. The future benefit must be large enough to make it worth taking the risk that the change might not produce the benefits anticipated. Since the investor has opportunities to invest money elsewhere, the expected return for this investment must be competitive with the returns for taking other comparable risks.

We may diagram the financial linkage as in Figure 5.2.

The manager, software engineer, and possibly the client negotiate these financial linkages.

Delegation

Although system changes sometimes involve just one person, software engineering usually concerns changes that are larger than can be made by one person. So the change is made by managing the work of many people. One definition of management is getting work done by other people. A project is a planned undertaking that requires focusing resources on achieving specified expectations.

To manage a project we must plan and assign work to subordinates. Planning involves defining tasks in terms of *what, when,* and *where* so that if the tasks are done as planned, the expectations placed on the manager will be satisfied.

Delegation involves dividing the job into tasks, then defining for subordinates the expectations of how they will do these tasks. Managers bear the risk that the tasks might not be done as expected, leaving them, as managers, in the position of not meeting the expectations that were placed on them. During the time between when the tasks are assigned and when it can be seen whether the expectations have been met, managers can become nervous. But once the tasks are finished, managers can compare their expectations with the actual performance. (See Figure 5.3.)

It is easier to coordinate decisions made within our own heads than to coordinate the decisions made in the heads of other people. Thus it is easier to make decisions ourselves than to delegate them. But managers must delegate because they cannot do all the work themselves. We might call the delegation of authority

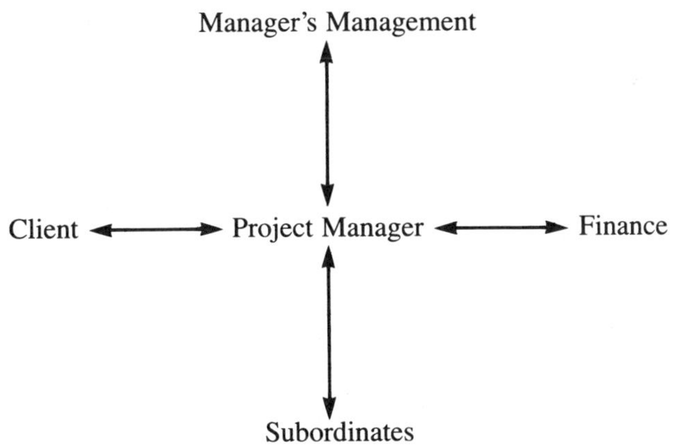

Figure 5.3 Project manager's linkages of expectations.

to make decisions **distributed thinking.** It is similar to the idea of distributed processing in computing and has similar problems.

Managers must motivate their subordinates to want to meet expectations. They motivate by clearly defining what they expect, and to do so, they rely largely on the subordinates' normal desire to do a good job. In a negative sense management may also motivate by spelling out the consequences of not meeting expectations. We'll describe some strategies for motivating people later in this chapter.

Expectations are passed to subordinates in the form of policies, procedures, and objectives. **Policies** are guidelines showing how to make decisions. **Procedures** are step-by-step descriptions of how to perform specific tasks. **Objectives** are specific things to be accomplished, such as a piece of work or a goal to be met.

Table 5.1 shows how managers pass down expectations to their subordinates. Management hopes that all the objectives can be met by their subordinates working within the guidelines management establishes. If not, exceptions must be passed back for the manager to rule on. If too many exceptions are passed back, the manager gains little release of his or her own time by delegating work to a subordinate. Thus managers would generally prefer not to have such exceptions.

One approach to management is called **Management by Objective.** Here manager and subordinate negotiate mutually agreeable expectations. A contract is then drawn up, stating these expectations explicitly. Whether or not the agreement is formally written, the subordinates must know clearly what is expected of them and what consequences will follow if expectations are or are not met.

Since managers are evaluated by the expectations their superiors place on them, they must accept responsibility for the work done for them by subordinates. One model of how managers sometimes delegate, which we offer as an example of what *not* to do, might be called the **Clone Model.** The Clone Model says that the manager feels most comfortable if that work is done precisely as he or she

Table 5.1 Expectations between manager and a subordinate

Manager		Subordinate
Define expectations: policies, procedures and objectives	→	Operates under these expectations
	←	Passes back exceptions to policies, or problems in meeting objectives—to be resolved by the manager
Rules on exceptions	→	
	←	Passes back reports of performance
Conveys acceptance or rewards and punishments based on comparing performance and expectations	→	
	←	Conveys information that goes beyond defined expectations (e.g., ideas for improvements, or information that indicates the system does not behave as the manager may have thought)
Presents manager with having to make decisions that were not expected and are not well structured		

would do it. Such a manager evaluates the work not by the objectives negotiated beforehand, but by how good a clone the subordinate is.

Skilled managers can collect around them people who have capabilities and approaches to problems that complement rather than clone their own capabilities and approaches. They must obtain the subordinates' commitment to management's expectations instead of to management's way of behavior. Managers must let subordinates perform their work their way. This requires unusual skill in evaluating and coordinating thc work of others.

Information and Decision Handling in the Organization

Let's consider some characteristics of how people handle information, ideas, and problems, with particular attention to how this affects the information flow and decision processes in an organization.

We all operate with an **agenda of concerns** that limits the number of ideas or problems we can consider at one time. When that agenda is full, a new problem cannot be considered until an old problem has been dealt with and removed from the agenda. This is similar to Miller's Principle, which concerns how much information we can handle at one time.

The best time to present a new idea or problem to a manager is when an old problem has just been resolved. A good way to be sure of being there at the right time is to have helped resolve an old problem.

All of us work with an extremely simple set of models in our heads that explain to us how things ought to behave and tell us how to make decisions. When we have a decision to make, if we have several models that seem to apply to the situation but give us different decisions, we feel uncomfortable. We then seek rationalizations as to why one model should apply so we can ignore the other conflicting models.

Perhaps we can tell ourselves that the information supporting one model came from reliable sources while the information supporting the others came from unreliable sources. Maybe we can argue that the circumstances do not apply to the other models. A battle goes on in our head. We may have to find reasons to ignore or reinterpret certain information to get one model to win and the others to surrender before we can stop the battle going on in our minds. Sometimes the reasons we use to stop the battle are not very logical, but they work. We call this **rationalization.** Rationalization is an important part of decision making.

We tend to want to keep things simple and orderly, and we often reject information that upsets our tidy view of the world. Thus a new idea presented to a manager may upset his or her simple, orderly structure. This fact of human nature makes it hard to get new ideas considered.

Sometimes people do not behave the way we would expect them to. We might even think their behavior is irrational. From the bottom of the organization looking up, we may often think managers behave irrationally. But what may be irrational from our perspective may be very rational if only we understood what models they were using and how they were rationalizing them. We refer to this concept of behavior as "rationalization—a rational explanation of irrational behavior."

From the preceding analysis we can begin to see why managers often tend to:

1. Avoid new information that requires them to change their view of how things behave
2. Avoid new ideas that upset their tidy views
3. Avoid risk
4. Avoid change, even though no change may cause a higher risk
5. Avoid delegating any more than what they absolutely need to.

This tendency may help explain why managers are often overloaded, have jammed communication lines, are not ready to handle exceptions and problems, and do not want to face new ideas.

Managers also feel uncomfortable when their subordinates use methods they do not understand. Managers will generally feel more comfortable with methods they themselves used when they were moving up through the organization, holding the positions now occupied by their subordinates. Management may be reluctant to see changes in the methods that have occurred since they did this type of work. They may feel particularly uncomfortable and even threatened by subordinates who are from a different discipline than the one they learned.

The introduction of software engineering methods particularly suffers from these problems. Since software engineering is relatively new, it is not part of many managers' experience. Often, though, these techniques can be used without the need for management consent. (The use of Trees may save enough time to convert them later into flowcharts if that is what the manager or client really wants.) But once the subordinates have used the new techniques themselves, they will require them when they become managers.

Managers who are reluctant to delegate tend to take on too much work themselves, leaving little or no reserve time or positions on their agenda of concerns to handle exceptions, ideas, or new views of reality that are passed up the communication chain from subordinates. This tendency produces an environment that rejects innovation or any change introduced as a consequence of suggestions from below. Communication is down, but not up.

The following poem was written by an executive of an international company (Drucker, 1974):

Along this tree
From root to crown
Ideas flow up
And vetoes down.

Information that reflects a good situation is preferred over information that reflects a bad situation. Good information seldom requires a decision or action. Information that confirms that the system works according to management's models is preferred over information that contradicts their models.

In this environment subordinates often learn to tell management what they believe management wants to hear. Subordinates are more comfortable if their managers are comfortable. The goal of keeping one's manager comfortable often conflicts with the project's goals. Getting caught between the manager and the project can cause a great deal of stress. It is usually best to resolve such conflicts in favor of what the manager wants.

We have developed a facetious "organizational chart" in Figure 5.4 to illustrate this phenomenon.

At the bottom of the organization the subordinates must confront and understand the problems. Their mouths are downturned. But knowing their manager's preference for good news and their desire to look better than the others at their same level, they report the most optimistic news they can. This good news prop-

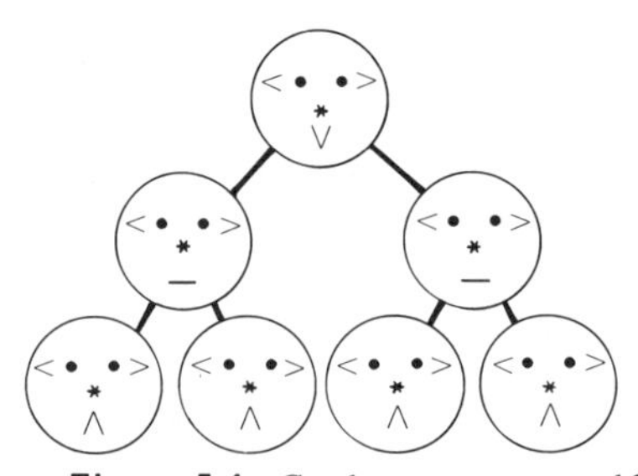

Figure 5.4 Good news goes up and bad news down.

agates to the top levels of management, who are unaware of problems and, consequently, very happy. An exaggeration, perhaps. But people with experience in business and industry will recognize the phenomenon.

Much knowledge about a company's operation can be learned from those at the bottom of the organization. Workers see how improvements can be made to the product, the process of making it, or relations with the customers. Initially, they may be concerned about how they and the company can do better. But if they are frustrated in trying to attract management's attention to possible improvements, they eventually lose respect for management and for the value of their own work. "If management doesn't care, why should I?" becomes their attitude. Their work deteriorates and the company is the loser.

This pattern will be reversed as more channels are provided for information to flow up as well as down. It has been a successful approach in some Japanese companies and some American high tech companies for several years. Many American companies, however, are only paying lip service to this concept; upward communication is still stopped at the higher levels of management.

Managers have limited time to consider new ideas, not only because of the problems discussed thus far, but because they have responsibility for many subordinates. Each subordinate can receive only a small fraction of the manager's attention, that fraction being smaller the higher in the organization the manager's position.

As one goes higher in most organizational ladders, more and more significant decisions are made with less and less information. This means that the information provided to management should be in a form that is brief and digestible.

When our work is reviewed by a manager, the life or death of the results of months of our hard effort will be determined in just a few minutes with very little information. It is in our interest to package that information so management can make the best possible decision. Our own success depends on our ability to communicate to management the essence of our work clearly and succinctly. Strategies for communicating are considered in Chapter 10.

Reception to Change

From the preceding section we can see why it is often difficult to get a fair hearing for a new idea.

In a complex system any changes are likely to have many consequences, some of which are not anticipated. So change to a complex system always involves a risk. Mature managers know the possibility of failure, but they also understand that a system that is not open to change may not improve, or may die because it cannot adapt to changing conditions and new needs.

It is easy to suggest a new idea, but it is somewhat more difficult to accept the responsibilities for the consequences of implementing an idea. Often people who suggest ideas make an implicit assumption that they will gladly take the credit for the idea if it works, but others should take the responsibilities for the consequences if it does not work.

A study of effective organizations has shown that in such organizations if there is no clear and compelling reason to reject an idea, the answer is, "yes, go ahead and do it." Less effective organizations tend to be biased toward finding reasons for saying no (Peters and Waterman, 1983).

There are many reasons for saying no to new ideas. "It can't be any good because if it were, we would have thought of it." This response is often called the "Not Invented Here" syndrome. Another reason is, "If this idea were any good, someone else would have done it years ago." One answer to this rejection is that just because the idea was thought of years ago and wasn't done does not mean it wasn't a good idea. Only a very few of many thousands of good ideas are ever implemented. And when an idea is first implemented, usually it is not the first time it was proposed.

Many good ideas—in information processing as well as elsewhere—were implemented only years after they were suggested. For example, in 1747 the Scottish naval surgeon James Lind demonstrated experimentally that citrus fruits in sailors' diets can prevent or cure scurvy. He proposed a method for preserving citrus fruits during extended voyages to make it practical to supplement the daily diet of the sailors. Thus he had both defined the problem and shown the solution. But it was not until 1795 that his idea was implemented. By one estimate more than 100,000 British sailors died of scurvy during the intervening forty-eight years, more than the number lost in combat during those same years. It was another seventy years and many more thousands of deaths before the idea was adopted by the merchant marine, and another forty-seven years after that before it was adopted by the American navy. Even Lind wasn't the first to propose the idea. The Danes had implemented it many years before it was suggested by Lind.

Other examples of rejected ideas include the case of the automobile engineer who warned management about the potentially deadly problem with the Pinto's fuel tank, which could explode on impact. Similarly, at least two engineers warned of the problem that later resulted in an accident at Three Mile Island and there were warnings before the Challenger tragedy. The people who understand the problems often are not able to convince those who have the power to make a change that a change is both necessary and possible.

Thomas Edison found that he had great difficulty selling his early inventions even when he could demonstrate their utility. Thereafter, once he had the basic idea, he set up the financing and marketing for his inventions before he bothered to work out the technical details. He designed to meet needs, building on the ideas of others and his own abilities to solve the practical problems that stood in the way of their success. He formed research teams to solve problems. His real strength was not just inventing, but getting the financial backing and putting together all the other ingredients necessary to make his inventions succeed.

Two American statisticians, W. Edwards Deming and Joseph M. Juran, failed to get a hearing for their methods of quality assurance in the United States. The

Japanese, who were recovering from World War II and were concerned about a bad reputation for low quality goods, did listen. Today, now that the Japanese have taken some markets away from us with their reputation for high quality, some American industries are also beginning to listen.

Developing the Organization

A manager must create an environment in which his or her subordinates can do the work. They need space and furniture, uninterrupted time, equipment such as personal computer workstations, help with administrative tasks such as accounting for attendance and planning time off, and so on. Much of a manager's time involves negotiating to obtain the resources and to produce the environment that his or her subordinates need to do their work. An essential part of the environment is an organization that indicates the natural pattern of authority and communication. An **organization** describes the management-subordinate structure that tells individuals whose expectations they are to meet, to whom they pass back exceptions, and by whom they will be evaluated.

The most common type of organization is hierarchical. Each person reports to only one manager, but he or she may have charge over several people in subordinate positions. The **hierarchical organization** has the structure of a tree, as in Figure 5.5.

People have training and experience that make them suited to do particular types of work such as programming or accounting. The company has a long-term responsibility for the development of each employee so that he or she is able to perform those functions. In a hierarchical organization each person works for a manager who is concerned with the function that employee performs and the application of that function to a project.

In a project-oriented organization you may report to both a manager responsible for your functional role and to one or more managers responsible for the projects to which you apply your function. Such an organization is called a **matrix** or **network organization** (see Figure 5.6). Here the rows and columns of the matrix correspond to function and project. The term is also used for any system in which one reports to more than one manager. This arrangement gives you the chance to be evaluated by several people instead of just one, thus reducing the problems

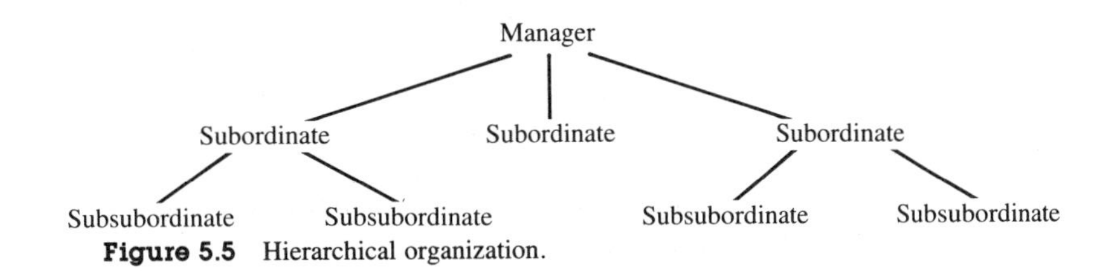

Figure 5.5 Hierarchical organization.

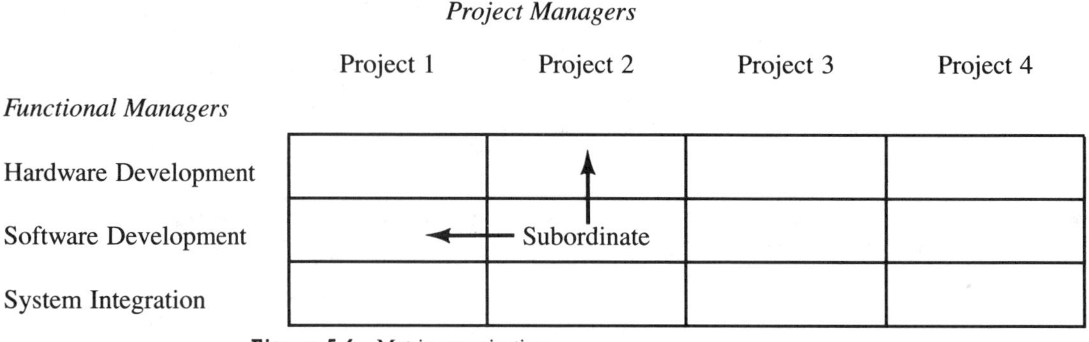

Figure 5.6 Matrix organization.

that may occur due to personality incompatibilities. On the other hand, you may become caught in the middle, not knowing whom to respond to as both managers negotiate over what you are to do with your time. You may also not know who is going to look after your interests when opportunities arise for training or promotion.

In a small project where the client for a project is within the company, the project may be coordinated by a **task force**. The task force is composed of the client and key people representing the various functions to be applied to the project. These people usually serve on the task force while continuing to maintain their usual responsibilities. The task force guides the day-to-day operation of the project. By having the client serve on the task force, the other members become better aware of his or her needs. Also the client becomes part of the work and thus is better prepared to accept the final product.

A **steering committee** of key managers may be set up in conjunction with a task force to provide longer term, month-by-month, high-level management guidance to the project. The steering committee maintains management commitment to the project, so the managers are already familiar with the project when their help is requested. Also they will be more inclined to support the final system because of their involvement in the project.

In a **chief programmer team,** an organization sometimes used for software development, everyone works around the chief programmer, similar to the way a surgical team works around the chief surgeon to improve his or her performance. The chief programmer does all the system and interface design, assigns the programming of modules to individual programmers, and coordinates the overall effort. A librarian handles all the running of programs and record keeping. An administrator may handle the interfaces with the client and the administrative duties such as time cards and vacation schedules. Others may contribute as experts in specific areas such as how to use certain tools. An assistant chief programmer works closely with the chief programmer. He or she may be responsible for integration and testing, and can assume the responsibilities of the chief programmer when necessary.

A chief programmer team is an attempt to retain as much coordination as possible within the head of one person. But if this person gets overloaded, the team can fall apart. The subordinates become frustrated because they don't know what to do and can see that their efforts are uncoordinated and wasted.

Management Skills

Many of the managerial tasks we've described thus far have in some way involved working with other people. In this last section we'll discuss some of the specific theories and tactics for assisting the subordinates in one's team to work together on a project.

Acquiring, Managing, and Motivating People

At the beginning of a project the project manager must acquire people with the capabilities needed for the project. They may be hired from the outside or reassigned from other projects that are coming to completion within the company. Often the choice of people brought into a new project has more to do with who is *available* as the other projects wind down than the actual personnel the project manager wants. Later, as the work of this new project winds down, the project manager will see people pulled off this project to work on another.

While these people report to the project manager, he or she must motivate them to be willing to commit themselves to the success of the project. And the manager should be concerned with helping them to develop and realize their career plans. This latter role may involve encouraging team members to take training.

Motivating people is an art. There are several theories on how people are motivated. These theories are constantly growing in number and being revised. We will briefly mention some of the now classic theories.

McGregor has proposed two theories (Rosenau, 1981). According to Theory *X*, people will do as little as they can get by with. They have to be watched constantly and must be motivated with rewards and punishments. Theory *Y* proposes that people are responsible and want to work to gain the satisfaction of accomplishment. They need only to be directed as to what work to do, then supported and encouraged. The more modern approaches lean toward Theory *Y*.

McGregor's models have now been extended with the addition of a new theory (Ouchi, 1981). Theory *Z* proposes that people gain satisfaction from the accomplishments of the group to which they belong. (Although called the Japanese model, many American companies such as IBM, Hewlett-Packard, and Intel use some of these same techniques.) Generally American industries tend to emphasize individuality and competition, while Japanese industries tend to emphasize satisfaction arising from the performance of the group.

Maslow's theory of the hierarchy of needs says that each successively higher level of need is sought only after needs at lower levels have been satisfied (Rosenau, 1981). Once the needs at one level are satisfied, greater satisfaction of them no longer has a motivating effect as one seeks to satisfy needs at the next higher level. The lowest level is physiological—the need to eat, sleep, have shelter, and so on. The second level concerns safety and security, the third is social, the fourth is esteem and ego, and the fifth or highest level involves self-actualization or fulfillment.

Hertzberg recognizes two major levels (Rosenau, 1981). The first, which he calls "hygiene factors," includes company policies, supervision, work conditions, and salary. Once provided, they are taken for granted and no longer serve as motivators. At the higher level employees are motivated by achievement, recognition, and pride in the work itself.

One factor that appears very important to motivation is feedback. Employees tend to interpret no feedback as an indication they are not doing well. A simple acknowledgment of what an employee has done can have a significant positive motivating effect. The theory of the *One Minute Manager* (Blanchard and Spencer, 1982) is that people should be told briefly but effectively when they have done something right, and similarly told briefly but effectively when they have done something wrong. This brief reprimand should focus on the wrong behavior but not reflect on any quality of the person. It should state clearly what the wrong behavior was and why it was wrong, then state how important the person is. The behavior is wrong, the person is good.

Power and Personal Traits

Power is the ability to control resources and people to achieve desired goals. The manager must learn how to motivate people to gain their commitment to the work. Often this motivation comes from the manager's leadership but sometimes he or she needs to exercise authority.

Managers must be able to wield the power to meet the expectations of their subordinates. This power allows them to make decisions that others will abide by. Power may be derived from the authority that goes with the position, or it may be authority primarily held by someone else and delegated to that particular manager. The authority that accompanies a position may be spelled out in a policy and procedures guide. For example, the policy and procedures guide may show that a manager has the authority to approve purchases up to a given maximum amount of money.

Power can take several forms. Managers can manage through their power to reward or punish, such as their ability to contribute to the determination of salaries, or to hire, fire, or reassign. Or they can manage through their ability to lead. Leadership requires maintaining the respect and loyalty of subordinates. It comes from their technical and managerial skills and their ability to provide subordinates with the expertise, guidance, and information they need to do their work.

Negotiation and Conflict

The expectations, constraints, and capabilities, which we have discussed, are subject to negotiation. A manager may negotiate with a supervisor or the client over the expectations and constraints they have expressed. Equally, a manager negotiates with subordinates over the expectations and constraints passed down to them.

Software engineers are involved in negotiations between the clients and the developers to obtain a match between the client's needs and the developer's means. They may also negotiate over alternative technical means to achieve the system objectives.

Negotiations are usually nonzero sums. The sum of the gains to all parties and their losses (when losses are taken as negative gains) add up to more than zero. Both parties then have more to gain by working together than by working individually.

We can do several things to make negotiations more productive and amicable. First, we can agree on the joint gains. Negotiations usually go more easily if the mutual gains are spelled out in the beginning. Then any other gains or losses are easier to accept. Second, we can delay judgments until the situation is well understood and all options are known. Dilemmas with no apparent answers often arise because we jump to conclusions so that some options are never considered. Judgments are difficult to change once they are made. Instead of being prepared to change a judgment, we tend to look for rationalizations that support it. Rationalizations can set bad decisions into concrete and keep us from looking for alternatives. If we don't make early judgments, we can more easily remain open to possibilities yet unrevealed that may resolve the problem.

We can learn from the techniques used by therapists. They enable people to make changes that will make their lives more comfortable. Therapists understand that people will not change if they are forced to defend and rationalize their present position. Many therapists, therefore, first empathize with the person, trying to understand the client's point of view. Then the therapists may reveal their own point of view to gain mutual trust. Only then do they confront the client with the consequences of the client's present actions and suggest how the client may feel more comfortable if he or she were to make a change. We may engage in a similar interaction with our client, our manager, or our subordinates.

Software engineers may find it helps to empathize with a client, a manager, or a subordinate before confronting them with the need for change. Don't you find it easier to accept someone's proposal if first they show they understand what you are trying to say?

It may be clear that we should do all of the above. But there is never enough time. Often we cannot negotiate; we must command. Skilled managers make time for negotiation but know when to command if they cannot negotiate.

Consultants hired to solve a problem seldom tell you to do something you did not know you should do. They help by telling you what priorities to set among all those things you already know you should do but don't have time to do.

Conflict is behavior that interferes with another's attaining his or her desired goals. Conflict can be beneficial if it presents new information that leads to better decisions. It becomes a problem if it results in poor decision making, delays, or costs that do not favorably affect the outcome of the project.

Conflict can occur over business decisions, technical questions, or personal matters. Often the less substance involved, the more hostile the conflict—perhaps because we substitute egos when we don't have substance.

Managers often need to resolve conflicts between their subordinates. Conflicts often arise from misunderstandings of expectations. The opportunity for such conflicts to arise can be reduced by making expectations clear.

Conflicts can also arise because of ego involvements concerning who contributed what and how much. It is important to provide a continual feedback to show subordinates that their contributions are understood and appreciated.

Sometimes conflicts can be resolved by making a decision about what is to be expected. Other times the conflict can be resolved by making someone feel comfortable with the decisions that have already been made.

We all like to receive our strokes. Regular strokes can reduce conflict.

Summary

A project is initiated when a client sees that certain expectations with the present system are not being met. This failure of expectations may occur because an earlier change did not do what it was intended to do, the system itself has changed, a new technology has created new expectations, or the system's goal has shifted. The project manager must estimate what the new system will be like, how much it will cost, and how long it will take to produce it. These estimates lead to new expectations.

Project managers do not work with unlimited freedom. They must direct their team within constraints, which say what cannot be done, and capabilities, which say what can be done. Striving for the highest possible performance, projects are limited by available resources of time and money, thus establishing what is possible and not possible.

Projects connect the expectations of all those involved. The expectations of the investor, who hopes to make money, are linked to the expectations of the client, who looks forward to getting a new system, and to the expectations of the project team, who anticipate being paid for their contribution. The project manager has to delegate tasks in such a way that subordinates do the work needed to satisfy the expectations put on the manager.

The human tendency to want to keep things simple and orderly leads managers to avoid new ideas and risk. Some managers may be unwilling to consider new methods suggested by subordinates or to delegate tasks to subordinates. As a result, subordinates often are reluctant to pass on realistic assessments and instead tell management what it wants to hear. Channels of information need to be open to flow up as well as down. Information for decision making should be presented in a brief, digestible form. Organizations need to establish a structure that allows new ideas to be heard and to encourage an atmosphere that promotes change.

An organization describes the management-subordinate structure. The most common organizational structure is hierarchical, with a treelike chain of command. In a matrix organization each subordinate may report to several managers. A chief programmer team organizes a group effort around an experienced programmer.

Once the project team has been assigned, the manager must strive to motivate team members to commit themselves to the work. Theories vary as to what inspires people to work and how much direction is necessary. Part of management's ability to motivate stems from its power over subordinates.

Another managerial role is negotiation and conflict resolution. Negotiations are more productive when the parties first agree on the gains they both can achieve by working together, and when judgments are delayed until all the facts and options have been considered. Conflicts may be beneficial because they allow new information to be heard, but they may also cause delays and added costs. Conflicts can be reduced by clarifying expectations and helping subordinates to see that their contributions are appreciated.

Exercises

1. Draw a diagram showing the linkages of expectations required to develop a project to make a change. Show by a different type of line the linkages that link the future to the present.
2. List qualities of a software product and characteristics of a project. Then state what expectations the client and the developer may have regarding these qualities and characteristics.

References

Baker, F. T. "Chief Programmer Team Management of Production Programming." *IBM Systems Journal*, Vol. 11, No. 1 (1972).

Blanchard, Kenneth, and Spencer Johnson. *The One Minute Manager.* New York: Berkley Books, 1982.

Brooks, Frederick P., Jr. *The Mythical Man-Month: Essays on Software Engineering.* Reading, Mass.: Addison-Wesley, 1974.

Cleland, David I., and William R. King. *Systems Analysis and Project Management.* New York: McGraw-Hill, 1975.

Cleland, David I., and William R. King. *Management: A Systems Approach.* New York: McGraw-Hill, 1972.

Donaldson, Hamish. *A Guide to the Successful Management of Computer Projects.* New York: John Wiley & Sons, 1978.

Drucker, Peter F. *Management: Tasks, Practices, Responsibilities*. New York: Harper and Row, 1974, p. 797.

Kerzner, Harold. *Project Management for Executives*. New York: Van Nostrand Reinhold, 1982.

Koontz, Harold, and Cyril O'Donnell. *Principles of Management, 3d ed*. New York: McGraw-Hill, 1955.

Koontz, Harold, Cyril O'Donnell, and Heinz Weihrick. *Essentials of Management*. New York: McGraw-Hill, 1982.

Metzger, Philip W. *Managing a Programming Project*. Englewood Cliffs, N.J.: Prentice-Hall, 1973.

Ouchi, William S. *Theory Z*. New York: Avon, 1981.

Peters, Thomas, and Robert H. Waterman, Jr. *In Search of Excellence: Lessons from America's Best Run Companies*. New York: Harper and Row, 1983.

Rieken, Bill, Jr. *Software Project Management*. Mountain View, Calif.: Association for Computing Education and Seminars, 1984.

Rosenau, Milton D., Jr. *Successful Project Management*. Belmont, Calif.: Lifetime Learning Publications, 1981.

Thamhain, Hans J., and David L. Wilemon. "Conflict Management in Project Life Cycles." *Sloan Management Review* (Summer 1975).

Thayer, Richard H., Arthur B. Pyster, and Roger C. Wood. "Major Issues in Software Engineering Project Management." *IEEE Trans. on Software Engineering*, vol. SE-7, no. 4, July 1981: 333–342.

Weinberg, Gerald M. *The Psychology of Computer Programming*. New York: Van Nostrand Reinhold, 1971.

6

Estimating the Project

- Accurate estimates are necessary because the client and project management must agree on the boundaries of cost, time, quality, and capability.
- Calculating the time required for a project involves more than figuring the number of people and the time needed for each task. In some cases adding more people decreases productivity and increases the time.
- The COCOMO method of estimation uses a set of parameters gathered from studies of earlier projects to predict how long a new project will take.
- The Putnam-Norton-Rayleigh curve and the tree growth model are ways of calculating the level of effort as a function of time required for a project.
- Estimates may be made from data gathered in your own past projects, but you may have difficulty collecting such data during a busy project schedule.
- Trees offer a way for planners to break down the elements of a project to assist in estimates. Such structures include the work breakdown tree, the requirements breakdown tree, and the parts breakdown tree.

Whether we undertake a project will depend on our estimates of how much it will cost and how long it will take. Once we decide to pursue the project, we must plan how we will use our resources during the project so that we can keep expenditures and schedules under control. Whether the project is considered a success will depend upon how well the actual costs and times compare with the expectations set up by our estimates. Good estimation is vital to the success of projects. But, at this point in the maturity of the art of software development, good estimation is very hard to come by. In this chapter we will discuss some of the current methods for estimation, and what needs to be done to improve models and data if there are to be better methods for estimation in the future.

Constraints on Management

Management is concerned with getting a specified job done while working under constraints that define what it can and cannot do. The primary constraints are pictured conceptually as a box in Figure 6.1.

The project team is constrained to stay within the box—not to exceed cost and time, yet to produce at least a minimum capability and quality. The expression 1/quality and 1/capability are used so that the higher values will lie within the box. Cost and time are resource constraints: capability and quality are performance constraints. This is an expansion of Figure 5.1, where we merged cost and time and merged quality and capability.

As the project proceeds, the manager may realize that these constraints cannot all be met. Then he or she must renegotiate the constraints with the client. Perhaps the project can stay within the original constraints of time and cost by producing a product that does not have all the capability or quality originally intended. Or the capability and quality may be maintained by increasing the time and cost and so on.

Another dimension not shown in this box is the efficiency of the process. Better methods—say top-down rather than unstructured development—may reduce

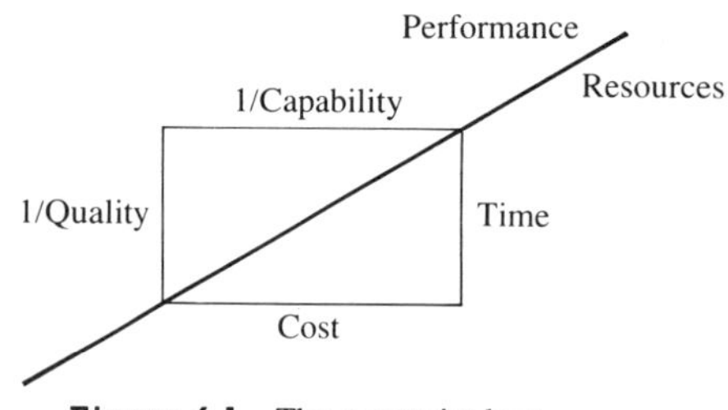

Figure 6.1 The constraint box.

the time and cost while producing the same capability and quality. Software engineering is concerned with using more efficient methods to obtain higher performance with fewer resources.

Mythical Person-Month

It has generally been assumed that the effort needed to get a particular piece of work done can be expressed as the product of the number of people and the time they take to do the work. According to this assumption, if the piece of work takes ten person-months, the project can be done equally well by one person in ten months, or by ten persons in one month. Apparently, therefore, the more people we use, the faster the work gets done.

Indeed, this relationship may be valid if we can assume that (1) what the people are to do has already been planned; (2) the workers do not need to communicate; and (3) the people do not get in each other's way. Brooks (1974) has shown how these assumptions are not true of a large software project. Each new person brought into the project needs unique instructions about what he or she is to do, which takes time from the people who provide the instructions. While the new person can relieve the others of some of their work effort, he or she also costs them additional communication effort.

How much effort is required due to this communication depends on how the work is organized. The worst case is if everyone has to communicate with everyone else. Then we have p people each communicating with $p - 1$ other people. Thus the amount of communication would be proportional to $p * (p - 1)$, or roughly p squared.

Figure 6.2 shows one curve representing the concept that the total effort is given by the product of people and time needed to do the work. On this curve the time is the effort (a constant) divided by the number of people. The other curve shows the added time due to the effort required to communicate. If e is the effort (people times time, or $p * t$) required without the communication burden, and $a * p * (p-1)$ is the additional effort due to communication (a is a constant), then $p * t = e + a * p * (p-1)$, or $t = e/p + a * (p-1)$. When these quantities are computed, they reveal that the time at first decreases but later rises as more people are added to the project.

Clearly, if a project is behind schedule and is beyond the point where the curve is beginning to rise, adding more people with the intention of speeding up the project will instead slow it down even further.

However, if the project is organized as a tree, each person communicates to just those people representing branches in his or her part of the tree, or about the same number of people irrespective of the total number of people working on the project. Then $p * t = e + b * p$, or $t = e/p + b$ (b is a constant). This curve drops as the number of people on the project increases, but it does not rise later

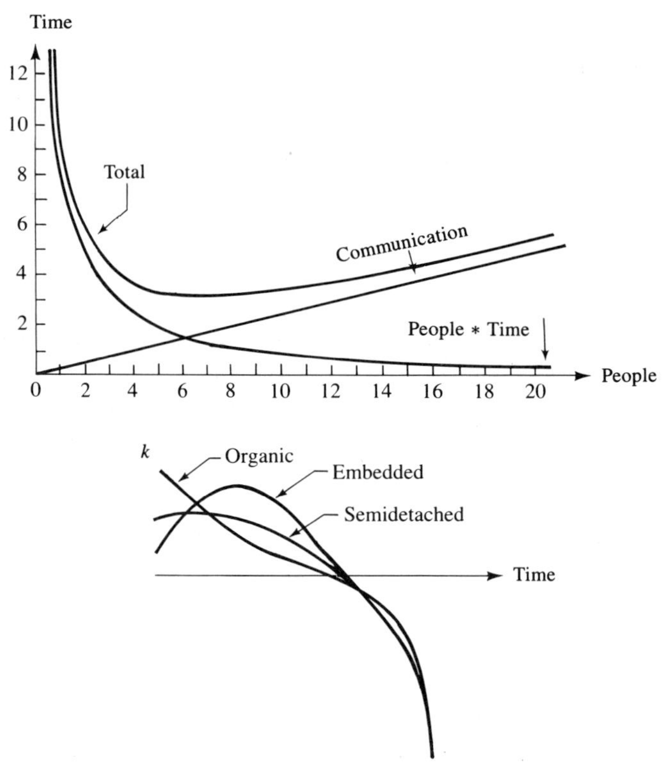

Figure 6.2 Mythical person-month.

as more people are added. Thus, if the project is organized as a hierarchy, communication effort can be saved, allowing management to speed up the project by adding more people, provided they plan their work by adding branches to maintain the hierarchy.

Estimation Methods

Several techniques have been created to calculate a software development project's length and required effort in person-months. We will discuss several of these models with their potentials and limitations.

COCOMO Model

The **COnstructive COst MOdel (COCOMO)** model is a procedure developed by Barry Boehm (1981) for estimating the effort in man-months and the schedule in months to develop a software system. The procedure is based on fitting data from sixty-three projects to a simple equation that can then be used to apply this experience to new projects. We will present here just an overview of Boehm's model. The complete model is the primary subject of his book, which is over 700 pages.

The COCOMO method is based on the premise that one can estimate *KDSI*, the number of thousand Delivered Source Instructions. This quantity is then used to compute an estimate for the effort and schedule. At first the method sounds as difficult as estimating the effort and schedule directly. It sounds a little like telling children they can catch a bird if they can put salt on its tail. However, if we can break down the system into modules, we may be able to estimate the *KDSI* for each module. Then we can sum the *KDSI* for the modules to get the *KDSI* for the whole system. Summing the *KDSI* over the modules first and then using this *KDSI* in the equation should give a better estimate than summing the effort and schedule for each module.

KDSI counts each line of multiple line statements, formats, data declarations, and the job control language. It does not count comments, code not developed by the project personnel, or code that is not delivered as part of the product, such as throwaway code used for testing.

KDSI may be estimated as follows:

1. Using a tree, break the system into modules small enough that they can be recognized as being similar to modules that have been programmed before and for which *KDSI* is known. Even though the system being developed may not be similar to one that has been programmed before, the modules may be.
2. Considering the similarities and differences in function and complexity between each new module and a similar module that has already been programmed, estimate the *KDSI* for the new modules.
3. Sum the *KDSI* over the modules to get the *KDSI* for the whole system. Do not count previously written code that is reused in this project.

The initial estimates for effort (*MM* in Man-Months) and scheduled Development Time (*TDEV* in months) are then obtained as follows:

$$MM = A\,(KDSI)^x$$
$$TDEV = B\,(MM)^y \text{ or } TDEV = C\,(KDSI)^z$$

where the parameters A, B, C, x, y and z depend on the type of system being developed. The coefficients in these equations can be obtained from Table 6.1.

Table 6.1 Coefficients of COCOMO model.

	A	*x*	*B*	*y*	*C*	*z*
COCOMO system types						
Organic	2.4	1.05	2.5	0.38	3.49	.399
Semidetached	3.0	1.12	2.5	0.35	3.67	.392
Embedded	3.6	1.20	2.5	0.32	3.77	.384

Table 6.1 and related material on pp. 146–148 are from Barry W. Boehm, *Software Engineering Economics*, © 1981, pp. 75, 90. Adapted by permission of Prentice-Hall, Inc., Englewood Cliffs, New Jersey.

The parameters in Table 6.1 are based on Boehm's COCOMO study.

The COCOMO types used in Table 6.1 are defined as follows:

Organic—A fast start is possible because the personnel are already familiar with similar projects and there is not a high initial learning and communications load. The constraints are not severe, so if there is difficulty in meeting them, there is often room to renegotiate.

Embedded—A slow start is required because the personnel are not familiar with similar projects and many technical interfaces have to be resolved. Much early planning and communication is required. There is little or no freedom to renegotiate requirements. (Considering the results the model predicts for the embedded project as compared to the organic project, we can assume greater effort is also involved in reviews and verifications in the embedded project.)

Semidetached—Intermediate between organic and embedded.

The general shape of these curves is shown in Figure 6.3.

Multipliers can be applied to the total project *MM* and *TDEV* to allocate effort and schedule to each phase of the project. The effort and schedule for each phase is given by:

$$MM_p = M_p\, MM$$
$$TDEV_p = S_p\, TDEV$$

where M_p and S_p are given by Table 6.2 and $_p$ is the phase.

The average number of Full-time Software Persons (*FSP*) for each phase is given by:

$$FSP_p = MM_p \,/\, TDEV_p$$

For example, given an embedded project with 128 *KDSI*:

$$MM = 3.6 * (128)^{1.20} = 1{,}216 \text{ man-months}$$
$$TDEV = 3.77 * (128)^{.384} = 24.3 \text{ months}$$

For the systems design phase, *sd*,

$$MM_{sd} = .18 * 1{,}216 = 219 \text{ man-months}$$
$$TDEV_{sd} = .36 * 24.3 = 8.7 \text{ months}$$
$$FSP_{sd} = 219/8.7 = 25.2 \text{ people}$$

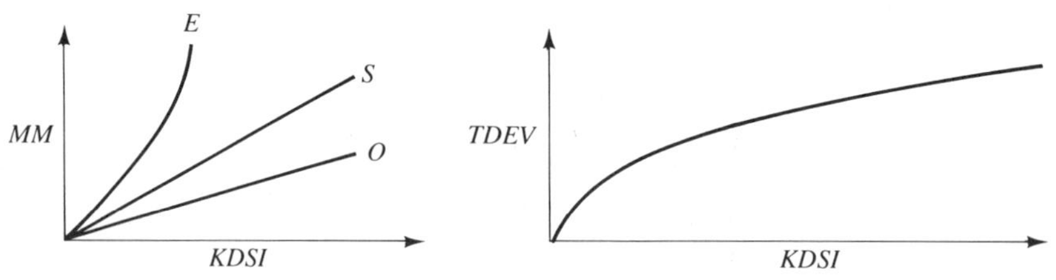

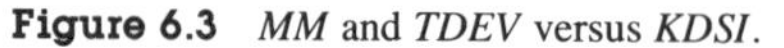

Figure 6.3 *MM* and *TDEV* versus *KDSI*.

Table 6.2 Fraction of total effort and schedule occurring in each phase.

	Organic				Semidetached				Embedded			
	Small 2 *KDSI*		Medium 32 *KDSI*		Medium 32 *KDSI*		Large 128 *KDSI*		Large 128 *KDSI*		Very large 512 *KDSI*	
PHASE	M_p	S_p	M_p	S_p	M_p	S_p	M_p	S_p	M_p	S_p	M_p	S_p
Plan & requirements	.06	.10	.06	.12	.07	.20	.07	.22	.08	.36	.08	.40
Systems design	.16	.19	.16	.19	.17	.26	.17	.27	.18	.36	.18	.38
Detail design	.26	.24	.24	.21	.25	.21	.24	.19	.25	.18	.24	.16
Module code & test	.42	.39	.38	.34	.33	.27	.31	.25	.26	.18	.24	.16
Integrate & test	.16	.18	.22	.26	.25	.26	.28	.29	.31	.28	.34	.30

More sophisticated versions of the model provide multipliers M_p and S_p, which not only depend on the size, type, and phase as in Table 6.2, but also correct the total *MM* and *TDEV* for each of the following fifteen attributes:[1]

Product Attributes
- Required reliability
- Data base size
- Object system complexity
- Severity of execution speed constraints
- Severity of storage constraints
- Underlying operating system, or data base management system, or application system used

Programming Environment Attributes
- Turnaround time
- Use of modern programming practices
- Use of automated tools

Personnel Attributes
- Capability and familiarity of analyst
- Experience with this application
- Capability of programmers
- Experience with the operating systems, data base management systems, and application systems
- Experience with the programming language

Project Attribute
- Schedule constraints—pressure to make tight schedule

These multipliers affect the total effort and schedule, and how the total is allocated to the phases. For example, if we were to develop a program that is difficult to pack into the memory size available, we would expect the extra time would show up in the Module Code and Test and Integrate and Test phases, where we will first know how much memory the code needs.

[1] Values for these multipliers appear in Boehm's book (1981).

We must be very careful using fits based on someone else's data unless we know all the circumstances of the projects on which the data were collected and how they relate to the project to be estimated. Boehm uses his COCOMO model at TRW. But, even though he has explained his work in a lengthy book, some other companies have had difficulty applying his model to their environment. We can make estimates using Boehm's equations and coefficients when no data of our own yet exist. As we get experience with our own projects, though, we should replace these coefficients with our own.

Rayleigh Curve for Effort versus Time

The number of people needed to work on a project changes as the project proceeds. It starts out low, rises, and then falls off. If we had the same number of people on the payroll throughout the project, we would have more people than we could manage at some times, and not enough people at other times (see Figure 6.4). We need a model to help us plan for the staffing and cash flow so we can plan how to pay and manage these people as they work.

Much work has been done to show how to obtain coefficients K and a so that the following equation will fit data on the level of effort (that is, number of people p) as a function of the time (t) into the project:

$$p = K * t * e^{-at^2}$$

The equation was originally used by Lord Rayleigh to represent processes in physics, which bear no obvious relation to the processes seen in project growth. More recently, P. V. Norton (1980) has suggested that this curve has the right shape for explaining project growth. It goes up, over the top, and then comes down asymptotically to the axis as project effort might be expected to. Lawrence Putnam (1980) has developed tables and curves to obtain the parameters K and a and has widely promoted the use of this equation. The curve has been called the **Rayleigh curve**, or **Putnam-Norton-Rayleigh (PNR)** curve (DeMarco, 1982).

This curve has been criticized because it rises too rapidly at the beginning as compared to what has been observed, and it does not represent the early planning phases of the project.

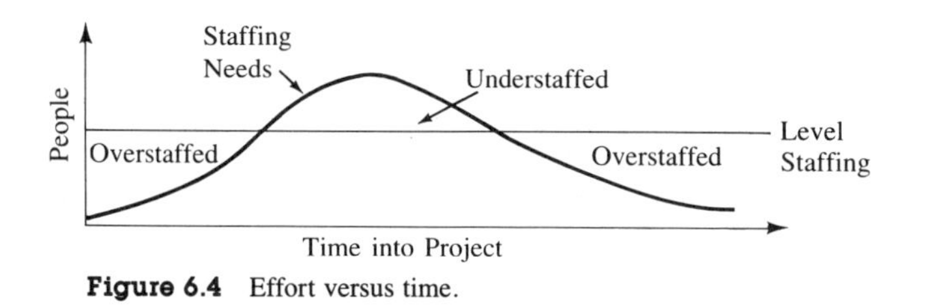

Figure 6.4 Effort versus time.

Improved Methods

To make the best possible estimates we need to know as much as possible about the project before starting and then break down the whole project into the details of its tasks and requirements. In this section we'll discuss some of the impediments to estimation and then we'll see how a model that resembles a tree can help to overcome some of these difficulties.

Problems in Making Good Estimates

Today the ability to estimate the effort and time to develop software is not very good. Better data must be collected today if better estimates are to be made tomorrow.

A major problem is that project managers are often asked to commit themselves to an estimate before they are told what they are expected to accomplish, what constraints will exist, or what resources will be available and when. Then they are expected to keep to their estimate even though there is a change in the expectations, constraints, and resources.

The expectations of what is to be done, what constraints are involved, and what resources are allowed should be spelled out before the estimate is made. Once the estimate is made, if these parameters change, the project management should be allowed to change the estimate accordingly.

But often the ability to estimate is so poor that project managers don't want the assumptions spelled out too clearly. They want to be able to wiggle out of the estimates later. It is very difficult to maintain simultaneously expectations on all dimensions of the constraint box (cost, time, capability, and quality). If a project is held to three dimensions, the manager may still be able to wiggle out of the fourth.

The fundamental law of project management is that as soon as we know we cannot meet our negotiated expectations, we should negotiate new expectations. The earlier we can renegotiate, the more time and freedom we have to replace the failed expectations with realizable ones. The longer we wait, the more trouble we will be in when we finally have to face up to not meeting the expectations.

To make estimates we need to have measures. An estimate is meaningless if we cannot later measure what we estimated. To define how to make a measurement, we need to have some model of the process we are measuring. Thus to develop a good data base for estimating projects in the future, we have to go through the following process now:

Develop a model.
Define the measures based on that model.
Collect data for these measures from past projects.

Then we may:

Estimate a new project.
Track and control that project.
Collect data based on the measures.
Compare this data to the estimates.
Revise the estimates.

After all this we may need to revise the model.

Thus we build estimates using our own or someone else's historical data about past performance. The data must show clearly what was done, what constraints were involved, what resources were available, how long it took, and what it cost. All the parameters, such as those we listed earlier with the COCOMO model, should be carefully recorded so that we can see how to apply this information to another project.

If we play the game of stating the expectations vaguely so we can wiggle out of them later, we are likely to lose our opportunity to collect the data we will need to make future estimates.

It is difficult to collect good data. Usually we are under pressure to complete the present project, and we know the data won't be of any value until some future project. Indeed if we don't survive the present project, there may not be a future project. One significant advantage of using automated aids in system development is that they can make it easier to collect the data needed for estimating future projects.

To make progress in improving our methods for estimation we should:

- Collect better historical data.
- Define exactly what is being estimated and what assumptions are being made.
- Use standardized methods of development, which not only reduce costs but produce historical data that can be more easily applied to other projects.
- Use automated tools, which both standardize and record the data.
- Work with better metrics. (Tree metrics might offer a better approach to measuring what we have done and what resources were used to do it.)
- Make better plans that better define what we are measuring.
- Control the work to the plan. Measure deviations from the plan. When deviations occur, seek an understanding of why they happened. Then either change the situation or change the plans and expectations. Do not let the project drift out of control.
- Use trees to make a hierarchical breakdown of the requirements, work structure, and system structure so that we know what it is we are measuring. The more detail shown in the breakdown, the better we will recognize the relation between the work we are estimating and work we have done before. These tree breakdowns will be discussed later in this chapter.
- Recognize that it is the details that are most overlooked and account for most of the hidden costs and errors in estimates. Either break down the work so that

these details are revealed, or add in an amount to account for the various classes of detailed and overhead activity (for example, meetings and interviews, thinking and planning, and so on).

- Do not overlook all the nonprogramming tasks such as manual writing, training, data conversion, and so on when estimating the whole software development project.

Tree Breakdowns

Trees can be used to show several related types of breakdowns. They can be used to show the breakdown of the requirements for the whole system into the requirements that must be met to satisfy them. This is the **requirements breakdown tree**. Trees can also be used to show the breakdown of all the work to be done into smaller work units and finally into well-defined tasks that can be assigned as the responsibility of specific people and scheduled. This is the **work breakdown tree**. And a tree can also be used to show the breakdown of the system into the physical parts that make it up. This may be called the **parts breakdown tree**, or the physical packaging tree or configuration breakdown tree.

Often these trees are so closely related that one can be derived from the other. As we break down the goals describing what the system is to do, we have a requirements tree. Certain requirements may be met by certain parts of the system, so this can be a parts breakdown. And there may be certain tasks needed to meet certain requirements, which implies a work breakdown tree. We will discuss requirements trees in Chapter 13 and work breakdown trees in this chapter.

The work breakdown structure is a way to define the tasks that must be done to produce the product. It takes the form of a tree, the work breakdown tree, and is based on the concept of defining goals, subgoals to achieve the goals, subsubgoals to achieve these subgoals, and so on until finally the subgoals can be identified as familiar tasks (see Figure 6.5).

Goals define what has to be done to finish the project. But just as importantly, they define what does not have to be done. Without clear goals much time can be wasted on work that is not necessary.

We assume that any new project, even though we have never done one like it before, can be broken down into tasks that are similar to tasks we have done before. Then we can make reasonably good estimates of the effort and costs

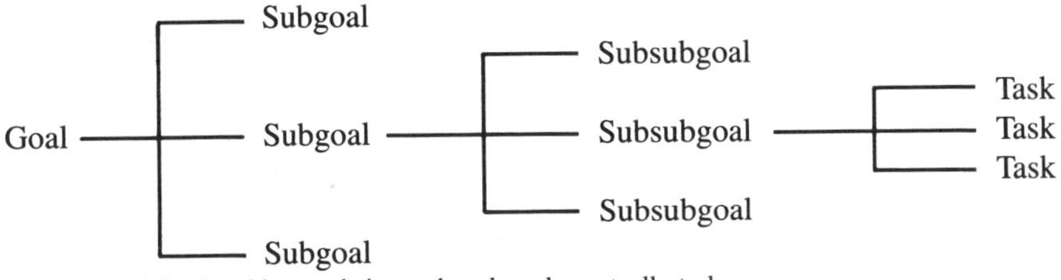

Figure 6.5 Breaking goals into subgoals and eventually tasks.

required to do these tasks. Add up these estimates for the individual tasks to get an estimate for the whole project. The work breakdown tree will also help to estimate and add in the planning and costs for integrating subsystems into larger systems. The planning and integrating occur on the branches.

For each of these small tasks we can estimate the effort and cost of doing the task. Effort is measured in person-hours—that is, the number of persons working multiplied by the time they work. Cost is the effort times the pay rate (dollars per hours) of the people doing the work.

This pay may be multiplied by an overhead factor to account for the costs that are not covered by specific billable tasks. For example, given an overhead factor of 1.4, the 1 would account for the direct labor and the .4 would provide for services such as payroll, purchasing, accounting, and the costs for facilities such as the building, heat, and light.

As a general rule the more we can break down the project, the better the estimates. There are two reasons for this rule:

1. Since the detailed tasks are smaller and more familiar, we can more easily visualize doing them and thus better estimate the effort and cost required to do them.
2. The principal errors in our estimates are usually due to overlooking the details. The effort and costs for overlooked details accumulate until they can represent a major portion of all the work, and thus a large error. It is easy to visualize the significant things to be done like writing a document. It is harder to remember all the other things we must do to write the document, like getting the information, organizing it, getting reviews, revising, having meetings, scheduling the meetings, and so on. These details are the part of the iceberg hidden below the water.

Overlooking details is a major reason for what is stated facetiously as the "90-90 rule," which says: "The first 90 percent of the work is done in 90 percent of the time, and the rest of the work requires another 90 percent of the time." A project can look nearly finished when all that remains is "clearing up the details." But the details can take a long time. Remember the saying:

> De-tails wag de dog.

A partial work breakdown tree showing the descriptions of the goals and tasks might look like Figure 6.6. We have intentionally expanded the "Write Requirements Specifications" in more detail than the rest of the tree to make a point below.

We could also represent the tree in indented form like an outline with a hierarchical decimal identification of tasks as in Figure 6.7.

If we tried to estimate the time to write a document without doing the tree, we might only have thought of the actual time we spend at our desk doing the writing. But when we start to break out all the details, we see there are many small tasks

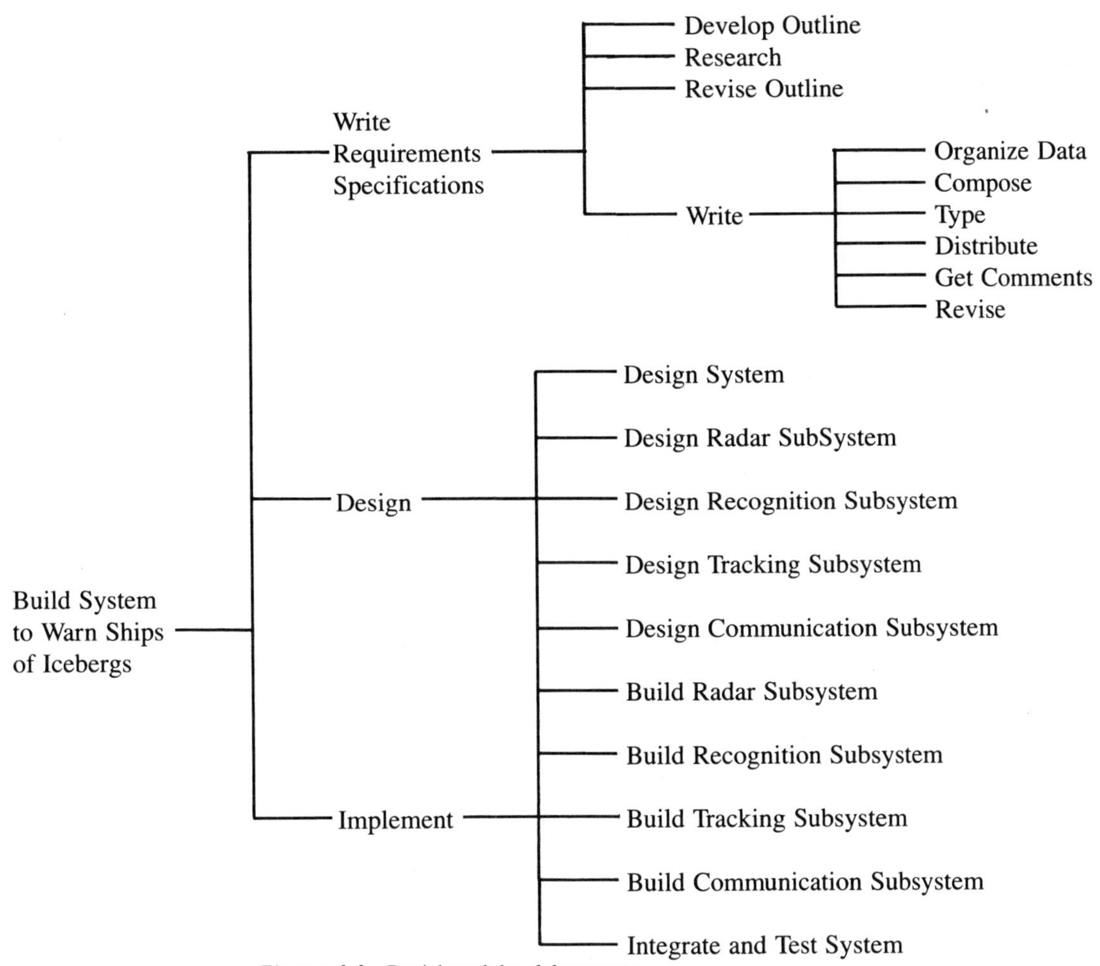

Figure 6.6 Partial work breakdown trees.

that together will add greatly to the effort. Without the discipline of the work breakdown tree, we might easily have overlooked this hidden effort.

What may have looked like a trivial task at one level is revealed as time- and cost-consuming detail when it is exploded into the lower levels. Such revelations are precisely why we must make a work breakdown tree; to discipline the process of expanding the work into sufficient detail to allow us to get good estimates.

Once we have defined the tasks with a work breakdown tree, we define for each task:

1. When can the task be started—that is, what other tasks must be done before this task can be started?

0.	Build System to Warn Ships of Icebergs
1.	Write Requirements Specifications
1.1	Develop Outline
1.2	Research
1.3	Revise Outline
1.4	Write
1.4.1	Organize Data
1.4.2	Compose
1.4.3	Type
1.4.4	Distribute
1.4.5	Get Comments
1.4.6	Revise
2.	Design
2.1	Design System
2.2	Design Radar Subsystem
2.3	Design Recognition Subsystem
2.4	Design Tracking Subsystem
2.5	Design Communication Subsystem
3.	Implement
3.1	Build Radar Subsystem
3.2	Build Recognition Subsystem
3.3	Build Tracking Subsystem
3.4	Build Communication Subsystem
3.5	Integrate and Test System

Figure 6.7 Outline form of work breakdown tree.

2. How do we determine when the task is completed?
3. How long will it take to do the task—the duration?
4. What are the expectations for how well it should be done?
5. What measures or tests are there to determine whether these expectations were met?
6. How can we tell how well the task is progressing before it is done?
7. Who will be responsible for completing this task satisfactorily?
8. What authorities will he or she need?
9. What resources will be needed to do the task?
   a. People with what skills?
   b. Facilities?
   c. Time?
   d. Money?
   e. Other?

This information will be used for developing a critical path schedule. Critical path scheduling will be discussed in the next chapter.

Tree Growth Model for Effort versus Time

The PNR curve has the right shape to explain the number of people on the project as a function of time—at least to the extent that it rises, goes over the top, and comes down approaching ever closer to the axis. But the data are not good enough to distinguish whether the PNR curve or any other curve that has this general property may be the better fit. In fact the PNR curve clearly does not give the right shape in the early period during analysis and requirements definition because it initially rises too fast. Thus the curve is usually not used to represent this period. Also the PNR curve cannot be developed as a consequence of a model that explains what we expect to be happening during the growth and decay of a software project.

The **tree growth model** is based on a model of what we might expect actually happens in software development to explain why the effort grows and decays during a project. Thus we can use it as a reference to compare the behavior of an actual project with what we might expect should be happening. The tree that we use to describe the growth and decay of effort is the same tree we discussed in Chapter 4, which we develop during the project to describe the requirements and design. Thus we can relate the requirements-design tree to the pattern of project growth and decay.

Before people can work, someone must plan what work they are to do. The project thus begins with one person or a small group of people who develop the initial plans. They are the root. After they work for a while, they can specify the work to be done by several more people. After these people work for a while, they can specify work to be done by yet more people, and so on. Thus the project grows like a tree. Branches develop as the work is planned and people are brought into the project to do the next level of planning or work. On the branches planning gets done. On the leaves code and reports are written and other work is done. Then the tree shrinks by absorbing its branches as the parts planned on those branches are now integrated into larger and larger subsystems. Eventually the whole system is integrated at the root.

Let us look at an illustration of tree growth and absorption (see Figure 6.8). Here we show the tree growth occurring from the root at the left and growing to the right. Then using a mirror image of the tree, we show its absorption as we move to the right.

Each branch is a task. The length of the task is shown by the length of the branch. The number on the branch is the level of effort for that task—that is, the number of person-months per month, or if you prefer the number of persons averaged over the month. As we move to the right, we see vertical limbs that represent branching as planning yields tasks to do more detailed planning or work. Eventually the planned work is done—modules are implemented. As we move further to the right, we see vertical limbs representing the integration of these modules into larger modules. When we finally return to the single branch on the far right, we have integrated the final modules into the whole system.

As we see in this illustration, not all the leaves occur at the same level nor are all the branches the same length. Thus, some branches are already being integrated while others are still growing. Dots in the figure show where some branches are shorter than other branches they must integrate with. (When we talk about

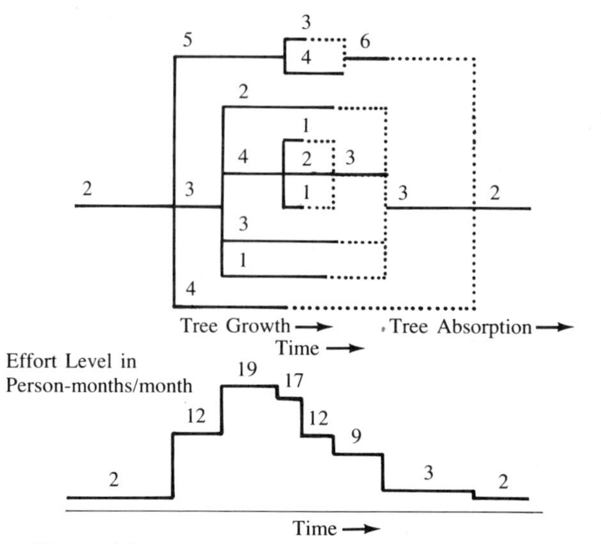

Figure 6.8 Profile of effort level as tree grows and decays.

critical path in Chapter 7, we will call this "slack.") Below this diagram another figure shows the effort level summed over all tasks occurring in each time period. This drawing gives us a profile of the effort level as a function of time.

We can make a simple mathematical model for the growth and decay of these project trees, which will help us to see how to gather project data on current projects that we can use to plan the level of manpower we need as a function of time for future projects.

The level of effort starts low during the planning stages and grows as the planning done by the initial people allows other people to assume their work. This relationship suggests that in the early stages the rate of change of the number of people on the project is proportional to the number of people already working on the project. This would give us the following simple differential equation:

$$\frac{dp}{dt} = k * p$$

If k were a constant, this would give us an unlimited exponential growth. So let k vary during the project. The quantity k is initially positive as the project grows, then decreases and goes negative as the project decays. If we look at the pattern of k with time for current projects, we may see what we can expect for future projects.

Exercise 2 shows that making the simple assumption that k starts positive and decreases linearly leads to a project effort curve that has the shape of a normal curve with part of the left tail chopped off. The normal curve has the shape for the initial rise of effort that we observe in actual projects, while the PNR curve

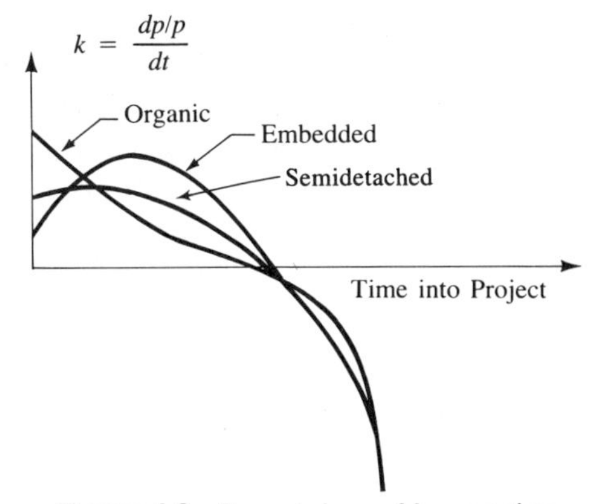

Figure 6.9 General shape of k versus time.

initially starts up too fast. Using actual data from projects rather than the linear assumption will give even better results. Figure 6.9 shows the general shape of the k curves for typical organic and embedded projects.

In an organic project k starts high, drops slowly, then drops off sharply, finally going negative. In an embedded project k starts low and rises, then drops off, finally going negative.

We might estimate the parameters to be used in the effort growth and decay model by considering the following:

- The length of the branches, which is how long it takes to plan before work can be branched off to other people
- The number of branches that form on the average with each branching
- The average number of levels required to define the tasks

The parameter k is proportional to the number of branches off each limb divided by the length of the branch. These parameters are related to the complexity of the product and can be estimated in the process of developing the work breakdown tree.

If we represent and develop our system as a Tree, we can count the branches as the project proceeds and relate this to the number of people needed at that time. This data can then be used when we are making estimates for later projects.

Using the Tree Model for Estimation

COCOMO is a macro approach, which might be useful if we have no data of our own and want to have a gross estimate just to determine whether the project is feasible. It may be used to make a proposal or bid on a project.

The tree model requires more detailed analysis of the project but should be able to give a more detailed estimate. As we collect data, we can relate the data to the parameters of the tree growth model. As we begin the early development of the tree for requirements, work breakdown, and system structure, we begin to get information that can be used to estimate how the rest of the tree will develop. The number of branches at each branching and the time interval between branchings are indications of the complexity and amount of communication, coordination, and integration required.

The tree model bases the estimate on information that is specific to the project being estimated instead of generic information from other people's projects as in the COCOMO approach. In the tree growth and self-absorption model, which we derived using differential equations, or in the COCOMO model, we were not looking at specific identifiable tasks. With the tree model we are.

The total effort and cost for the whole project can be estimated by summing the efforts and costs for each of the tasks. But we must also consider the effort and costs associated with defining and planning these tasks before they are done, and with integrating them into the whole system after they are done. The more tasks there are, the more levels there should be in the work breakdown tree. As we sum the effort and cost of each task, we will also add in the additional effort and cost at each branch to cover definition and planning (as the tree grows) and integration (as the tree contracts). At each task or branch we must consider the type of employee who would do that work and apply the appropriate pay rate. Figure 6.10 is a work breakdown tree showing this process of summation without the descriptions.

This estimation tree includes both the costs that arise during the definition and planning when we go out through the branches, and those that arise when we integrate the work we had planned as we come back through the same branches.

This figure suggests an approach to how we can estimate a project. We carry the estimation tree out as far as we can afford to. On the leaves we put the best estimates for all thc work that lies beyond that has not yet been expanded as a

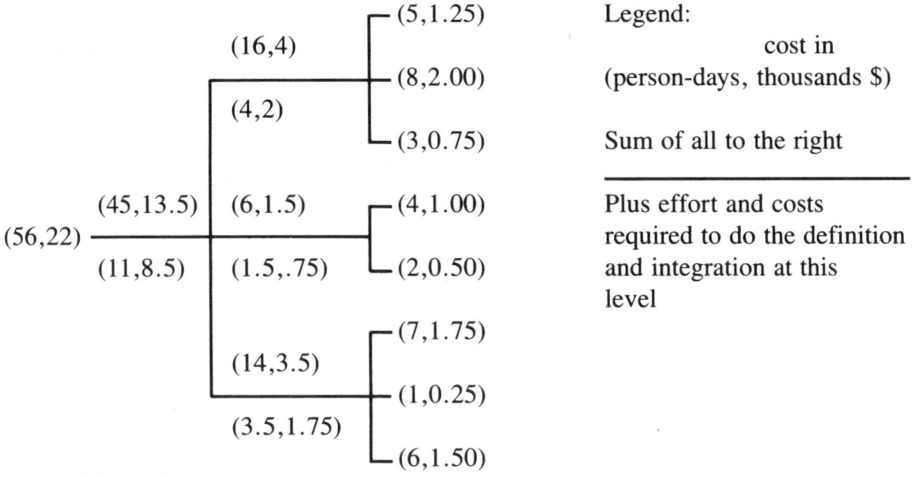

(56,22)
(45,13.5)
(11,8.5)
(16,4)
(4,2)
(5,1.25)
(8,2.00)
(3,0.75)
(6,1.5)
(1.5,.75)
(4,1.00)
(2,0.50)
(14,3.5)
(3.5,1.75)
(7,1.75)
(1,0.25)
(6,1.50)
Legend:
cost in
(person-days, thousands $)
Sum of all to the right
Plus effort and costs required to do the definition and integration at this level

Figure 6.10 Estimation tree.

tree. The sum back to the root is then our best estimate for the whole project to date. As the project proceeds, we extend the tree further to obtain better estimates.

If we represent the functional requirements as a tree on a computer and develop this tree through the design and implementation, we can collect cost and time data as the branches are developed and integrated. In this way we can generate the costs of the subtrees. Then when we have a new project, we can break it down until we can recognize a similarity to familiar subtrees for which we have cost and time data.

Once we are committed to do the project, we must develop a plan at the level of detail such that we can schedule individual tasks. We can make such a detailed schedule using critical path techniques, which we will discuss in the next chapter.

Summary

Before a project actually begins the client and management will want to know the costs and time required to complete all the work needed to produce a product with a specified quality and capability.

Planners sometimes assume that adding people will decrease the time to do the job. But, depending on the organizational structure, adding more people can actually reduce productivity.

The COnstructive COst MOdel method estimates the time and manpower required to complete a planned project. The method uses data gathered from earlier projects to establish a set of parameters, which are then applied to a new project. COCOMO's premise is that we can estimate the time and effort to complete a program provided we can estimate the number of thousand Delivered Source Instructions (*KDSI*). The calculations characterize systems as organic, embedded, or semidetached. The method also provides corrections for fifteen attributes of the planned system and the process that produces it.

The Putnam-Norton-Rayleigh curve and the tree growth model show the shape of project growth with time.

Good estimates are difficult to make because software developers are often asked to commit themselves to projects before all the expectations have been spelled out. As a result, managers sometimes try to wiggle out of their estimates. Estimation methods can be improved by collecting one's own data from past projects and using measurements based on good models. Details must not be overlooked when collecting data because such small details can add up to significant effects on the time and cost.

Trees are a good means of breaking down a project into all the detailed components. A requirements breakdown tree shows the requirements for the whole system. The work breakdown tree describes the tasks that combine to produce the system. A parts breakdown tree portrays the physical constituents of the system. A work breakdown tree is a starting point for developing a project schedule and cost estimate.

Exercises

1. Given the development of an organic type system of forty thousand delivered source lines of code, use the COCOMO model to estimate the level of effort and time required for each phase.
2. Assume that the value of k in the differential equation for project growth and decay is as follows:

$$k = \frac{u - t}{\sigma^2}$$

where u and σ^2 are constants. This is an equation for a decreasing straight line. Show that this equation has a solution that can be written in the following two forms:

$$p = K_1 * \frac{e^{qt} * e^{-t^2/2\sigma^2}}{\sqrt{2\pi\sigma^2}} \text{ where } q = u/\sigma^2$$

$$= K_2 * \frac{e^{-(t-u)^2/2\sigma^2}}{\sqrt{2\pi\sigma^2}}$$

Note that the first form is an exponential rise, which we would get if we used k as a constant, multiplied by a decay function, which is the right side of the normal distribution. The second form is a normal with the t axis moved to the right by an amount u, cutting off a piece on the left end of the curve.

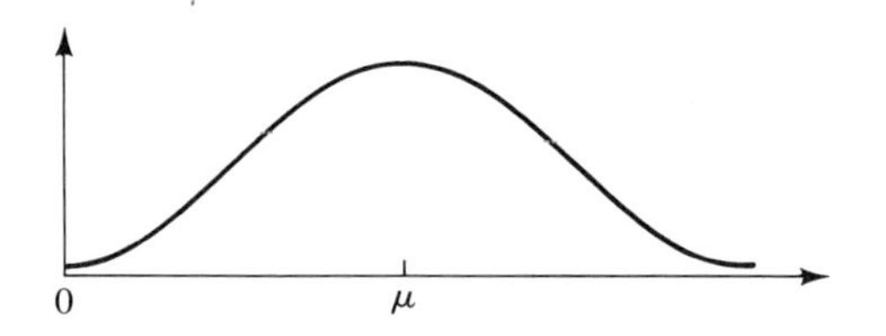

3. Try choosing several typical values of *KDSI* (say 32, 50, 128, and 512) and use the COCOMO equations for either the organic or embedded models to obtain *TDEV*. Now try making a fit of this data to the equation:

$$TDEV = a * \ln(KDSI)$$

Note that to fit a straight line of the form

$$y = a * x$$

through points such that the sum of squares of the distances of the points off the line is a minimum, we compute a as:

$$a = \text{sum } x * y \,/\, \text{sum } x^2$$

where the sum is over all the points. Here we use $\ln(KDSI)$ for x.

Plot this equation and the data points. Does this fit make sense in the light of the tree growth model?

References

Arthur, Lowell Jay. *Measuring Programmer Productivity and Software Quality*. New York: John Wiley & Sons, 1985.

Basili, Victor R. *Tutorial on Models and Metrics for Software Management and Engineering*. Los Alamitos, CA.: IEEE Computer Society Press, Catalog No. EHO-167-7, 1980.

Boehm, Barry W. *Software Engineering Economics*. Englewood Cliffs, N.J.: Prentice-Hall, 1981.

Brooks, Frederick P., Jr. *The Mythical Man-Month: Essays on Software Engineering*. Reading, Mass.: Addison-Wesley, 1974.

Conte, S.D., H.E. Dunsmore, and V.Y. Shen. *Software Engineering Metrics and Models*. Menlo Park, Calif.: Benjamin/Cummings, 1986.

DeMarco, Tom. *Controlling Software Projects: Management, Measurement and Estimation*. New York: Yourdon Press, 1982.

Norton, P. "Useful Tools for Project Management." In *Software Cost Estimating and Life Cycle Control,* edited by L. Putnam, New York: IEEE Computer Society Press, 1980, 216–225.

Putnam, L., ed. *Software Cost Estimating and Life Cycle Control*. New York: IEEE Computer Society Press, 1980.

7

Scheduling and Controlling the Project

- Once you know the project begin time, have a scheduling network showing what tasks precede which others, and have estimates for the task durations, you can calculate the earliest and latest begin and finish times for each task. Slack is the latest begin time minus earliest begin time.
- The critical path is the longest path through the schedule from first to last task.
- Two methods can be used to represent a critical path network: the precedence network and the IJ network.
- A graph is a set of points and the lines between them. A directed graph used for scheduling is often called a network.
- Graphs representing information flow in the design of a system contain circuits but graphs showing the building of a system do not.
- Worksheets keep track of the time spent during the course of a project.
- Bar charts help managers follow the project schedule. They can show tasks, resources, and progress to date.
- Reviews try to identify problems as early as possible. Types of reviews include the walkthrough, an inspection, and a formal management review.
- Change control procedures monitor the changes suggested by the client, the government, or the project team in the course of the project.

The last chapter described how to define the tasks to be done and estimate the time and effort required to do them and integrate them. In this chapter we'll turn to the development of schedules that show when each task should be done. The discussion will explain how to track whether the work has been done as scheduled, and how to control the project and bring it into conformance with the plan, or when and how to make a new plan.

A project plan is a reference from which to make changes. The manager starts with a plan so everyone will know what to do and how to do it with an expectation of completing the project within the schedule. Without a plan a project would drift and never reach its goal. But, as the work proceeds, events occur that require the plan to be changed. Thus the plan has to be kept up to date. In this chapter we'll look at ways to maintain control over the inevitable changes in the plan and the product as it is developed.

Graphs, Trees, and Networks

Many of the explanations in the following discussions rely on a type of diagram called a graph. To understand these graphs, we need to define a number of related terms. A tree, such as we discussed in Chapter 4, is a type of graph. Next we will describe other graph concepts that will be useful in scheduling.

When we hear the term *graph,* we may think of y plotted as a function of x. Here we mean something different. A **graph** is a set of points, called **nodes**, and lines between them. If the lines do not have a direction, they are referred to as **edges**, and the graph is known as an **undirected graph**. If the lines have a direction, they are called **arcs**, as we said in Chapter 3, and the graph is a **directed graph**, or digraph. A sequence of lines from node to node (in the direction of the arcs if it is a directed graph) is considered a **path**. If the path comes back to the node where it started, the path is known as a **circuit**.

Figure 7.1 is an example of a graph containing a circuit.

A tree is a special kind of undirected graph that has one node called a root; every other node has just one path leading to it from the root. In Chapter 6 we used a tree to break down a project into the required tasks. In this chapter we'll use a directed graph to show the order in which these tasks can be completed. The graph used for scheduling is often called a **network.**

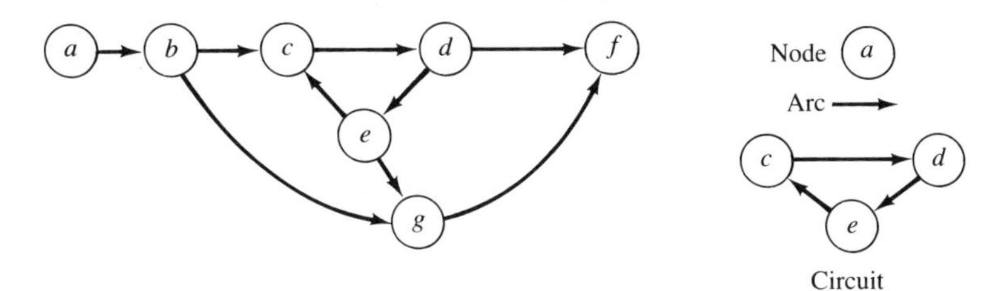

Figure 7.1 Arcs, graphs, and circuits.

If task *a* must be done before task *b*, we say that *a* is a **predecessor** to *b*, or that *b* is a **successor** to *a*. This relationship can be represented in a graph by drawing an arc from *a* to *b*.

Critical Path Scheduling

To develop a schedule, a project is first broken down into tasks, as we did in Chapter 6 when we developed the work breakdown tree. The tasks were the leaves of this tree. The next step is to consider the following for each task:

1. What other tasks must be completed before this task can be started (the predecessor tasks)?
2. How long will it take to perform this task (the duration) once these predecessors have been completed?

The plan may be described in a precedence table (Table 7.1) or as a graph (Figure 7.2). The table is more compact and can be used as computer input. The graph makes it easier to visualize.

Given a project begin time, we can compute the Earliest Begin (*EB*) and Earliest Finish (*EF*) times for each task as follows:

1. If a task is not preceded by any other task, its earliest begin time is the project begin time.
2. The earliest finish time is the earliest begin time plus the duration.
3. The earliest begin time of a task that has predecessors is the largest of the earliest finish times of its predecessors.
4. The computed project finish time is the largest of the earliest finish times of those tasks that have no successor.

Note that before we can compute the earliest begin time for an activity we must have already computed the times for the tasks that precede it. A numbering of tasks for which each task has a higher number than those that precede it is called a **topological order**. Thus we compute the earliest times for the tasks following

Table 7.1 Precedence table.

Task	Predecessors (/ means no predecessor)	Duration
a	/	9
b	/	3
c	*a*	4
d	*a,b*	6
e	*c,d*	3
f	*e*	8
g	d	6
h	*f,g*	6
i	*g*	3

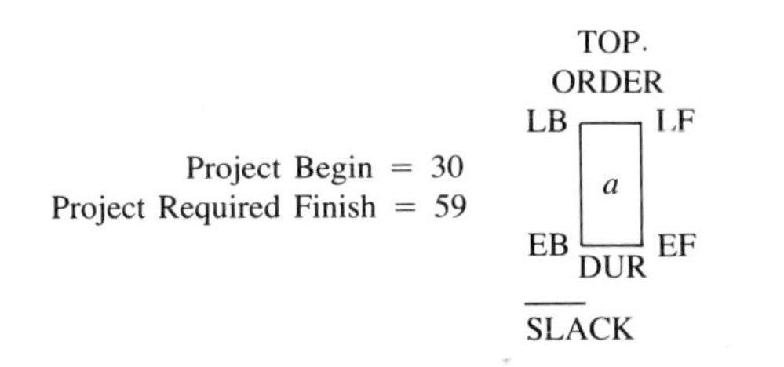

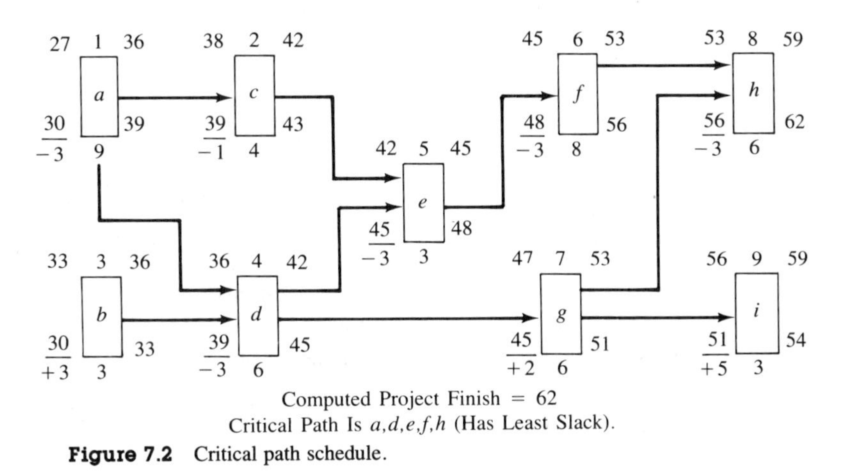

Figure 7.2 Critical path schedule.

the topological order. (The numbers above the tasks in Figure 7.2 show a topological order.)

It may be specified that the project must be finished by a given required project finish time. If this has not been specified, we use the computed project finish time for the required project finish time.

We can compute the very latest times that each task must be finished (*LF*) and begun (*LB*) so that the project will be finished by the required project finish time as follows:

1. If a task is not succeeded by any other task, its latest finish time is the required project finish time.
2. The latest begin time of a task is its latest finish time minus its duration.
3. The latest finish time for a task that has successors is the smallest latest begin time of all its successors.

Thus, we compute the latest times in reverse topological order.

We can also compute a quantity called the **slack**, which indicates how much freedom we have between the earliest we can begin a task such that all its predecessors are finished and the latest we can begin it and still meet the required project finish time. The slack (*S*) is the latest begin time minus the earliest begin time (see Figure 7.3).

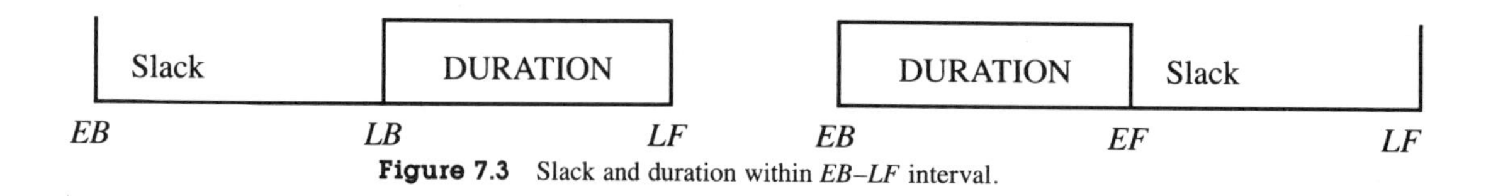

Figure 7.3 Slack and duration within *EB–LF* interval.

One of the properties of these rules is that there will be at least one path from a task with no predecessor to a task with no successor such that every task along this path has a slack equal to the required project finish time minus the computed project finish time. This path is called the **critical path**. It is the longest path; the sequence of tasks that determines the length of the project. The slack along the critical path is the smallest slack that occurs anywhere in the schedule. If the project is to be shortened, these are the tasks that must be shortened. If the project is not to be delayed, these are the tasks that must be controlled so that they do not slip.

Obviously, if the required project finish is earlier than the computed project finish, we have a problem. The slacks along the critical path will then be negative. Although this fact signifies that this schedule is impossible, it shows where we must shorten the durations to make a new schedule that does meet the required project finish.

If we must shorten the durations to satisfy the required project finish, we must be sure we know how to obtain the shorter durations. We may be able to shorten the durations by adding more people to do the task. But, as we saw in Chapter 6, sometimes adding people can increase rather than decrease the time required. We may also shorten the project by planning different tasks having different predecessors.

In Figure 7.2 the required project finish time is 59, but the computed project finish time is 62. The difference is -3. There is a path of tasks with a slack of -3 from an activity with no predecessor (a) to an activity with no successor (h). This the critical path (a,d,e,f,h).

Precedence and IJ Networks

Two methods are now commonly used to represent the critical path network. The method that we just used in the preceding section, which puts the activities on the nodes, is often called a **precedence network**. The older method, which puts the activities on the lines instead of the nodes, is often called an **IJ Network**.

To distinguish and compare these methods, let's consider the simple example in Figure 7.4.

The input to a critical path program expressing just the predecessors in this network is shown in Table 7.2.

If we later find that we need to add another task as a predecessor to B, we would change the input as shown in Table 7.3 (with the changes underlined).

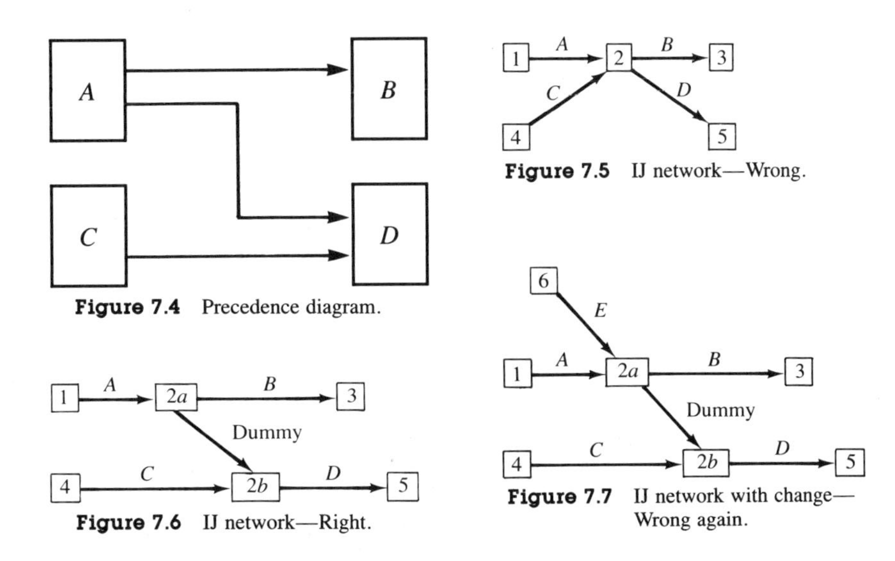

Figure 7.4 Precedence diagram.

Figure 7.5 IJ network—Wrong.

Figure 7.6 IJ network—Right.

Figure 7.7 IJ network with change—Wrong again.

To diagram this as an IJ network, we must put each activity between two nodes called **events**. (An activity is specified by the I and J numbers of the beginning and ending events, thus the name IJ network.) The precedences between activities are shown by making the end event of the first the same as the beginning event of the second. Thus we might be inclined to design an IJ network for Table 7.2 as shown in Figure 7.5. To describe these precedences as an IJ network to a critical path scheduling program, we would use the input in Table 7.4. However, we see that where event 2 ties *A* as a predecessor to *B* and *D*, and where event 2 ties *C* as a predecessor to *D*, event 2 also has the effect of tying *C* as a predecessor to *B*, which we do not intend.

This problem can be avoided by splitting event 2 into two events with a dummy activity that has zero duration between them, as in Figure 7.6. Now the input to an IJ network critical path program would be as in Table 7.5.

If we need to add another activity as a predecessor to *B*, we might be inclined to change our network as in Figure 7.7. But we would now have *E* preceding *D* as well as *B*. We would have to change the network by introducing another dummy, as in Figure 7.8. And the new input would read as in Table 7.6.

Table 7.2 Precedence table.

Task	**Predecessor (/ means no predecessor)**
A	/
B	*A*
C	/
D	*A*,*C*

Table 7.3 Precedence table with change.

Task	**Predecessor**
A	/
B	*A*,*E*
C	/
D	*A*,*C*
E	/

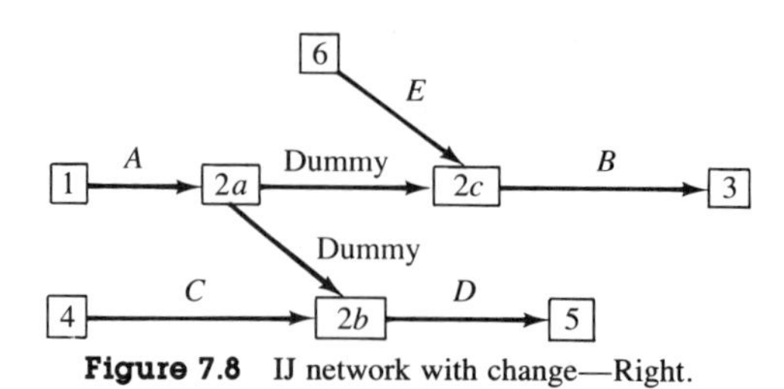

Figure 7.8 IJ network with change—Right.

The problems of maintaining the IJ network diagram and input are sufficiently burdensome that the input is often not kept up to date as changes occur in how the project is planned. If the plan is not updated, the usefulness of critical path scheduling is lost.[1] Using the precedence scheme it is not necessary to draw and maintain a diagram to find where dummy activities are required before input can be developed for the computer. Thus, with the precedence scheme it is easier to develop the initial network and maintain it as changes occur, making it more likely that it will be properly kept up to date.

Table 7.4 Input for IJ network—Wrong.

Task	IJ
A	1,2
B	2,3
C	4,2
D	2,5

Table 7.5 Input for IJ network—Right.

Task	IJ
A	1,2*a*
B	2*a*,3
C	4,2*b*
D	2*b*,5
Dummy	2*a*,2*b*

Table 7.6 Input to IJ network—Finally right.

Task	Predecessor
A	1,2*a*
B	2*c*,3
C	4,2*b*
D	2*b*,5
E	6,2*c*
Dummy	2*a*,2*b*
Dummy	2*a*,2*c*

Design Structure System

The critical path calculation discussed thus far works when we plan and schedule the tasks required to *build* a system. It does not work quite this way if we are planning and scheduling the tasks to *design* a system. There's a good reason for the difference. When we look at the graph of the precedences between tasks to build a system, there are no circuits. But when we schedule the tasks to design a system, there are circuits. Let's see why this is true.

[1]When the precedence method was proposed by the author in 1962, he pointed out that the old IJ method used by PERT required dummy activities and people to maintain the network, while the precedence method got rid of all the dummies. This remark was considered impertinent, so the new method was naturally called IMPERT.

Torn Circuits

When we are planning the design of a system, the tasks are the work we do to determine the parameters that describe how to build it. The precedences arise because of the information from one task needed to do another. These tasks often depend on each other such that neither can proceed without knowing information produced by the other. Once the system is built, the parts must fit or work together—for example, the left part must fit with the right part, and the right part must fit with the left part. So the design of each depends on knowing the design of the other. This relationship appears in the design graph as circuits (see Figure 7.9).

How do we deal with these circuits in the design graph? A circuit would seem to imply that we cannot start a task until that task has already been completed, which is a logical absurdity. There are two approaches to this apparent dilemma:

1. We can sometimes do all the tasks in a circuit together.
2. We can start by using an assumption, either by ignoring or guessing at the information from the predecessor. Then once we have done all the tasks in the circuit, we make a review to see if we made a good assumption. If not, we make a new assumption and go around again. This process of repeating is called **iteration**.

If these tasks are done by the same person and they are simple, they can often be done together. If they are done by different people, or are too complex for one person to deal with all at one time, they are iterated.

Circuits imply iteration or tasks that must be done together.

Once these parameters are determined during the design, then the parts can be made one at a time from the design specifications so they fit together properly when they are built. If we did our design well, when the system is built, there will be no need to iterate. Any iterations should be done during the design with paper and pencil.

Thus the graph representing the design of a system will have circuits so that the graph representing the building of the system will not have circuits.

We show where we make an assumption about a predecessor (ignore or guess) by drawing a line through the arc. This is a **tear**. We must make enough assumptions to break all the circuits. Having used an assumption at the arrow end of the torn arc, we then proceed through the tasks in sequence until we come to the tail

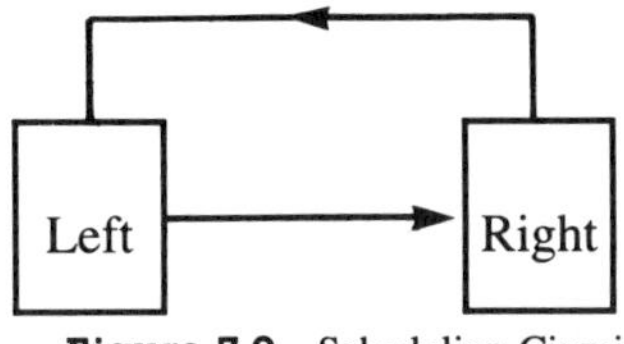

Figure 7.9 Scheduling Circuit

end of that same arc. Then we have a review to see whether our assumption was a good one. If so, we go on. If not, we make a better assumption and iterate around the circuit again.

Planning the design of a system involves defining the tasks, determining what tasks require information from what other tasks, tearing arcs to break all the circuits, and then estimating how many iterations there will be. Once these steps have been taken, the circuits can be opened up at the tears and laid out end to end to represent each iteration. Then we can assign times to each iteration of each task and develop a critical path schedule as shown earlier in this chapter.

The circuits are torn and unwrapped to form a finite number of distinct iterations.

Let's consider the example of the design of an order entry system. We will initially look at just two design variables: file organization and response time. (Later in this section we will look at these variables as part of a larger set of variables.) We assume that for each design variable there is a corresponding task to determine that variable.

Given the file organization, we can compute the response time. Given the response time, we can decide whether the file organization is adequate or whether another file organization should be sought. The nature of the connection is illustrated in Figure 7.10.

Clearly this is a circuit. As it stands, there is no place to start that we don't first need something we don't yet have. To resolve this dilemma, we may assume a file organization and then determine the response time; or we can simply ignore the response time so we can start by determining first the file organization.

Let's do the latter. We can show this by tearing the arc from response time to file organization (shown with a dashed line across the arc) (see Figure 7.11). Then we make a topological ordering of these variables so that all arcs except the tear go from a lower number to a higher number (shown by the numbers above the variables).

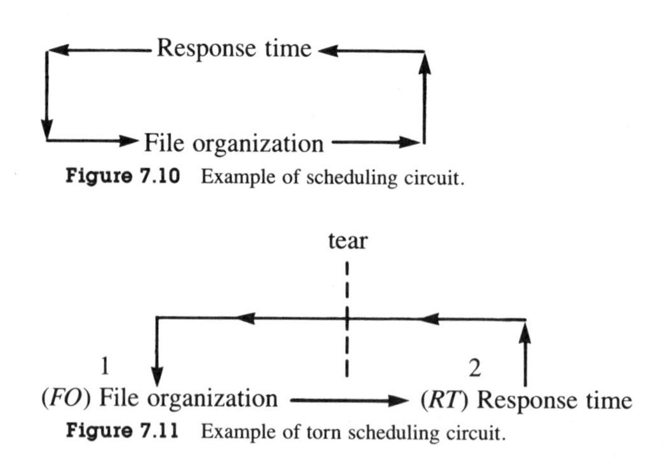

Figure 7.10 Example of scheduling circuit.

Figure 7.11 Example of torn scheduling circuit.

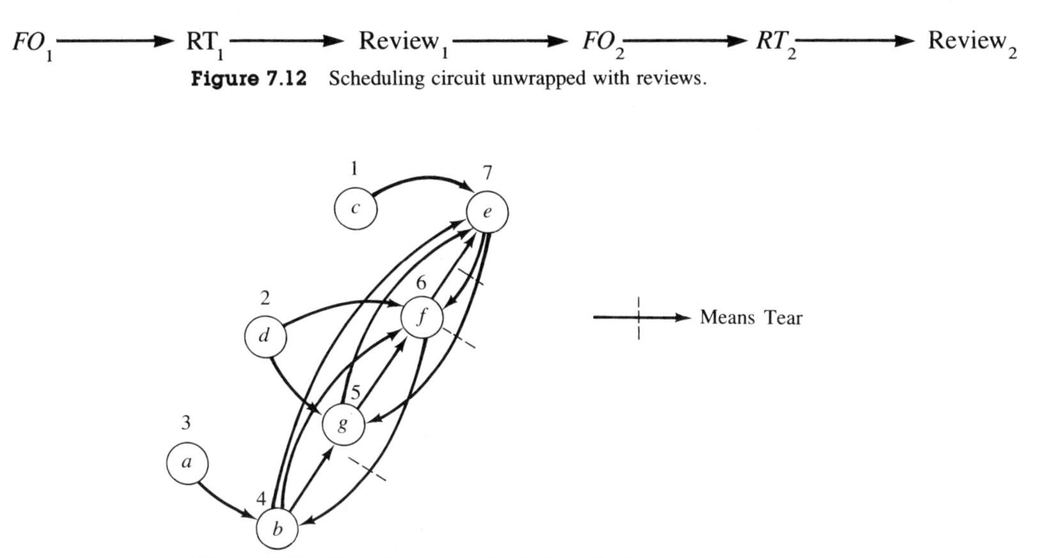

Figure 7.12 Scheduling circuit unwrapped with reviews.

Figure 7.13 Precedence graph for design of order entry system.

We then unwrap this circuit into iterations with reviews, as in Figure 7.12. After adding durations, we can compute a critical path schedule.

Now let's consider the system of seven tasks from the design of an order entry system, which includes the tasks we just considered. Table 7.7 shows for each task what other tasks produce information needed to do this task. These information requirements establish the precedence constraints. (See Figure 7.13.)

The torn arcs show where we make assumptions about the predecessors by ignoring them or using guesses. The numbers above the nodes show the order in which we can do the tasks on the first iteration. Note that the nodes have been ordered so that each node is preceded only by nodes of lower numbers, or by tears. So all arcs go from lower numbers to higher numbers, except the tears, which go from higher numbers to lower numbers. We can now do the tasks in the sequence of these numbers, ignoring or using guesses about higher number tasks we have not gotten to yet.

Table 7.7 Precedence table for design of order entry system.

Task	Predecessors (/ means no predecessor)
a. Size of inventory	/
b. File size	*a,f*
c. No. orders / day	/
d. Desired response time	/
e. Resulting response time	*b,c,f,g*
f. File organization	*b,d,e,g*
g. File Hardware	*b,d,e*

The graph shows what is connected to what, which is called the **structure**. It does not show what these connections mean, which is called the **semantics**. The structure shows whether the tearing we choose breaks all the circuits. But there are many ways to tear so as to break all the circuits. The semantics is used to judge whether a choice of tearing makes sense—whether the guesses the tears call for are easy or difficult to make. The following is a procedure to assign order numbers to the nodes so that each node is preceded only by nodes with lower numbers or tears. This helps us determine the order in which we will do the tasks when we later schedule them.

Ordering Procedure

1. Circle each set of nodes so that each set is the largest set such that there is a path from every node in the set to every other node in the same set. These sets of nodes are called **blocks**.
2. Assign order numbers to the blocks so that every arc between nodes in different blocks goes from a node in a lower numbered block to a node in a higher numbered block.
3. Tear sufficient arcs within each block so that no circuits remain. There may be many ways to tear. Of these possible ways, we choose one way so that it is easy to make guesses where the tears call for guesses to be made.
4. Assign order numbers to the nodes so that:
   a. Within the blocks the nodes have contiguous numbers, and
   b. Only torn arcs go from higher numbered nodes to lower numbered nodes.

Figure 7.14 illustrates this ordering procedure.

Methods are available for analyzing these graphs to find a set of tears that break all circuits; these methods allow us to interact with the computer program as we consider the semantics to choose tears where it is practical to make guesses (Steward, 1981a,b).

Precedence Matrices

A graph can also be represented by a matrix. (A matrix is more compact than a graph, is a clearer representation for a large system, and can be input to a com-

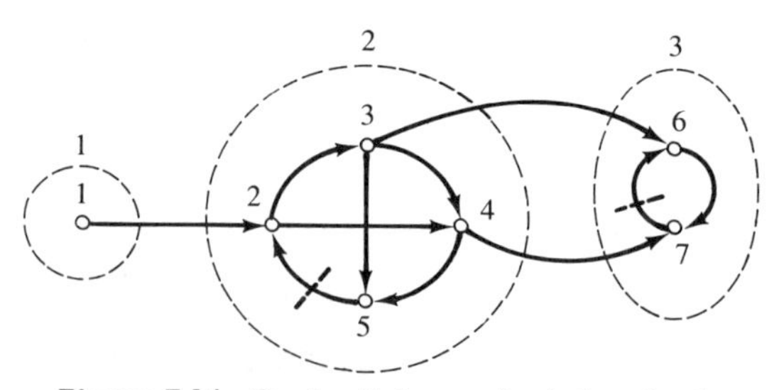

Figure 7.14 Graph with tears and ordering of nodes.

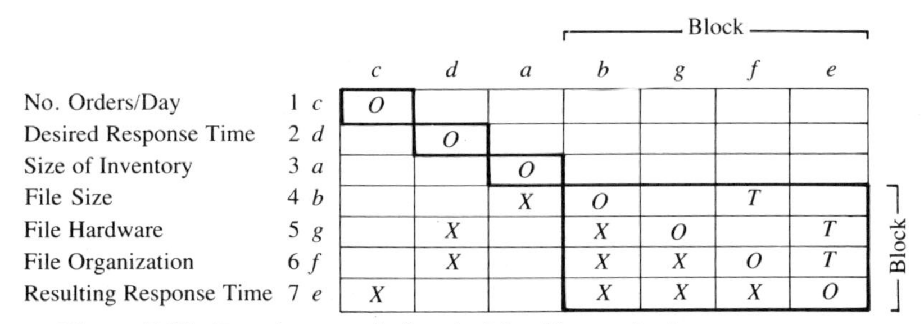

		c	*d*	*a*	*b*	*g*	*f*	*e*
No. Orders/Day	1 *c*	O						
Desired Response Time	2 *d*		O					
Size of Inventory	3 *a*			O				
File Size	4 *b*			X	O		T	
File Hardware	5 *g*		X		X	O		T
File Organization	6 *f*		X		X	X	O	T
Resulting Response Time	7 *e*	X			X	X	X	O

Figure 7.15 Precedence matrix for schedule with torn circuits.

puter program. The graph is a good visual representation for a small system, but if there are too many nodes it can be quite confusing.) The rows and columns in the matrix represent the nodes in the graph; the off-diagonal marks represent the arcs. The columns in the matrix must always be in the same order as the rows. In the matrix in Figure 7.15 the rows are in the order established in Figure 7.13.

O's always mark the diagonal. An *X* in row *f* column *b* means that the task to determine *f* (file organization) requires information from the task that determines *b* (file size).

When we put the rows and columns in the sequence of the numbers we assigned the nodes in the graph, we see that the marks corresponding to the torn arcs (marked with *T*'s) are from higher to lower numbers and thus occur above the diagonal. All the other marks are on or below the diagonal.

Note that *b*, *f*, *g*, and *e* are all in a common block. A block is defined by noting that there is a path from any node in the block to any other node in the same block. For example, if we choose *e* and *b* in the block, the path from *e* to *b* is $e \rightarrow f \rightarrow b$, and the path from *b* to *e* is $b \rightarrow e$. The block is marked off by a line above the columns in the block and by a line to the right of the rows in the block. Note that when the rows and columns are ordered this way, the marks appear only within a block, on or below the diagonal.

A design review should be made after completing the tasks in a block to see whether the assumptions were valid, and thus decide whether to make another iteration or to move on. The *T*'s above the diagonal show the assumptions that were made that need to be checked in the review.

If the system is small enough, all these tasks might be done by one person. But he or she may have to consider them one at a time because of the limits on how much information one person can handle at any time (Miller's Principle). Most projects require many people working together to do the design. Thus, the tasks may be assigned to different people. The choice of where tears are made will determine the order in which these people do their tasks, how the information flows from one task to another, where design reviews are required, and who should communicate their results to whom. Thus, the choice of tears leads to a design plan.

Once the tears are made, the circuits can be broken and the tasks in the circuit laid out as distinct iterations. Durations can then be given for each iteration of

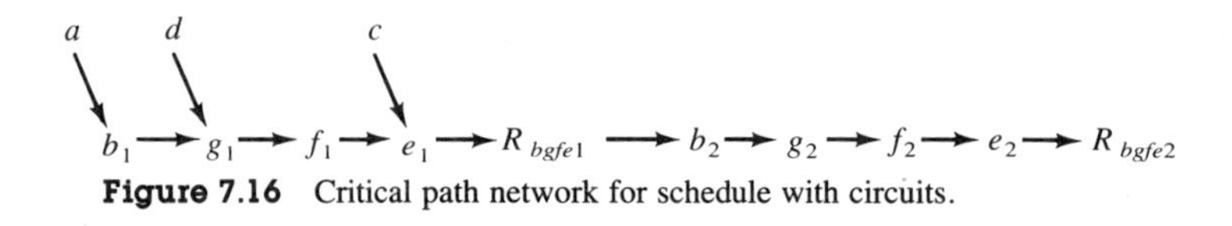

Figure 7.16 Critical path network for schedule with circuits.

each task, and a critical path schedule can be computed. The graph for this schedule without the durations would then look like Figure 7.16 where b_2 represents iteration 2 of task *b*, and R_{bgfe2} represents the review of tasks *b*, *g*, *f*, and *e* at the end of iteration 2, and so on. It has been assumed here that two iterations would be adequate. Predecessors such as $d \rightarrow f$, which are implied by other predecessors (that is, $d \rightarrow g$ and $g \rightarrow f$) have been deleted to make the diagram less cluttered. If in the review it is decided that fewer or more iterations are required, the schedule is changed accordingly.

Project Tracking and Control

Once the critical path has been drawn and the project begins, project managers must actually follow the work in progress to see how closely it is matching the schedule. It is often necessary to reschedule as changes occur.

Keeping the Schedule Updated

When we have negative slack as in Figure 7.2, we have to seek a way of replanning so we can achieve the desired project required finish. To fix the schedule so that it can be completed by the project required finish, assume it is found that *f* can bc shortened to 7 and *d* can be broken into two tasks: *d*1 of length 3, which precedes *d*2 and e, and *d*2 of length 4, which precedes *g*. Now we recompute the times as in Figure 7.17 and get a completion time of 59 as required.

As the project proceeds, circumstances always occur that were not planned: material is not delivered as expected, clients change their minds, resources are not available when needed, or some tasks take longer than expected. As actual dates occur, we plug them in and compute all the following dates. New plans may have to be made to determine how we will get from where we are to as close as possible to finishing the project as required. We then update our plans and expectations.

Assume we are now in week 42. Task *c* has just finished, and task *d*1 did not get started until time 40. How does this affect the computed project finish, the early times of each task, and the critical path?

We may also be able to make better estimates of some of the durations of tasks that lie ahead. The new network and computations may change what tasks are on the critical path and when we will need certain resources or parts to be delivered. This could affect the plans of other people who should be informed as soon as our plans have changed.

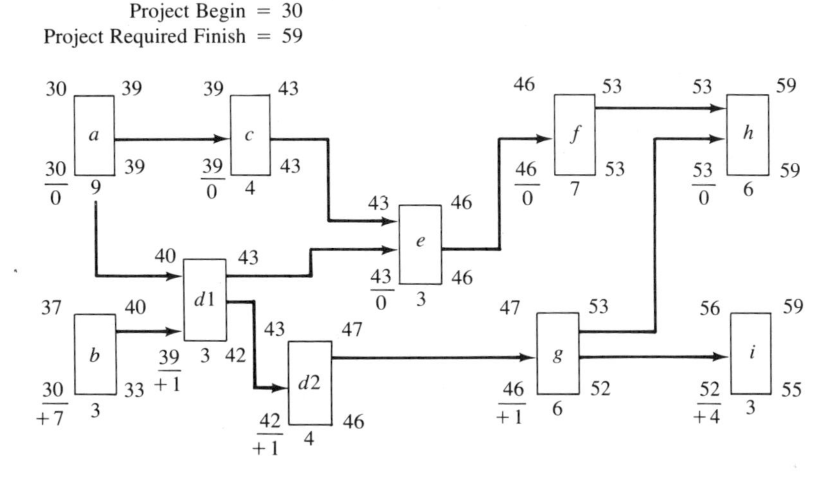

Computed Project Finish = 59
New Critical Path is *a,c,e,f,h*

Figure 7.17 Updated critical path schedule.

Time and Cost Accounting

In many organizations people work on more than one project each week. Managers must keep track of the amount of time each person is charging to each project, because the clients will be charged for this time. Managers may also want to know how much each task costs so that this information can be used to make estimates for the cost to do similar tasks in future projects. Finally, management often needs to monitor how the work is going to anticipate and forestall any problems in the project.

Many different time tracking systems can do the job. The following scheme illustrates what might be done. Table 7.8 shows all the calculations and data. Line *a* is entered when the plan is made. Lines *b* and *c* are entered by the employee each week. (The table might be set up so that the data are entered each day.) Lines *d*, *e*, *f*, *g*, and *h* are computed to help make inferences about how the work is going as compared to plan.

The employee would not see a full worksheet like this. He or she would enter only his or her data—either on a time sheet, which someone else keys into the computer, or directly on a personal microcomputer workstation. A manager tracking the work may look at this data on a printout or on a personal microcumputer workstation.

The vertical stroke in line *a* shows when the task is planned to be completed. The vertical stroke in line *c* indicates, for comparison, when it was actually completed. Line *f* tells us how the work spent compares with the estimate. The underlined figure on line *f* is the final tally of how much effort was used compared to plan. Line *g* shows the current best estimate for how long it will take to

Table 7.8 Tracking and projecting actual times against plan*.

TASK A

Start	wk1	wk2	wk3	wk4	wk5	
						PLAN DATA ENTERED
0	10	10	10	10	10	a. Hours planned this week
						WEEKLY DATA ENTERED
0	12	15	20	20	10	b. Hours actually spent this week
40	30	22	12	5	0	c. Estimated hours yet to complete
						COMPUTED DATA
0	10	20	30	40	50	d. Total hours planned to date $[d(-1)+a]$
0	12	27	47	67	77	e. Total hours spent to date $[e(-1)+b]$
0	+2	+7	+17	+27	+27	f. Hours spent + or − from plan so far $[e-d]$
40	42	49	59	72	77	g. Est. total time to complete $(c+e)$
0	+2	+9	+19	+32	+37	h. Est. hours to complete + or − from plan $[g-g(0)]$

TASK B

Start	wk1	wk2	wk3	wk4	wk5	
						PLAN DATA ENTERED
0	20	20	10	0	0	a. Hours planned this week
						WEEKLY DATA ENTERED
0	10	10	10	10	10	b. Hours actually spent this week
50	40	30	20	10	0	c. Estimated hours yet to complete
						COMPUTED DATA
0	20	40	50	50	50	d. Total hours planned to date $[d(-1)+a]$
0	10	20	30	40	50	e. Total hours spent to date $[e(-1)+b]$
0	−10	−20	−20	−10	0	f. Hours spent + or − from plan so far $[e-d]$
50	50	50	50	50	50	g. Est. total time to complete $(c+e)$
0	0	0	0	0	0	h. Est. hours to complete + or − from plan $[g-g(0)]$

* − 1 refers to previous week
0 refers to start

complete. Line *h* compares this new estimate with the original estimate. Lines *d* and *e* are intermediate calculations to compute *f*, *g*, and *h*.

Looking at lines *f* and *g* for these two tasks, we can see that Task *A* was taking much more effort than had been expected but was finished on time. Not as much effort was spent each week on Task *B* as had been planned and it was not finished on time.

Such a table can be used to see problems as they occur so they can be corrected before the task is finished. It will also record the times for this project and can be used as data for estimating future projects.

Displaying the Schedule with Bar Charts

We can display the critical path schedule with a **bar chart**, also called a **Gantt chart**. For each task a bar shows when the task is scheduled to begin and finish; a label describes the task (see Figure 7.18).

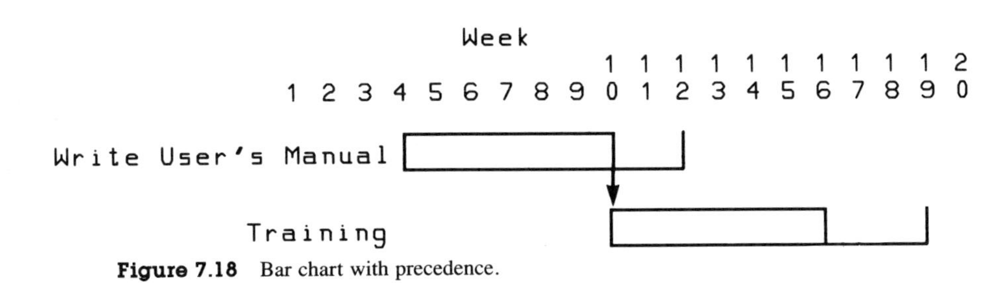

Figure 7.18 Bar chart with precedence.

This Figure would imply that the task "Write User's Manual" is scheduled for an earliest begin in week 4, an earliest finish in week 10, and a latest finish in week 12. "Training" would begin in week 10, be finished by week 16, with a latest finish in week 20. The downward arrow can be used to show the precedence relation between tasks. However, using arrows to show all the precedences can clutter the chart, so we use this device with discretion. The primary purpose of the bar chart is to display the time when tasks are to be done, not the precedences.

Resource Planning

For each task the amount of each resource used can also be shown on the bar. Then a plot can be made of the amount of that resource used over time (see Figure 7.19). Such plotting shows how many people are needed and when they are needed if this plan is to be followed.

Now we may want to level the plan so that there is not too much variation in how many people are needed at different times. It is difficult and expensive to move people into and out of the project so that they are ready just when they are needed and not on the payroll when they aren't. We may also feel some moral obligation not to make the lives of these people difficult by hiring and firing or

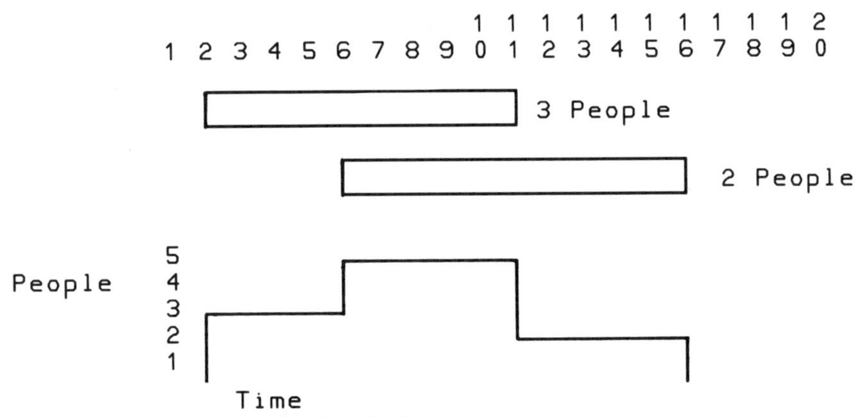

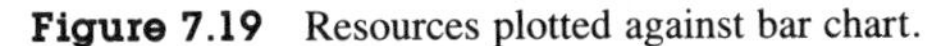

Figure 7.19 Resources plotted against bar chart.

moving them back and forth between projects. There may be some freedom to move a task back and forth between its earliest and latest times. We can use this freedom to move a task back and forth between its earliest and latest times to lower the peaks and fill in the valleys. The resource curve can also be flattened by changing the level of resources assigned, which would lengthen or shorten the bars. If necessary, we could also delay the project to flatten the resource curve.

Tracking Progress

As the project proceeds, we need to monitor what is happening and when it happens to see whether the project is going according to plan. When all expectations are not being met, which is quite frequently, we must understand why and make changes to how the work is being done and how future work is planned so that we have the best chance of still completing the project as required. If we find we cannot complete it when originally expected, we must renegotiate these new expectations.

A bar chart can be used to keep track of progress (see Figure 7.20). The bar is filled in to show how much of the task has been completed. When showing the progress, we should keep in mind the tendency for the details at the end of a task to take a long time and thus be careful not to overstate this progress. A symbol such as the V may be used to show the new time when the task is expected to be complete; another symbol, such as a solid triangle, can be used to show when the task was actually completed.

A variation on using the bar to record progress is called the **grief-joy chart**. Only what happened that was different than expected is shown. In Figure 7.21 a NOW line indicates the date the information on the chart was updated. If progress is better than expected, the bar is filled with the progress only to the right of the NOW line to show just the unexpected progress. This filled part of the bar is called "joy." If progress is behind what is expected, the bar is filled to the left of NOW, showing only the progress that was expected but didn't happen. This filled part of the bar is called "grief." No fill-in indicates that the progress is as expected.

The first bar in Figure 7.21 shows how the same progress indicated in Figure 7.20 would be displayed on a grief-joy chart. The second bar assumes that progress was behind what was expected, completed only up through week 3 when it should be up through NOW, which is week 6. The bars are plotted against their latest times so that the grief or joy displayed is real, not hidden by any slack.

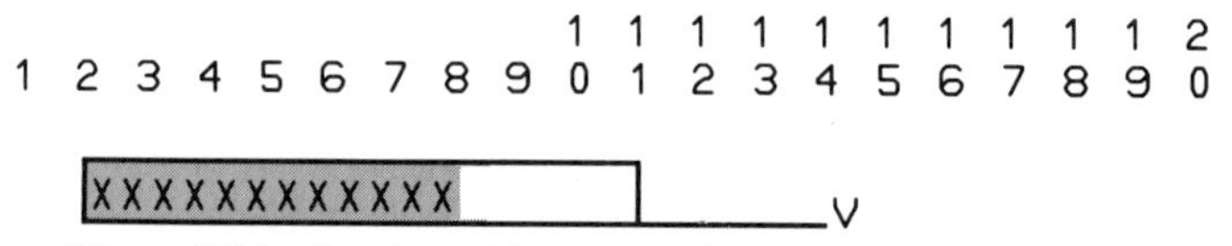

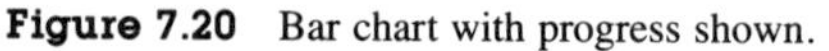

Figure 7.20 Bar chart with progress shown.

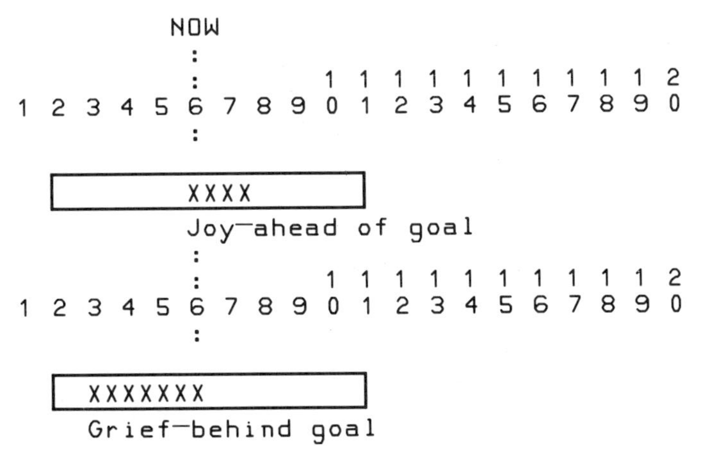

Figure 7.21 Grief-joy chart.

Metrics

Taking the time for paperwork first, before actually working on the product, can pay off significantly in cutting the costs of changing the product late in the process. But how much investment in paper should be made? Is the paperwork productive? Is it going well? Can we measure progress by how much paper has been generated? Can we tell how much more effort and time will be required? Is paper being generated for paper's sake?

To make the "paper first—product later" approach work, we must be able to provide management with measures of the costs of producing the paper and its value later on. Until this is done—and it isn't easy—the "paper first—product later" approach will have to be justified to management on the basis of reason and faith. Reason and faith do not last long under pressure to get the product out.

Management is used to looking at the number of lines of completed source statements as a measure of progress. If a Tree is used, the unit of measurement is each branch of Tree generated. This system offers a measure that encompasses the whole process from front end planning through the back end code generation.

Traceability

It should be possible to trace every step taken in building the product back to the statement in the design that made it necessary. Similarly, it should be possible to trace every statement in the design back to the statement in the requirements specification that made that design statement necessary. Otherwise, we may build a product that is not as specified in the requirements specification, or we may do work on building a product that is not required by the design, which only increases the cost and time.

If a tree is used to represent the product from top to bottom, from the user requirements to code, the tree provides this traceability in the description of the product. This is a prime reason for having one method that can be used consistently from top to bottom and from beginning to end so that traceability is not lost when switching from one method of description to another.

The work breakdown tree offers another dimension in traceability. Every task can be traced through the subgoals that make it necessary, and the subgoals to their subgoals and finally to the overall goal.

Reviews

As we have already pointed out, the longer the interval between when an error is introduced and when it is found, the greater the cost of fixing it. All the work based on the error may need to be redone. Thus it is in our interest to catch errors as soon as possible after they are introduced.

In the later stages of a project the system can be seen and errors can be found by running tests. But errors should be caught early in the process when the system exists only on paper. At this point the system cannot be run in order to test it. The paper must be reviewed by humans.

Reviews have many advantages over waiting to run a test. Reviews can find problems long before there is anything to test. A review can often catch errors in logic that would require many test cases to find. A test may determine only whether the system does or does not run correctly for that one particular case. When a test finds an error, one must still find the error's cause. A review may find the error at its cause. Furthermore, a review can check the system directly to determine whether it conforms to documents such as the requirements specifications.

There are several kinds of review. A **walkthrough** is a review by peers looking for errors or potential errors in the system. This may be informal. A programmer may go over the program with a fellow programmer. Often, by explaining to someone else what he or she is doing, the programmer can spot errors. When one programmer reviews another's work, both can learn new techniques. Also, if programmers working on different parts of the system review each other's programs, they can resolve problems about how the parts should fit together or communicate.

Reviews can also be very formal. A **formal management review** may take days to prepare and rehearse so that the right information is conveyed quickly and effectively. Presentations sometimes include visual aids such as flip charts or slides. Formal reviews present information that management needs to make decisions about whether to

- Continue the project
- Change the plans, schedule, or resources
- Redo work
- Review changes with the client

An **inspection** is somewhere between an informal walkthrough with peers and a more formal review by management. It is a formal review by peers. The inspection may be done by a team made up of a moderator, the author of the work to be inspected, and several technical peers of the author who act as inspectors. They may work from checklists, which help focus their attention on likely sources of errors, such as data declaration errors, data reference errors, and so on (Myers, 1979). The moderator sets up the inspection by distributing the materials to be inspected, scheduling and moderating the meetings, maintaining a log of all errors or concerns that develop, and making plans for follow-up action. The inspectors must have the material to be inspected in plenty of time to go over it before the meeting.

The atmosphere of the inspection should be such that people will look for faults or potential faults in the system, not faults in the author. The author should not feel on the defensive. He or she should be as interested as everyone else in finding any errors or potential problems that may exist. Thus managers who may have to make personnel evaluations of the author should probably not be present, and reports of the meeting should not reflect on the author personally.

The emphasis of the inspection should be on finding faults, not in determining how to fix them. It is easy to have a group of people agree that some matter is in error or should be looked at further. But discussions of how to fix it can go on endlessly, consuming all the time scheduled for the meeting.

An inspection can be used to resolve the technical problems before making a review before management. Then the management review can concentrate on management problems such as costs, schedules, resource allocation, and so on.

Configuration and Change Control

As we explained in Chapter 2, the configuration of a system is the current definition of what the system is or is planned to be. The definition includes the system's parts and how they go together, and it takes into account the assumptions that have been made about the system. As the system is developed, its configuration changes. If we do not know what the configuration of the system is, we do not know what we are working with. And if we do not know what we are working with, we do not know what we are doing. Keeping track of the configuration of a complex system is an onerous, but a vital, job.

An error or change may affect many parts of the system and many documents. Some errors or changes may affect whether the product will meet the requirements specifications. If so, the modifications should be reviewed with the client. Errors or changes can affect whether work must be redone, how plans for work yet to be done must be changed, and whether documents must be reviewed to determine whether work is conforming to them. The change must be documented, and all documents affected directly or by the effects of these changes must be corrected accordingly. All the people whose work is affected need to be notified.

Change control is the process of seeing that the effects of errors and changes are properly taken care of. Large projects may have a Change Control Board to review all errors and proposed changes, to decide what action is required, and to initiate that action.

Some errors found in the system, or even deviations from the controlling documents, may be acceptable. Not every error should be fixed. But the people responsible for these documents must review the consequences of the errors and decide whether a fix is necessary.

It is very difficult to work on systems that seem to be constantly changing. All sorts of people have some idea about how they would like to see the system changed. Perhaps the client wants things differently than he or she had originally specified, or the government has changed its regulations, or people working on the system see a better way to do it. With many changes affecting many parts of the system there can be continual chaos with no progress. Often these changes are in principle good, but their consequences and costs may be too great. All changes should be proposed through some form of documentation, perhaps called a change proposal, addressed to the change control board.

The graphs and matrices made for the design structure system are helpful in determining what is affected by a change. Going down the column corresponding to a changed task will show which other tasks are directly affected. For example, in the matrix for the order entry system design (see Figure 7.15), if we change the desired response time (*d*), we can go down that column and see that it would directly affect file hardware (*g*) and file organization (*f*). These tasks or documents should be reviewed. If they change, we need to go down those columns to see what is changed next, and so on.

Once we have determined what tasks are affected, the change control board can estimate the time and cost to make the change and the consequences of not making it, then decide whether the change will be made.

Project management may decide to freeze documents periodically, not allowing any changes to be made to them. During the period of this freeze the change control board has time to review new change proposals and people have time to get work done unimpeded by changes. Then those changes can be issued all at once as change notices distributed to the people responsible for the affected tasks. A new set of documents is issued reflecting the new freeze, and people work to these documents until the next freeze cycle.

Summary

To compute a project schedule we use a network, which is a special kind of graph. A graph is a set of points, called nodes, and the lines between certain pairs of nodes. If the lines have directions, they are called arcs and the graph is a directed graph. If the lines do not have directions, they are called edges and the graph is an undirected graph. A sequence of arcs from node to node in a directed graph is a path. If the path returns to where it started, it is a circuit. A directed graph used for scheduling is a network.

Once the tasks have been defined with the help of a work breakdown tree (Chapter 6), a network can be drawn with nodes representing tasks and arcs between nodes to show where one task is a predecessor of another, that is, one task must be finished before the next can begin. Given such a network, an estimated duration for each task, and a start date for the project, we can compute the earliest and latest time each task can begin and finish, and the earliest time the project can be completed. The difference between the latest and earliest begin time of a task is its slack, the amount of time a task can slip before delaying the project. The longest path through the network, which defines how long the project takes, is called the critical path.

Generally it can be expected that some things will not happen as planned. There has to be constant vigilance to find these deviations and plan what to do about them. Thus, during the course of the project, the network and its plan will be changed many times.

The older way of drawing scheduling networks, called IJ networks, involves putting the tasks on the arcs, but this requires the use of dummy activities, which makes the networks harder to draw. The newer way, called precedence networks, puts the tasks on the nodes and is easier to draw and maintain as the plan changes.

If we draw a graph for the tasks to design a system with the arcs showing where one design task depends on another, the graph will have circuits. This is because some of the tasks will be interdependent, that is, task *a* depends on task *b* and task *b* depends on task *a*. Such a graph cannot be scheduled until we decide where we will tear arcs to represent guesses. These tears break the circuits, which can then be unwrapped to show design iterations. Then we can have a network for which we can compute a schedule.

Periodic reviews can catch errors early in the paper stage when the cost of fixing the errors is small. Reviews should focus on the faults in the work, not in the people. Walkthroughs are reviews by peers. Inspections are more formal reviews with follow-ups to see that any problems are taken care of. Usually technical reviews should occur first so that management can focus their reviews on management issues.

It is next to impossible to manage a project when changes are being made continually unless the changes are well-managed. Change control involves evaluating proposed changes for their effects on the product, schedule, and cost, then notifying people whose work is affected of approved changes, and seeing that all documents reflect the changes to show the current concept of the product.

Exercises

1. Given the following tasks, draw the critical path network and compute the earliest and latest begin and finish times for each task.

Task	Duration	Predecessors
A	3	*C*
B	4	*A*
C	8	/
D	6	*C*,*E*
E	3	/

Project Begin = 8
Project Required Finish = 256

2. Assume the schedule shown in Exercise 1 and that it is now week 16. *C* actually finished at week 15, and *E* will take another two weeks to finish. Now what does the schedule look like?
3. Starting again with the schedule in Exercise 1, assume that there is an external requirement that task *A* be completed by time 19. How does this change the latest times and slack?
4. Draw the bar chart for the schedule in Exercise 1. Show on the bar the earliest begin, earliest finish, and latest finish.

   Assume the following number of units of resources for each task.

   A 2 units of resource *r*
   B 5 units of resource *s*
   C 3 units of resource *t*
   E 4 units of resource *t*
   D 2 units of resource *t*

   For each resource, plot the resource level versus time with all the tasks at their earliest times, and with all the tasks at their latest times.

   Now plot the resource levels for resource *t* with *C* at its earliest time, *D* at its latest time, and *E* between them. What is the effect of this resource leveling on the maximum number of people required?
5. In the following graph:

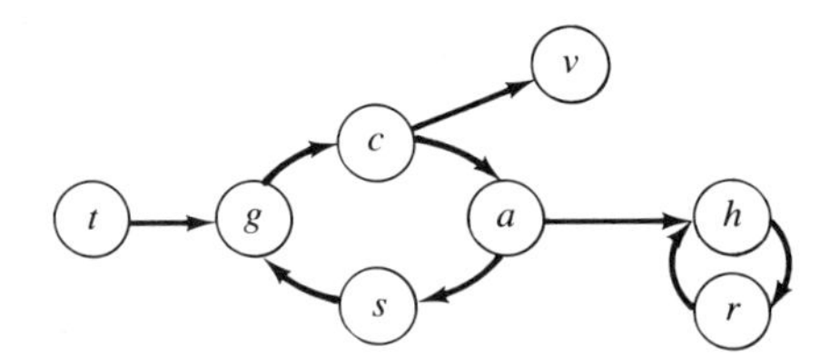

   a. Circle the blocks.
   b. Mark one tear in each block.
   c. Number the nodes such that all untorn arcs go from a low number to a high number, and the nodes in each block are numbered contiguously.
   d. Find the bad pun.
6. Given the following precedence matrix:
   a. Draw the graph.
   b. Circle the blocks.

	a	*b*	*c*	*d*	*e*	*f*	*g*
a	O					X	
b		O		X		X	X
c			O		X		
d		X	X	O			
e	X		X		O		
f			X		X	O	
g				X			O

c. Mark tears in each block to break all circuits.
d. Number the nodes such that all untorn arcs go from a low number to a high number, and the nodes in each block are numbered contiguously.
e. Make up a new matrix with the rows and columns in the sequence of this numbering.
f. Outline with dark lines the blocks on the diagonal.

References

Bersoff, Edward H., Vilas D. Hendserson, and Stanley G. Siegel. *Software Configuration Management: An Investment in Product Integrity*. Englewood Cliffs, N.J.: Prentice-Hall, 1980.

Boehm, Barry W. *Software Engineering Economics*. Englewood Cliffs, N.J.: Prentice-Hall, 1981.

DeMarco, Tom. *Controlling Software Projects: Management, Measurement and Estimation*. New York: Yourdon Press, 1982.

Gunther, Richard C. *Management Methodology for Software Product Engineering*. New York: Wiley Interscience, 1978.

Horowitz, E. *Practical Strategies for Developing Large Software Systems*. Reading, Mass.: Addison-Wesley, 1978.

McClure, Carma. *Managing Software Development and Maintenance*. New York: Van Nostrand Reinhold, 1981.

Myers, Glenford J. *The Art of Software Testing*. New York: Wiley Interscience, 1979.

Reifer, D. J. *Software Management*, 2nd rev. ed. Los Alamitos, CA.: IEEE Computer Society Press, 1981.

Steward, Donald V. "The Design Structure System: A Method for Managing the Design of Complex Systems. *IEEE Transactions on Engineering Management*, August 1981a.

Steward, Donald V. *Systems Analysis and Management: Structure, Strategy and Design*. Petrocelli Books, Princeton, N.J., 1981b.

Yourdon, Edward. *Managing the Structured Techniques*. 2d ed. Englewood Cliffs, N.J.: Prentice-Hall, 1979.

Yourdon, Edward N. *Structured Walkthroughs*, 2nd. ed. Englewood Cliffs, N.J.: Prentice-Hall, 1980.

Applying the Technology

Software engineering is concerned with using computer technologies to produce products that solve problems. The software engineer is responsible for understanding these technologies. Many of them such as data structures, file design, and communications are the subject of other courses and books, and thus need not be covered here. However, some topics typically slip through the cracks. In some cases there is not time to devote whole courses to them, and in other cases there is not enough material to warrant a whole course.

Part 4 is a brief survey of some of the technologies of concern to the software engineer. These topics are background to Chapters 12–15. If they are covered in other courses, this part may be skipped or considered a review.

8

Technical Considerations in Analysis and Design

- Data base management systems are organized around their data so many processes and departments are able to share the same data.
- The capability of having many processors, often distributed in many places, all working on the same problems, offers opportunities and challenges to systems analysis and design.
- Another analysis and design challenge is the integration of computing technology into the electronic office.
- People using the new innovations in computers and communications can work when and where they choose—no longer being obliged to come together in central offices or to meet at certain times.
- Electronic mail gives users the ability to send and receive messages with neither sender nor receiver having to wait for the other to participate at the same time.
- Expert systems put the knowledge of one or more experts in a computer data base so that knowledge is made available to those who may need it.
- Artificial intelligence is the field of creating computer systems to do the tasks that usually require human intelligence.
- Data processing must guard against the dangers of errors, loss, mischief, fraud, and hazards such as fire or floods.
- Systems are judged by how well they perform in terms of speed, capacity, reliability, and availability.

This chapter looks briefly at some technologies that software engineers need to be familiar with when they analyze the weaknesses and strengths of an existing system (Chapter 12) and develop the requirements, feasibility and design of a new system (Chapters 13–15). The technologies are presented in an historical perspective. Because of its survey nature, the material is arranged in the form of a catalog of pertinent technologies with very short descriptions.

After the survey of technologies, we'll discuss some of those features of systems that must be considered in their analysis and design, such as the use of controls and security. Finally we'll give a short overview of the analysis of performance. Performance analysis is important in determining why existing systems are not behaving as we would like, or in evaluating the expected performance of proposed designs to establish whether they will satisfy the needed requirements.

Technologies Used in Information Processing

Due to the fast pace of the development of information technologies, we are likely to encounter today the full historical span of computing systems—manual systems that have only recently been converted to computers as computer prices have come down, computer systems still using twenty-years-old technologies, and the very latest systems using innovations just now appearing in the journals. While some people have been conservative or have never heard of the new approaches, others are anxious to use state-of-the-art methods. Thus, in systems development we will encounter existing systems from a wide range of technologies. We will consider where computing has been and where it is going as it affects analysis, requirements, and design.

Processing Modes

When computers were very expensive, we were concerned with how to get the data to the computer to make the computer most efficient. Now, as computer costs have plummeted, we are concerned with how to get the computing power to us to make us most efficient. We have moved from an era in which the large, centralized computer made the most sense into an era in which it makes more sense, in many cases, to use decentralized computers. When we have problems requiring a lot of computing power, we can use computers that have many processors working simultaneously.

Single-Process Mainframe Computers

When stored program electronic computers first came into commercial use in the early 1950s, a user would have exclusive access to the machine for a short period. He or she would go onto the machine and run a problem; when one user was finished, another user would run the machine. Users paid $500 or more per hour for machine time, and their jobs often ran several hours. These early machines were very expensive and not very fast. To make them economic, users ran them twenty-four hours a day. Only problems that justified their high cost were ever run, and mistakes were costly. But while the user was running a program, the relationship was one person to one machine.

Multiprogramming

Grosch's Law observed that doubling the cost of a machine could increase its computing power by a factor of four. Thus it became cost effective to buy larger machines provided this additional power could be used. But if only one program ran on the machine at a time, the computing unit was often idle while IO was being done. By using separate computing, IO processors, and an interrupt mechanism, it became possible to write operating systems that switched the computing and IO processors between programs. Then computing for one program could occur at the same time as the IO for another. This is **multiprogramming.**

Batch Processing

Multiprogramming allows jobs to be stacked with their operating instructions so that a number of jobs can be loaded into the machine at one time. This capability leads to **batch processing.** Here work is accumulated for the machine until it is ready to be run as a batch. Then the batch is sent to the machine where someone else runs it when they are ready, and the output is returned to us when it is done. Batch computing is done at the convenience of the computer rather than of the user.

Timesharing

Another approach to loading the machine with many programs is **timesharing.** If communication processors are added to the multiprogramming capability, many people can run their programs by remote access. Each person runs his or her own program at a pace controlled by how fast he or she can provide input and respond to output. This again restores the regime of one person to one apparent machine. Users do not have to wait to accumulate a batch of data and wait again for someone else to process it. But because inputting data through a keyboard is so time consuming, the data are retained at the central machine so they will not have to be input again. Data backup and security are thus in the hands of the computer center.

On-Line Processing

In **on-line processing** the computer operates as the process occurs, but the speed of the process may be affected by the response time of the computer. An example is an on-line teller machine, where the user may wait a moment to get a response from the machine. On-line processes are usually timeshared.

Real-Time Processing

Real-time processing takes place when the computer must respond to interrupts to respond to a process without delaying the process. As computers have gotten faster, they have become able to keep up with faster processes, leading to more applications of real-time processing. In closed-loop systems the computer reads data generated by the process, makes calculations, and operates controls that change the process in real time.

Multiprocessing

Multiprocessing has been used to increase computing speed by parceling out pieces of the work to various units that may work simultaneously. Often each of these units has a special purpose, like communications processors or channels on input-output devices. But there are now also systems of many similar computing units working simultaneously on a common problem. With a system of many computing units, instead of the system being either up (operating) or down (not operating), it might, for example, be 6 percent down because some of the computing units are not operating while others are.

One of the important current challenges has become the development of operating systems and compilers, or specially designed computers that will automatically parcel out the work to the various units so that as much work as possible can be done simultaneously. Data flow computers are designed to assign computing units to data once the data are ready to be processed. The methods of handling asynchronous processes discussed in Chapters 3 and 4 allow us to represent systems that can use this multiprocessing capability.

Embedded Computers

Processors on chips with extensive computing capabilities have become inexpensive. They now appear as parts of automobile controls and displays, TV sets, airplane flight controls, laboratory instruments, and microwave ovens. When such processors are considered primarily not as computers but only as a part of something else, they are called **embedded computers.** Many embedded computer systems use several processors that are multiprocessed.

Trends

The problems faced in society and industry have become more complex and the need for information processing has become more demanding. As the use of computers has spread from the working environment of experts and enthusiasts to the working environment of most people, computer systems have had to become easier to use.

Complexity and the Burden on Information Processing

At the same time that the cost of computing is coming down, the need to compute is going up. A larger fraction of our work force is now involved with handling information rather than handling actual material items.

The problems of our society are becoming more complex because the increasing interaction between problems means that we cannot solve problems one at a time independently. What people used to be able to do without affecting others now has effects on other people, who demand to be parties to the decision. These connections make life more difficult and require more people to have more information in order to make decisions.

Not many years ago someone could cut trees on his or her own land for fuel. The smoke would dissipate before it reached the edge of the property. Now we have more people packed closer together. We reach halfway around the world to get much of our energy, and the smoke carries into other countries, where they object to the acid rain. Although we once acted as though resources were infinite and each person used his or her own private resources, now we recognize that many resources are limited and we must interact to arrange how to share them. To resolve the problems arising from interactions, we must store and process more information than we have ever dealt with before.

User Friendliness

Early machines were used solely by computer professionals. Little attention was paid to making systems easy to use. Indeed professionals often took pride in how hard the machines were to use. Today everyone uses computers. Systems have to be easy to use by people who are not prepared to spend a career learning how to handle them. Unlike computer scientists, average users these days do not consider the operation of computers the focal point of their lives. Programs have to describe themselves to the user. Many programs are now being written and sold, but the ones that sell the most are those that users can most comfortably apply to their work. User interaction will be discussed further in Chapter 9.

Communications and the Personalization of Computing

When everyone used the same large, centralized computer, that computer could be used as the focal point for communications and sharing data. Now that much of our computing is becoming distributed among smaller computers, those computers must communicate. Communications and personal computers lead to the ability to create, edit, communicate, store, and share data in an electronic office.

Communications

As people want to use the same common data but do not use the same computer, computers are being tied together with communications. Communications occur not only between terminals and computers, but between computers and other computers. As the load of communications becomes heavier, special communications processing computers have been introduced to handle communications. These input-output multiplexers themselves are multiprogrammed to handle many communications lines.

Personal Computer Workstations and Local Area Networks

Today the cost of microcomputers is so low and the capability so great that for many applications it is practical to have one's own computer with the power of machines that once had to be shared with others. The cost to the user of running just one of these jobs on the old machines would today allow the user to buy his or her own machine. Once an individual has run the jobs that justify the machine's purchase, the computer is still available the rest of the time at no additional cost to run any other jobs.

Since individuals now each have their own personal computers, the advantage of different departments such as payroll and personnel maintaining and using the same data would be lost unless the personal computers could communicate. **Local area networking (LAN)** allows separate personal computers to communicate with each other and with machines that maintain files (file servers) to share common data.

Desynchronized Space and Time

The combination of computers and communications is having a significant effect on where and when we can do our work. In an agricultural society people worked at their own pace on farms that they shared with their families. In the industrial society of the nineteenth and twentieth centuries workers traveled each day to central factories or offices to work together paced by the same clock. Workers needed to be together to work on the same materials or information. Now we are entering a new form of society. Since many of us work with information rather than material items, instead of transporting ourselves on highways to some location where we can work together on the information, the information can be transmitted to us over wires. Our work is desynchronized in space (Toffler, 1980).

A principal means of communication has been the telephone. The use of the telephone is time synchronized; both parties must be on the phone at the same time to have a conversation. Electronic mail allows us to leave a message without having to synchronize with anyone. Studies have shown that in business communication it takes an average of three or four phone calls before the caller and called are time synchronized and can talk. The other calls result in busy signals, no answer, or someone taking a message to return a call. Attempting to return a call brings the same frustrations and just continues this game of telephone tag.

Electronic Mail

Electronic mail allows for a time-desynchronized communication—that is, we can leave messages without having to synchronize with anyone. A message can be typed and kept on a data store until the receiver is ready to look at it. Whenever they choose, receivers can check to see what messages are waiting for them. They need not be interrupted by a phone call. Senders type their message when they think about it, without having to think about it again to make a call later.

This same process can be done with recorded voice messages. The computer calls back until the message is delivered. This saves us from having to disturb our work to think about making the call again.

By using electronic mail we can do our work where we please. That may be at home. Or it might be at a nearby center, where we can work with friends. If we want to work with different friends, we can walk to a different work center. The people we work with need not be people working for the same organization or with the same information.

If we work on the West Coast, we don't need to be in the office early to handle business with people on the East Coast. People can work together on the same problem, although they are in different cities and working at different times. When it is necessary to communicate in time synchronization, electronic mail can be used to set up a time when all parties are available to talk.

The wired society can be very important to us at a time when energy resources are becoming more obviously limited. A significant amount of energy can be saved by not having to drive to work or go to the store as often. Store items can be ordered by electronic mail and delivered to the door more efficiently than if all the customers make their tours to all the stores. Before electronic mail could become a reality, enough people had to have the appropriate equipment to send and receive messages. Word processing got equipment into a large number of offices so that the machines could be used to talk to other word processors. This growth of word processors set the stage for electronic mail, and electronic mail sets the stage for coordinating office functions with computerized scheduling and task management, which leads to the electronic office.

Electronic Office Information Systems

Word processing was the thin edge that led us into the automated office. Entering text on a keyboard converts the text to computer-readable form. We can then modify, file, move, retrieve, and print the text as we wish. We can process graphic information in the same manner.

Key entry requires a conversion from human form to computer form, which is expensive. Printing is a conversion from computer form to human form, which is very inexpensive. Once the expensive key entry has been paid for, we should think about what else could be done with that data in its computer form.

Once in computer form, data can be transmitted by electronic mail and displayed or printed out at some other location, allowing a space desynchronization, or placed in a data store and held to be made available when convenient, providing a time desynchronization.

The arrangement is not really an "automated office," although this term is commonly used. It might better be called an **electronic office.** The functions of the office are not automated in the sense that they are taken over by the computer. The computer and the people in the office work together. The electronic devices form a system to help people do their work and make decisions.

The growth of electronic offices has been somewhat hampered because the independent development of networking techniques by several companies has produced a number of incompatible communications protocols. Machines with different protocols need translators to talk to each other.

There are two possible solutions to this incompatibility problem. One is to standardize protocols so that everyone uses the same one. The other is to develop electronic equipment, called gateways, that will translate from one protocol to another.

Once people learn how to use electronic mail to organize their work, they have begun transforming their environment into the electronic office. The most difficult part of the evolution of the electronic office, however, is rethinking of office functions in light of these new capabilities.

Most of the increases in industrial productivity in the past have come from introducing machinery in the factory. Now we can look for an increase in productivity in the office through the introduction of electronic office equipment and procedures.

The early use of electronics in the office improved the productivity of clerical workers. But the largest cost in the office is not the wages of clericals but the wages of managers and professional workers. Thus the greater potential for savings from the electronic office is the increased productivity of managers and professionals.

Fourth-Generation Approaches

We are moving from languages with which we describe how to solve problems to languages we use to describe what the problem is, leaving it to the system to work out the how. These fourth-generation languages usually work within some domain of problem types that they are able to solve. One of the first domains to yield useful fourth-generation systems is classical data processing and data base inquiry. Communications allows us to distribute our processing and data storage to many different sites. One very pressing current problem is how to design systems in which processing and storage are distributed.

Data Base Management Systems

Initially files were made specifically for the program that used them. If the same data were used by several programs, each program had its own file specifically formatted for its use by that program. The program was dominant; the file was its adjunct.

When many people shared the use of a common machine, they could share the data maintained on that machine as well. Instead of running separate programs, each with its own data, people found they could integrate applications with several programs using the same data base. Payroll and personnel didn't need their own files on each employee, so that each department had to change its files whenever the data changed. They could share one file and make the change in one place.

The effort saved by not having to keep up two files could be used to make extra sure that the one file was correct.

The data in a **data base management system** must describe itself by a pattern (schema) so that it can be understood by all the programs using it. This led to the dominance of the data base over the program. Now we frequently think of the program as something attached to the data base rather than vice versa.

Distributed Systems

Many new systems have their data bases and computing capabilities distributed among several locations. A bank may have many branches. Each branch may store and process the accounts of the customers of that branch. The branches may periodically feed information into a central office to compute the total assets and liabilities of the combined bank. Customers at one branch are able to make transactions on accounts maintained at another branch.

The design of such systems must consider where the people who use the data are, where various data are maintained, what transactions are made on what data, where these transactions are processed, how the integrity of the data is maintained when it is used by different machines and people, and how this affects the communications between sites.

Processing can also be distributed between parts of a job done on a mainframe and other parts done on personal computer workstations. Decision support systems may be considered yet another dimension of a distributed system, where the data and processing are distributed between the user and the computer. We will discuss decision support systems shortly.

Standard Data Processing Patterns

Data processing has a common pattern to it: (1) generate a transaction, and then (2) process that transaction.

Generating a transaction involves matching data about the same entity from two or more different sources. Different types of processing are distinguished by how the transaction is generated. Let's look at several examples.

Payroll is an example of batch processing. For each employee there is a record on a master file showing the employee number, name and other identifying information, and pay rate. Every pay period the employee makes out a time card showing the number of hours he or she worked. To determine how much the worker is to be paid, a match must be made between the pay rate on the master file and the number of hours the employee worked on the transaction file, which has the information from the time cards.

In batch processing, since all the data are accumulated before processing begins, this matching can be done very efficiently by having the master file sorted by employee number, and then having the time card records sorted by the same employee number. Then the computing process can proceed down both files in sequence, either finding the matching records for the same employee, or finding records on one file that don't exist on the other. This technique is sequential processing. It is associated with batch processing and with matching records from different sources using sort and match.

An example of on-line processing is a person walking into a bank to make a deposit or withdrawal. Before the transaction can be made, a match must be made between a record identifying the customer and his or her transaction and the master record for that customer that shows the balance. On-line processing is associated with direct access or look-up using a directory.

Once the transaction is generated, the next steps involve computing, retrieving, updating, and outputting. Computing is required to produce output data from input data. Retrieving may be done to check some input or result against stored information. Updating may be done to revise stored information. Outputting presents information in human readable form. Data processing applications can be designed using these basic processes.

Application Generators and Prototyping

Application generators make it easy to develop some types of data processing applications quickly by modifying procedures already written for these standard data processing steps.

Application generators operate at a level higher than compilers. While a compiler is a program that uses a language to describe and program procedures, an **application generator** is a program that uses a language to adapt already programmed procedures to a specific application.

Many so-called data base management systems have the capability not only of managing data bases, but also of generating input forms, formatting output reports, and defining calculations to process the data. These are more than just data base management systems. They might properly be called application generators.

Application generators can be used to develop a program very quickly that does most if not all of what we want. We can experiment with it, change it, and carry it as far as the applications generator will allow us to go. For many applications this may be enough. But we may want more speed or a capacity to handle more data, or to perform functions we cannot do with the applications generator. If so, we can write a special program for our application. But having run the applications generator version first and been able to experiment with it, we now know exactly what the requirements for that program should be. This is one approach to prototyping. Prototyping will be discussed further in Chapter 9.

Fifth-Generation Approaches

Fifth-generation approaches are associated with artificial intelligence. Exactly what is meant by artificial intelligence, and what techniques it does or does not include, is a matter of some controversy. Generally, artificial intelligence involves getting computers to do things that typically we have thought machines were not able to do, such as recognizing patterns and working out how to solve problems. After many years of hard research and high hopes, some subareas have broken loose with commercially-practical techniques, such as character recognition and expert systems.

Decision Support and Expert Systems

Sometimes it is not practical to dump into the machine all the vast background of available information that the computer would need to solve a problem. Instead it is more appropriate for us to work with the computer as partners, pooling our knowledge and its computing abilities. The user is only asked for such information as may be needed as the problem solving progresses. Such systems are called **decision support systems.** (The term decision support system has also been associated with what used to be called "management information systems," that is, any systems that supply management with information to make decisions.) They have been used for such problems as helping a group of experts who understand various aspects of city problems to think together to make a plan for attracting investment to the city (Sprague and Carlson, 1982; Steward, 1981).

Another approach is to find someone who is an expert on a class of problems, debrief the expert, and code his or her knowledge into a computer data base. Then with this knowledge base and a system of rules, the computer can act like the expert. Such arrangements, called **expert systems,** have been developed for such problems as analyzing geological data to suggest where to mine for certain minerals. One test of such a system found a significant deposit of molybdenum (Feigenbaum and McCorduck, 1983). Other expert systems have been able to analyze symptoms of human patients to suggest medical diagnoses and medications and show their reasoning to doctors who can then either accept or reject these suggestions (Shortliffe, 1976).

Artificial Intelligence

Artificial intelligence (AI) is the attempt to program computers to do things that we would consider intelligent if people were doing them. Expert systems are one of the early commercial successes of artificial intelligence. However, some would say that expert systems are not part of artificial intelligence.

After many years of struggle, significant progress is being made in solving a number of artificial intelligence problems on a useful scale. Optical character recognition, which is now well established, is an example of pattern recognition, which is often considered part of artificial intelligence. Speech recognition systems with limited vocabularies are becoming commercially available. Researchers have created translating systems able to do the bulk of a translation from one language to another, leaving the rest to be done by human translators. And data bases that can be queried in natural languages are now being used.

Thinkware

A new generation of programs are acting as thinking aids to help us increase our effectiveness in organizing our thoughts, gaining insights, solving problems, and making decisions. Outline editors, for instance, which assist writers in organizing their thoughts into hierarchical structures are a pre-fourth generation example of a thinking aid. Other sophisticated artificial intelligence-based programs can help an executive develop and evaluate the criteria for a decision, learn and use decision rules, and complete a complex decision-making process.

Transitions

When people did their own operations manually, they could maintain direct control over their own data, its processing, and when that processing was done. When systems were converted to batch processing, the users lost much of that control. With timesharing, users gained some control over their processing and when it was done, but they were still at the mercy of the computer center to maintain the integrity of their data. Personal computers tied together with networks again allow users to maintain control over both their data and its processing.

Some users made each of these transitions as they came along, from manual to batch, to timeshare, to minicomputer, and to personal microcomputers with networks. The hardest part of these transitions was adjusting the rest of the operation to changes in the timing between when data were generated and when the results were available. Other users waited and missed some of these transitions, going directly from manual to timesharing or to an individual minicomputer or microcomputer.

Thus, many characteristics of data processing have come full cycle. Systems have returned to a one-on-one relation between a user and a machine, but they are doing a higher volume of processing, and accomplishing it faster and less expensively. For many applications we are so hooked on the computer and have become used to such increased volumes of information handling that we could not go back to the manual methods.

Security and Control

Security and control are needed to protect computer systems from the hostilities of the outside world.

If we invest time and money to obtain something, such as data, computational results, software, or hardware, we may also find it appropriate to invest some time and money to protect it. Obviously we must balance the cost to protect data against the cost to generate it again.

Processing Perils

Among the risks facing data processing systems are errors, loss, mischief, fraud, and hazard.

Errors

People can work with a great deal of error in their communication. Using redundancy in the data and reasoning from their own knowledge, they can usually resolve most of this error. But until we develop artificial intelligence algorithms so that computer programs can do this type of reasoning, we must develop computer programs to reject data it does not know how to handle correctly. By making programs easy to use (which we'll discuss in Chapter 9), we can also make it less likely that the program will get bad data.

For computer programs to resolve errors in data,the program developer must anticipate all the types of errors that might occur. This forethought requires careful preparation in developing the program and often a large number of lines of code to handle all those potential sources of error.

To know whether something is wrong, the system needs some form of reference. The various methods of error resolution can be categorized by how this reference is made.

The reference can come from redundancy in the data entered. For example, the data may be key-entered twice and compared. Or errors might be detected from redundancy embedded in the input, such as an identification code with one character that can be computed from the other characters.

The reference may be made by comparing input information with information already on file—for instance, comparing an identification code and name to see that it corresponds with someone in the file. Or a program may make a calculation or comparison to determine whether the input is acceptable, such as comparing the input to a range to determine whether it is acceptable. Input may be rejected if its format cannot be interpreted, such as an alphabetic character in a field that should be numeric.

What happens when input data are rejected? It might be possible to enter it again. With on-line systems the person entering the data is present to reenter the correct data. But a batch system is run when the person responsible for the data is not present. Usually that person must be consulted, and the correction inserted either into a special run or held for the next run.

It must be determined whether each item of data can be checked and corrected individually, or whether once the data are corrected the whole process must be rerun. Sometimes the wrong value can be flagged as it is processed. Later, when the correct data are known, the entry can be reversed. For example, if a wrong value has been added to a sum, the incorrect value can be subtracted and the correct value added. This is called either reversing or backing out the error.

Loss of Data

Data can be lost by physical damage to the media it is recorded on, or by carelessness such as accidentally causing data to be recorded over data that was to be saved. Or, as with paperwork, diskettes or tapes can be mislaid.

The usual ways of protecting against loss of data are to retain copies, or retain other data from which the required data can be regenerated.

In batch processing an old master file and the transactions are used to produce a new file. By saving the old file and the transactions, the new file can be regenerated if it is lost.

Protecting data in on-line systems is more difficult. The new data usually write over the old data so that the old is no longer available if it is necessary to regenerate the new. We must, therefore, back up this data by making a periodic dump of the file and saving all the transactions since that dump. Another approach is to save the transactions, and if one of them is found to be in error, back it out and enter the correct value, thus avoiding the regeneration of the whole file.

Mischief

We must design systems to protect them against the mischief of people who destroy equipment or information, obtain access to information they should not have, or use facilities they don't pay for.

One way to protect against mischief is to control access. The two types of access are physical and logical. Physical access is the power to affect things by actually being there. Logical access is the power to affect things through an information connection by wires. We control physical access with a lock and key or with guards. We control logical access by passwords. When phone access is involved, it is often a good practice to have the system call back the person at the phone number where the computer expects the individual to be.

We may give different people permission to have different types of access to the system depending upon their responsibilities and needs. These types of access include: creating a new file, appending data to an existing file, changing data on an existing file, reading a file, deleting a file, or executing a program. We may also restrict people to having certain types of access to only certain subsets of the data.

Restricted access can protect data not only from people with mischief in mind but also from the carelessness of people not properly trained to care for it.

Fraud

Fraud concerns changing information to deceive or misrepresent, to cause someone to part with something of value, or to cause them to give up a right. We audit systems to check whether they are being misused. In an audit we check to see whether people have tampered with the data or with the programs.

With billions of dollars moving every day as bits across telephone lines, the opportunities for crime are tremendous unless careful precautions are taken.

We can check to see whether programs have been tampered with by running reference data through them to see that the expected answers are produced. We can often detect data tampering by periodically inspecting the transactions and their results.

One of the most effective ways to control against fraud is to control the people capable of perpetrating the fraud. They can be denied the access they would need. Or we can make it more difficult for them by designing the system so that several people are needed to commit a crime. The crime is therefore less lucrative to any one person because more people must divide the loot. And there is a greater probability of discovering the crime because there are more people who may give themselves away. Thus systems are purposefully built to require the independent action of more than one person to execute a transaction. For example, the system should not allow the same person to issue a purchase order and process the receipt of the material without knowledge of at least one other person.

Hazard

The standard hazards are acts of God and careless people such as fire, wind, and flood, and the acts of the power company such as interruptions, surges, or spikes in the power supply.

An insurance ad tells us that companies that lose their vital records in a fire have a much higher chance of going out of business than those whose records survive. The message should be clear for protecting vital computer files.

Some companies make periodic copies of all their vital computer files and store them in fireproof vaults or vaults located at a remote site.

Data and equipment can be wiped out by floods caused by human error or the weather. Damage to computer centers amounting to millions of dollars has been caused by such acts as faulty plumbing in a rest room, or a careless chemist who left the water running in leaky equipment.

Control Measures

In addition to the security precautions necessary to protect our data, we also need monitoring procedures so that data are not used to invade an individual's privacy, and so that we can trace who used a system at a particular time.

Privacy

When systems are created, the designers must consider how to protect against threats to privacy arising as side effects from the use of computerized data. Violations of privacy can arise from acts as simple as failing to demagnetize a payroll tape before it is released for use by another application. Control of access to the data and logging who accessed it are important methods for maintaining privacy.

Audit Trails and Logs

It may be necessary to maintain records of various types to protect the security of the system. Audit trails recording transactions can be used to trace tampering.

Logs of who accessed which systems or data at what time can be used later to trace who was using the system when the tampering occurred.

Audit trails can have other uses such as regenerating data that are lost or determining exactly how certain results were arrived at.

Performance Analysis

When specifying the requirements for a system, client and provider define the system's function and its performance. **Function** refers to what the system is to do. **Performance** concerns how well the system does it. Earlier in this chapter we discussed the functions of various categories of information systems such as batch, on-line, and so on. Now we will discuss four categories of performance: speed, capacity, reliability, and availability.

Speed

Speed in an information system can have many different meanings. It could mean throughput—that is, the number of transactions the system can handle in an hour or a day. In an interactive system speed could mean the response time, or the elapsed period from when an input is finished being entered until the output starts to print out.

Often a simple number like an average, a maximum, or a minimum does not provide enough information. We may need to know how the system will respond to various distributions of the arrival of transactions over time.

If over a long period the average rate at which transactions arrive is greater than the average rate at which they can be processed, the system will lose transactions. But we may still be able to handle transactions arriving for a short period at a rate faster than we can process them if we can hold them in a queue until they can be processed later. In this case we need to know, given a particular distribution of arrival times, how long must the storage for the queue be so as not to lose any transactions, or how long will a transaction be delayed while waiting in the queue.

To answer such questions where the distributions in time of the arrivals are drawn randomly from statistical classes of distributions (such as the Poisson distribution, which will be discussed later), we can use the mathematical techniques of queuing theory. Or, given a specific distribution of arrivals with time represented by a histogram, we can make a simulation to determine the maximum queue length and delay in handling a transaction (Martin, 1967). We have the space here to consider only a simple queuing simulation problem to introduce some of the concepts. A book on queuing theory or computer performance analysis should be consulted for more information (Ferrari, 1978).

Simulation of Queuing

Let's assume we have a histogram showing the number of transactions that can be served by our system in each time period. (In queuing analysis we often speak of the system as "serving" rather than "processing" transactions.) The transactions arrive randomly. When transactions arrive faster than the system can serve them, the unserved transactions are put into a first-in first-out queue.

Figure 8.1 is a histogram showing the distribution of transactions arriving in each time interval. It also shows that in each time interval two transactions can be served. As you can see, in some intervals more transactions arrive than can be served, and in others fewer. Below the diagram is written the given number of transactions that enter the system during each interval (taken from the histogram), and the values that can be calculated for the number of transactions served in the interval, the number that remain in the queue waiting to be served, and the longest that any item entering the queue in that interval must wait before being served. For this particular distribution the maximum queue length is shown to be five, and the maximum wait is three intervals. The following relations show how the numbers in Figure 8.1 are computed.

Transactions arriving = given data

Served during interval = lesser of (service rate) or (the number arriving in this interval plus the number in the queue at the end of the last interval)

Remaining in queue = sum of the number arriving in this interval and the number in the queue in the last interval, minus the number served in this interval

Figure 8.2 is a diagram showing the total number of arrivals and services since the queue last became nonzero. We can use this diagram to see the length of the queue and the maximum length of time a transaction arriving in a particular time interval must wait before it is served. The number that must be held in the queue is the total number that have arrived minus the number that have been served.

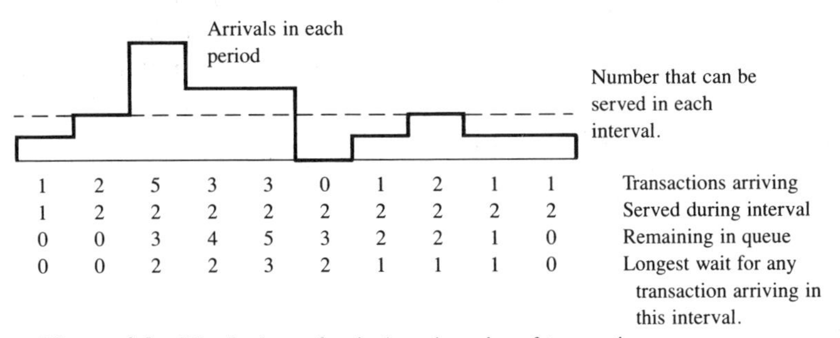

Figure 8.1 Distributions of arrivals and service of transactions.

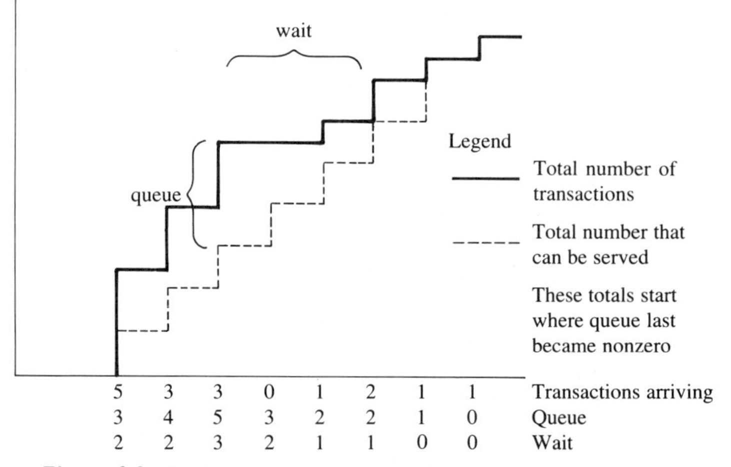

Figure 8.2 Total number of arrivals and services.

By looking at the line for the number that have arrived and moving over to the line for the number served, we can count the number of time intervals before transactions arriving in that interval are finally served. For this particular distribution of arrivals, Figure 8.2 indicates that the maximum queue length is five (as we computed earlier) and the maximum delay is three. Thus, for this particular distribution of arrivals, if we did not provide for a queue length of at least five, we would lose transactions.

Statistical Analysis of Queuing

Generally we do not know the precise distribution of the arrival times of the transactions as we assumed earlier. Instead we assume that they can be characterized by some statistical function. We will show some examples of what assumptions can be made and the conclusions that can be drawn using probability and queuing theory. This will only touch the surface, but we hope it will entice the reader to pursue the topic elsewhere (Martin, 1967).

The most common assumption is that the arrivals are uniform—that is, the probability of an arrival during any interval of time is a function of the length of the interval but independent of when that interval occurs.

How many arrivals occur within a given interval for a uniform arrival rate can be determined from the Poisson distribution, which is commonly treated in elementary courses on probability and statistics.

Let a = arrival rate in arrivals per second
t_a = mean arrival time in seconds = $1/a$
$m = a * t$ = mean number of arrivals during time t

Then by the Poisson formula the probability of having exactly n arrivals during the time interval t is:

$$P(n) = \frac{e^{-m} * m^n}{n!}$$

where $n!$ is n factorial $= n * (n - 1) * (n - 2) \ldots 3 * 2 * 1$

We are usually interested in the probability that the number arriving during an interval is greater than some number N, which may be the size we have allowed for a buffer. Tables of the Poisson distribution giving $P(n)$ and the sum of $P(n)$ for n larger than N (that is, $P(n > N)$ are common in elementary probability and statistics books or books discussing performance analysis.

This same assumption about the distribution of arrival times leads to the ability to estimate the probability that the time between arrivals t is less than some value T. This is given by the following equation:

$$P(t \leqslant T) = 1 - e^{-T/ta}$$

When we are concerned with the time between events, we call this an exponential distribution.

If we assume that the arrival distribution is uniform and the next transaction to be processed is not chosen based upon knowing the time required to process it, then we can estimate a number of interesting average values from the Khintchine-Polloczek formula as follows:

Let:

r = service rate—that is, the average number that can be served per second

s = average time to service a transaction $= 1/r$

w = the average number of transactions waiting to be served

q = the average number either waiting or being served

t_w = mean time waiting to be served

t_q = mean time while waiting and being served

then:

$$w = k * \frac{\rho^2}{(1 - \rho)}$$

where:

ρ = utilization $= a/r = a * s$, which is the ratio of the rate of arrivals to the rate they can be served. ρ must be less than one if the queue is not to grow indefinitely.

k = 1/2 if the service times are constant—that is, every transaction can be processed in the same amount of time, or

= 1 if the service times are exponentially distributed.
Usually we can assume that k will be between 1/2 and 1, where k = 1 gives a high, conservative estimate for queue size and times.

We can compute the other quantities from the following relations:

$$q = a * t_q = a * (t_w + s)$$

$$q = w + \rho$$

If the service time is also exponentially distributed (which gives a pessimistic, high value for queue length), then the probability of the queue length exceeding the value N (which could be the size of the buffer we provide for the queue) is:

$$P(q \geqslant N) = \sum_{q \geqslant N} (1 - \rho) * \rho^q$$

This explanation only touches the surface, but it indicates what can be done using probabilistic and statistical assumptions to estimate such items as how large a buffer should be so that the probability of losing transactions is less than some given value. Refer to Martin, 1967 for more information.

Capacity

Another performance issue is the **capacity** of the program—that is, how large a problem it can handle. For example, when writing a program to schedule tasks, we are concerned with how many tasks the program can handle without becoming excessively slow. This measure is a concern in deciding on the hardware, data structures, and algorithms to be used. The size of problem that can be handled comfortably depends on the size of the internal and peripheral memory, how much of the memory is used by the program itself, and the data structures. The choice of data structures is intimately related to the choice of algorithms that work with them and whether the data is to be stored in primary or secondary memory, or split between primary and secondary memory. It is usually quite difficult to know how much memory will be used by the program until the program is written. Thus, the size of the problem that can be handled is often settled very late in the development process. One should look forward to this problem when choosing the algorithms and data structures.

Reliability and Availability

Two other important performance measures are reliability and availability. **Reliability** is the probability that a system will continue to perform correctly for some given time. **Availability** is the probability that it will be available to perform when needed.

A system may fail to perform correctly either because there has been some change to the system, or because it becomes exposed to new input or a new environment that it is not able to handle.

Definitions that are used in evaluating reliability are:

- *Mean Time to Failure*—average time from when the system first performs correctly to when it fails.
- *Mean Time to Repair*—average time from when the system fails until it is restored to correct performance.
- *Mean Time between Failures*—Mean Time to Fail plus Mean Time to Repair

Availability is measured as the fraction of the total time that the system is needed that it is performing as needed.

$$\text{Availability} = \frac{\text{Amount of time available when needed}}{\text{Amount of time needed}}$$

The early machines were not very reliable by today's standards. The UNIVAC I had completely duplicated arithmetic-logic circuits, and buses were compared after each operation; memory was periodically scanned and checked for parity. As machines became more reliable, much of this self-checking in the machine was eliminated. Tapes were still a major source of error, so data on tapes were recorded with parity bits, which were tested when the data were read.

Newer machines are returning to the use of parity checking in primary memory. Some systems are now designed with duplicated computing units and data storage so that they can run without interruption. This is being done even though computers are now much more reliable because computing has become much more vital. We have become so dependent upon the machine that there are many things we can no longer do manually if the machine fails.

We are using computers to do things like controlling space vehicles that could not be controlled quickly enough by hand and are too costly to afford an abort due to a machine error. People's finances are controlled by banking machines, and their very lives may depend upon computers in medical machines. The next war may be commanded by machines because people cannot react and make decisions fast enough for the modern war machine. Who, if anyone, survives is likely to be determined not by whose machines are most capable but by whose systems are most reliable.

Not only are machines being duplicated for reliability, but the programs for the different machines are being written by different people to avoid the different machines failing in the same way because of using the same software with the same errors.

Summary

As we analyze old systems and develop new ones, we will come across many different technologies. Information processing began with extremely expensive machines, operated by one user, running one problem at a time. When it was recognized that the larger the machine the more computing could be done for the dollar, the emphasis went to larger machines with various methods devised to use their greater computing powers. Data were submitted in batches so the machine could be given several programs to work on simultaneously, computing on one program while an IO processor was doing the input/output for another (multiprogramming). Communication processors were attached so that several users could use the machine at the same time, giving the appearance that each user had his or her own machine (timesharing). As long as many users used the same machine, it was easy for them to share data.

Finally, computers became so inexpensive that it was practical to give each user his or her own personal computer. At that point, local area networks became necessary so users could communicate with each other and share data. Personal computers linked by local area networks to create, edit, communicate, store, and share data can be integrated into an automated office. Processors have become so inexpensive that it is practical to use many processors running simultaneously to solve large problems, provided we can determine how to program them.

Early systems centered on programs that each had their own unique data file. It is now becoming more common to have systems that are centered on data bases that many programs can use.

Using fourth-generation languages, we can state what we want done without having to declare how to do it. Fifth-generation systems help us think through complex problems, or use the stored knowledge of experts to do that thinking for us.

When we analyze existing systems or devise new ones, we should consider how the system is protected from human error, loss of data, mischief, fraud, and hazard and how we can ensure privacy.

When we analyze the performance of existing systems or the proposed designs of new systems, we may wish to use the simulation or statistical queuing theory to analyze for systems speed and required buffer sizes.

Exercises

1. Given the following distribution of arrivals, answer the following questions for the processing rate of two, and for the processing rate of three:
   a. What is the minimum queue storage required so as not to lose any transactions?
   b. What is the maximum delay in processing a transaction?

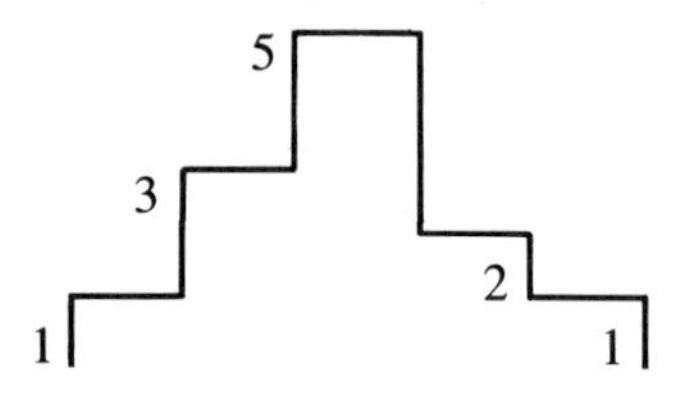

2. Assuming uniform distributions for the arrival of transactions and for service times, with an average arrival rate of eight per second and an average service rate of ten per second:
   a. What is the probability of having exactly ten arrivals in a one-second interval?
   b. What is the utilization?
   c. What is the average number of transactions waiting or being served?
   d. What is the average time a transaction spends waiting in the queue or being served?
   e. Estimate the probability that the queue will be longer than ten.

References

Computer (Issue on Data Flow Machines, six articles) Vol. 15, No. 2, February 1982.

Feigenbaum, Edward A., and Pamela McCorduck. *The Fifth Generation: Artificial Intelligence and Japan's Computer Challenge to the World*. Reading, MA: Addison Wesley, 1983.

Ferrari, Domenico. *Computer Systems Performance Evaluation*. Englewood Cliffs, N.J.: Prentice-Hall, 1978.

Ferrari, Domenico, G. Serazzi, and A. Zeigner. *Measurement and Tuning of Computer Systems*. Englewood Cliffs, N.J.: Prentice-Hall, 1983.

MacNair, Edward A., and Charles H. Sauer. *Elements of Practical Performance Modeling*. Englewood Cliffs, N.J.: Prentice-Hall, 1985.

Martin, James. *Design of Real-Time Computer Systems*. Englewood Cliffs, N.J.: Prentice-Hall, 1967.

Shortliffe, E. *Computer Based Medical Consultations: MYCIN*. American Elsevier, 1976.

Sprague, Ralph H., Jr., and Eric D. Carlson. *Building Effective Decision Support Systems*. Englewood Cliffs, N.J.: Prentice-Hall, 1982.

Steward, Donald V. *Systems Analysis and Management: Structure, Strategy and Design*. Princeton, N.J.: Petrocelli Books, 1981.

Toffler, Alvin. *The Third Wave*. New York: Morrow, 1980.

9

User Interfaces

- Friendly software communicates well. Designing user-computer interfaces is a communication skill.
- Programs should respond to users in ways that resemble as closely as possible a polite human reply.
- For an effective interface, programs must work as partners with the user, helping the user to understand rather than faulting the user for making errors.
- Users should be allowed to change their minds and to take risks when they explore a program.
- The friendliest software learns the user's habits and adapts to them.
- The design of good programs can combine the simplicity needed by the novice with the power required by the more experienced user. The same program may offer both menus and commands.
- Programs need to be consistent in their behavior but, equally, need not be boring or lose the user's attention.
- Rapid prototyping allows the software engineer to spend a small amount of time developing enough of the system for the client to visualize the final product.

Computers once were used primarily by computer professionals who were prepared to devote their lives to learning the idiosyncrasies of computer systems. They even relished their superior abilities to deal with difficult programs. Software writers had little motivation to design systems that easily communicated with the novice user. Now computer systems are being used by people who are interested only in getting the results, not in the mystique of using the computer. They have other careers. The computer is a means to an end, not the end in itself. To satisfy these users, we need to develop programs that are so friendly that people do not have to struggle to use them. In this chapter we will discuss several principles for making programs easier to use so that, as software engineers, we will be able to analyze and design effective user-program interfaces.

We will also consider how to write prototype programs, which behave superficially like the final system, yet take only a small fraction of the time and cost to develop. Using such a prototype is an excellent way of first exploring the interface between the user and the program. Prototyping also allows clients to visualize the final product; thus at an early stage clients and developers can see if they are on the same track. It is a supplement to the abstract methods of visualization discussed in Chapters 3 and 4.

Principles for Developing User Friendly Interfaces

A tremendous number of computer programs are being written today for microcomputers. Less than 1 percent of these programs account for more than half the total sales. (This is an example of Pareto's law, which says that a small fraction of the actors account for the major fraction of the effect.) One distinguishing feature of most of these top-selling programs is that they are easy to use.

Friendly software is software that communicates well, just as friendly people tend to be people who communicate well. The design of person-computer interfaces is a communication skill, like writing, not an engineering skill (Heckel, 1984). When we design these interfaces, we put away our engineering heads and bring out our communication heads.

Most of the following suggestions for designing user friendly programs make sense if we think of ourselves in the role of the program. So:

Suggestion 1

Think of yourself as the program and how you would talk to the people who are using it.

To get into the spirit of this suggestion, we will write from now on as though you were the program.

Imagine you're talking to someone who is unsure of himself or herself. This person doesn't understand what is happening, is uncertain how to address you, is afraid of being shown up by doing something stupid, and doesn't know what is expected. Would you be so crude as to reply to his or her hesitating attempts to start a conversation by saying:

```
ERROR 27
```

Certainly not. So why should you as a computer program respond in this way? Wouldn't it be nicer to say:

```
I DON'T UNDERSTAND WHAT YOU SAID.
THESE ARE SOME ANSWERS TO MY QUESTION THAT I UNDERSTAND
     Y or YES
     N or NO
THANK YOU.
```

After all, this person has probably spoken to you clearly with a message that would be understood by millions of people. Only you, a naive program, don't understand the message. In fact you don't understand very much of the rich language that people use, so he or she has to talk down to you in your own unique language. You make people learn your language with all its peculiarities because you are incapable of learning their language. And then you want to make them feel like the stupid ones (Heckel, 1984)? If the program can't understand, it is the program, not the user, that is incapable.

Suggestion 2

Be polite. Don't put down the user.

Remember, your programmer doesn't make money unless the user buys you. So treat the user like a friend so he or she will like you and tell friends what a great guy and fabulous conversationalist you are. Then other friends will buy you and there will be more of you. That means more money for the person who wrote and sold you. So:

Suggestion 3

Constantly sell yourself to the user.

Users may only use you occasionally and may use many other programs as well. They can easily forget how to talk your special language. It's not like talking to people who all share the same language. So:

Suggestion 4

Picture the user as one who has just returned from vacation and has not used the program in a month, or as someone who has lost the user's manual, which he or she probably didn't read anyway. Or imagine the user has just had a three martini lunch or is a first-time user full of FUD (Fear, Uncertainty, and Doubt) (Heckel, 1984).

You and your user are to work together to solve a problem. So:

Suggestion 5

Think of your user as a partner. Work together with him or her. Offer support when it's needed.

You may wish to give a tutorial to get the user started. Start with easy specific illustrations before getting into generalizations. But, as you work with the illustration, explain what you are doing and why. Then the user will be ready to understand the generalizations and take off on a solo experience.

Suggestion 6

Don't punish. Teach. If the user enters something you don't understand, help get the conversation back on track.

Show users how to get from where they are, having just said something you don't understand (note we didn't say "having just made an error") to where you can pick up the conversation again. Don't send the user back to the operating system without a clue. Offer some choices, and even let the user change a response if either of you don't like it.

Suggestion 7

Check quickly whether you understand the user's response. If you don't, get the matter straightened out before you and the user get too involved in a misunderstanding. Give the user a chance to change the response instead of dumping him or her with cement boots into the operating system.

Suggestion 8

Make it easy for the user to change his or her mind.

Children learn by exploring. During these explorations their parents try to prevent them from coming to any harm. In some sense you have to be a parent as you train your user.

Suggestion 9

Allow users to explore without risk and anxiety. Allow them to do and undo without harm. Help them cope (Heckel, 1984).

The user may not be like you. So don't use the Golden Rule, which says: "Do unto others as you would have them do unto you." Don't treat the user as a machine. Instead use the silver rule, which says:

Suggestion 10

Do unto your users as they would have you do unto them.

After a while working with someone else you should begin to learn how he or she operates. That makes it easier to converse. So:

Suggestion 11

Try to learn the user's habits, and adapt to them.

In the simplest form this suggestion means allowing users to make up a profile file that describes how they want all the default answers to questions set. But give users a chance to override these defaults. This allows them to adapt you to their habits. In the ultimate this could become an area of artificial intelligence. Then you would keep track of user responses and learn how to set up their profile accordingly.

Suggestion 12

Organize the structure of the interaction to flatter the organization of the user, not follow the organization of the program (Heckel, 1984).

Remember Miller's Principle. It applies here too. Even though your user may be very intelligent, everyone has limits on how much they can keep track of at one time. Try to organize yourself so that the user does not have too much to track. Don't make attempts to talk with you into a three-ring circus. This advice leads to the cardinal suggestion:

Suggestion 13

(KISS) Keep It Simple Simon.

Minimize how much users must remember about how to talk with you and about what is going on when they are talking with you.

Have you ever talked to someone and out of the blue he or she says something completely unexpected? It's hard to continue talking after that. Indeed such a response is usually a very good method to bring a conversation to a close.

Suggestion 14

Be consistent. Don't do the unexpected. Avoid anything that might be confusing or leave your user not knowing how to respond.

Ralph Waldo Emerson said, "foolish consistency is the hobgoblin of little minds." But what did he know about talking with computers? Yes, sometimes consistency, even dealing with programs, can be boring. But a certain consistency is necessary to understand what is going on in a conversation. Only once we have achieved clarity can we move on to novelty.

Suggestion 15

Make sure what you ask users to do has meaning to them. Avoid the arbitrary.

Suggestion 16

If you do the same type of thing again, do it the same way. Keep the rules users must remember and understand to an absolute minimum.

Remember, users are trying to learn how you behave. It is difficult for them if just when they think they understand who you are, you become someone else. Don't be a moving target.

A popular product today is "integrated software." With these packages users can do many different types of operations in much the same way—that is, using the same editing commands to edit a document, a line in a spreadsheet, or a line in a data base. Not surprisingly, integrated software packages are among the best selling programs. Users appreciate being able to use a limited number of rules to do a variety of tasks.

Suggestion 17

Make simple things simple, then more complicated things more complicated only if they need be.

Reviews of programs sometimes say that some programs are good for the novice while others are more powerful but should be left to the experienced user. But can't a program be designed to be easy for the novice, yet powerful for the experienced? Why not? Do simple things simply, but require more of the user who wants to do more. Remember, today's experienced user was once a novice. If you don't attract and hold a user as a novice, you may never have the opportunity to work with him or her as an experienced user.

Menus are good for novices who need to be shown what their options are. Let them select by pointing to what they want. Instead of having them make one difficult choice, give them a series of menus that lets them make a sequence of simple choices, each adding more detail to what they want. A set of menus could be organized like a tree, as in Figure 9.1.

Commands are better for experienced users who know what they want to do and don't want to waste time going through menus. Let them type a cryptic key sequence that gets them immediately what they want.

Menus and commands can be used in the same program. To open a file called "Text" and move to the fourth line and delete it, inexperienced users could go through the menus. The shorthand commands can be shown on the menu and learned while using the menus. Once they have learned the shorthand, they could

```
                             ┌O—New File (N)———┼—— "Give old file
 ┌O—Open File (O)———————————┤                       name"
 │                           └O—Existing File (E)——┼—— "Give
 │                                                       new file
 │                            ┌O—Up (U)                  name"
 ├O—Move cursor (M)——————————┤
 │                            └O—Down (D)
─┤
 │                        ┌O—Add line (A)           Here the O means
 ├O—Edit (E)——————————————┼O—Change line (C)        option.
 │                        └O—Delete line (D)
 ├O—Close  File (C)
 └O—Remove File (R)
```

Figure 9.1 Tree representing the structure of menu input.

just type:

```
O"Text"4MD
```

One approach to displaying menus is to have the options appear as a string of keywords at the bottom of the screen. Users can move the highlight to a keyword and see displayed a description of what that option does. They can then select the option if they want it. As they highlight each keyword, they will also be shown a one-letter command that they can learn to use directly.

The menu tree is a model of your behavior. Your behavior model should be simple enough that your user can figure out how you will behave. Thus we have the suggestion:

Suggestion 18

Develop a simple model of how the program will behave in all instances. Be sure that users can learn to understand this model as they use the program.

The next is another cardinal suggestion:

Suggestion 19

Provide feedback so the user will constantly know: What is happening and why. Where he or she is in the menu or command structure. What options are available at this point. What is expected of him or her. Should the user wait or type something?

Prompt users to show what you want. If the computer is tied up and not ready to accept the user input, tell them so with a message like:

```
I AM BUSY NOW or PLEASE WAIT or READING DATA.
```

Don't let users look at a blank screen wondering whether the machine is lost or not working, or whether it is waiting for them to do something. Repeat and update the message every few seconds to show you haven't forgotten about them.

Suggestion 20

Keep a neat screen. Put things together that go together and separate those that don't. Place things where users will expect them.

Don't clutter the screen. Show users where in the process or set of menus they are. Put the same type of information in the same place on different screens so users know where to look for it.

On the other hand, remember that users can become so accustomed to seeing the same thing in the same place that they ignore it and may fail to see it change. If users must be alerted, present the information in a way that will catch their attention. You can use a number of different features available on most screens: highlighting, blinking, reverse video, underlining, change of color, or beeping.

Suggestion 21

Keep users' attention. Respond quickly enough that their minds don't wander and have trouble picking up again with what they were doing.

Suggestion 22

Given a choice, have them point rather than say.

Next time you go to a cafeteria, try pointing to one thing and asking for another. See which is more effective.

A mouse is better than cursor keys for pointing.

Suggestion 23

Appeal to what is already understood. Use analogies such as icons.

You can interpret this as "Use graphics." The new generation of software has users point to icons that look like something they are familiar with—like a trash can to throw away a file. This way when users want to get rid of something, they don't have to remember whether they should type ERASE, DELETE, SCRATCH, or ERA or SCR. And if they are to use several systems, don't require users to say ERASE for one and DELETE for another.

Suggestion 24

Don't be boring.

I once enjoyed using a system that, each time I signed on, would present me with a different piece of wisdom for the day, such as:

```
TODAY IS A BAD DAY TO PLAY LEAP FROG WITH A UNICORN.
```

The next time it would say:

```
TODAY IS A GOOD DAY TO TAKE THE PLUNGE.
```

and the following day say:

```
WILL THE PERSON WHO TOOK THE PLUNGE YESTERDAY PLEASE
RETURN IT.
```

A little humor can be helpful, but be very careful to use good taste.

Seeing the same thing said the same way time after time is boring. Soon the user ignores what you say. Granted that earlier we said be consistent. But don't be so consistent that you lose your user's attention. After you have made sure that you can communicate clearly, consider how you can give your message a little variety. Variety makes your information more compelling provided it does not interfere with clarity.

Suggestion 25

Be concerned about the amount of information the user needs from one screen while looking at another.

It is hard for a user to remember information from one screen to use when working later on another screen. Try to present on each screen the information users may require while working on that screen. This may mean repeating the same information on several screens so it is available when they need it.

Many systems today have the ability to copy part of the screen onto a "notepad," a file that can be displayed on the screen even while in the middle of another operation. The note overlays what is on the screen, then disappears revealing what it had overlayed once the user is finished with it.

Suggestion 26

Have someone else run the program and observe the problems he or she has.

No matter how good you think you are, it is always a smart idea to have someone else run you and make a critique. Such a test user is likely to discover many oversights; correcting these will help make you more friendly. There are limits to how well designers can judge their own work. They may not realize how much they know about the program, so that they build in assumptions that the user is not told about.

Ideally there should be a procedure as early as possible in the development process whereby someone runs the program to work out any problems. But how can such a test run be made until the program is finished? In the next section we'll explore the technique of rapid prototyping, which does give clients and developers an early look at the system.

Rapid Prototyping

Rapid prototyping is an approach to obtaining an early visualization of what the product will look like by building a system that appears to the user like the system to be developed. That visualization gives clients a chance to decide whether what the software engineer has in mind is what they want, and it gives the project team a chance to change the design as soon as possible. The approach is particularly useful for studying the interface between the user and the program, which is why we discuss it in this chapter.

Rapid prototyping depends on our ability to put together quickly and inexpensively a system that will look to the customer something like the system he or she can expect to get. It raises a question. Doesn't it take almost as much work to make a prototype as to make the final system?

The answer lies in the observation that sometimes a small amount of effort can produce much of what is wanted, but it still takes a lot of additional effort to produce exactly what is wanted. This observation is **Pareto's law**. It is generally possible to obtain a large portion of the most important capability of the system after implementing only a small part of that system (see Figure 9.2). The key to prototyping lies in determining this small part and then figuring how to handle its interfaces to the parts not yet developed.

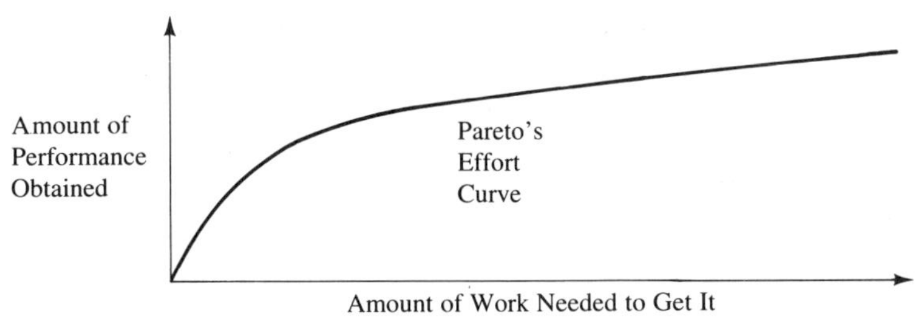

Figure 9.2 The curve of Pareto's law.

Thus, if we know what work obtains the greatest initial return, we can achieve a large fraction of what we want quickly and inexpensively. With a prototype clients can use their imagination to bridge the gap to see what they want. But it may take a great amount of additional work on the system to bridge the gap from the performance of the prototype to the performance of the final product.

To create a quick prototype, we can sometimes compromise performance characteristics of the system such as speed, size of the largest problem that can be handled, accuracy, security, or completeness of error handling. What we get will be enough for clients to extrapolate in their imagination to what they want. But to get the full speed or size of the problem or proper error handling may require much additional work.

Generally those aspects of the system that pose the greatest risk should be prototyped first. This risk may occur because no one knows whether an algorithm will work or will be fast enough. Before we make a big investment developing other parts of the system on the assumption that the algorithm will be satisfactory, we want to try it out.

Often the first risk to be resolved is the user interface. The developer wants the user to have a chance to see how the system will operate as soon as possible in order to get the client's feedback.

Given that we want to try out certain parts of the system, our next chore is to determine how to satisfy the interfaces between the parts we implement and those we don't implement. In some sense we need to "trick" the implemented parts into working properly, because we are investing considerably less work in producing these temporary interfaces now than it would take to write the real interfaces later.

If the interface is the input of data, we can often compute by hand the interface data and put it on a file to pass it to the module. If the interface is an output, we may have to write a program to read the data and print it out to be checked by someone reading it, or write a program to compare it to manually precomputed data. These special programs to trick the interfaces are usually thrown away once they are used and are not delivered with the final product.

Instead of emphasizing particular modules, the prototype may emphasize particular functions of a module. Other functions intended for the final system may be left unprogrammed for the prototype demonstration. This approach may produce several versions of the same module.

Significant savings may be made in developing a prototype by maintaining a data base only large enough for the demonstration. We may maintain this data by hand for the prototype demonstration and write the full data-handling routine later.

The code to check the validity of input data is usually very time consuming to write. By being extra careful in preparing this data, we may be able to demonstrate the module without having to write the code to do this data checking. Clearly this code must still be written before the system can be turned over to users because they will not have the time to be as careful in developing their input as we are.

One important question arises when using the prototyping approach. Once the prototype has been used to clarify what the final product should be, should the prototype then be extended to become the final product, or should the prototype be used only to help write requirements for the final program and then thrown away? Extending the prototype to become the final program can often be just an excuse for writing a sloppy program that cannot be maintained. Most often the methods used to write prototypes to produce a running program quickly and cheaply do not result in programs that can be maintained.

One approach to rapid prototyping says that once we have developed a prototype that illustrates how the system will operate and we have received our feedback from the user, we should throw the prototype away, write requirements specifications using what was learned from the prototype, and write a new program to these specifications.

Application generators allow us to build prototypes quickly to get user feedback. Such prototypes may or may not provide adequate performance. If they do, they become the final product. If not, it may be necessary to write a more efficient program to these same specifications in a high level language. As application generators become more efficient, more applications can be developed using applications generators without recourse to rewriting them in another language.

Summary

Since computers are now used by many people who are not professionals in the field, software must aim to communicate in a fashion that does not require a lot of knowledge about how the system works. Designing friendly software is a matter of creating programs that communicate well with novice as well as experienced users.

One way of imagining the interface between user and software is to think of yourself as a program. Think of how you must respond to make yourself understood by a user. In this sense the program should be a partner, willing to work together with the user.

For the sake of flexibility users ought to be allowed to change their minds and to explore without fear of failure or risk. Friendly programs adapt to the user rather than the other way around. They learn the user's habits and try to adjust to these.

By being consistent and performing the same task the same way in a variety of contexts, programs—such as integrated software packages—can enable inexperienced users to accomplish a great deal. Programs may also offer a choice of methods, such as menus and commands, to accommodate both the novice and the more skilled user.

To keep users constantly apprised of what they are to do, software must offer feedback, indicating exactly what the user is to do next. The same parts of the program should always appear in the same part of the screen so that users don't get lost. But, equally, the features of the program must arouse attention so that steps are not overlooked.

Allowing someone else besides the developer to run the program is often an effective way to spot oversights.

Rapid prototyping creates enough of a system to give the client and the developer a sense of the final product. Software engineers can determine if a particularly risky algorithm is going to work before going ahead with the full program. Clients can see what the user interface will be like.

Exercises

1. Run a commercially available program, and critique it based on this chapter's suggestions and such other ideas as you may have about how to make a program friendly. Do this with one of your own programs, or better yet, have someone else run and critique one of your programs.
2. Draw up menu trees and screens for improving the program you have critiqued.
3. Develop a user's manual for a program you would like to write. Show all the screens and the menu tree, and describe how to use any commands. (An outline for writing a user's manual appears in Chapter 13).

References

ACM Software Engineering Notes: Special Issue on Rapid Prototyping. Association for Computing Machinery, *ACM SIGSOFT SEN*, vol 7, no 5 (December 1982), entire issue.

Bennett, John L. "The Concept of Architecture Applied to User Interfaces in Interactive Computer Systems." *INTERACT*, 1984.

Boehm, Barry W., Terence E. Gray, and Thomas Seewaldt. "Prototyping Versus Specifying: A Multiproject Experiment." *IEEE Transactions on Software Engineering,* SE-10, 3 (May 1984), 290–302.

Crawford, Chris. "The ATARI Tutorial (Part 10: Human Engineering)."*Byte,* 7, June 1982, 302–318.

Heckel, Paul. *The Elements of Friendly Software Design*. New York: Warner Books, 1984.

Rutkowski, Chris. "An Introduction to the Human Applications Standard Computer Interface." *Byte,* 7, October 1982, 291–310 and November 1982, 379–390.

Simpson, Henry. "A Human-Factors Style Guide for Program Design." *Byte,* April 1982, 108.

Spencer, Richard H. *Computer Usability Testing and Evaluation*. Englewood Cliffs, N.J.: Prentice-Hall, 1985.

Stepping through the Process

In Parts 1 through 4 we have built the foundations. Now in Part 5 we use those foundations as we step through the phases of the software development process. These chapters can be shorter than they might otherwise be because we do not need to introduce new methods for each phase. Instead the integrated methods we've described thus far are used through all the phases.

Each phase has as its output one or more documents. A good place to begin the planning for the work to be done in a phase is the document outline. Much of the value of the work done in a phase is captured in its documents. Thus the documents must be written so that they can be read, understood, and used for the purpose intended. So we begin Part 5 with Chapter 10, "Documenting the Project," in which we discuss the general principles of how to write project documents.

Each project organization should have a standards and procedures guide, which describes the basis for how things are to be done in all projects. This guide should show outlines for all the standard project documents. The standards and procedures guide is a good starting place for developing the project management plan, which describes what is to be done for a specific project and how it is to be done. In Chapter 11, "Planning the Project," we discuss what a standards and procedures guide contains and how the project management plan is developed for a specific project.

Chapters 12 through 17 are presented in the order in which the work of the phases tends to be done. We say "tends to" because the work does not strictly follow this sequence. There is often an overlap of work in the various phases and sometimes work must be repeated. Chapter 12 describes how to analyze the current system and its expectations. However, these same methods of analysis are used again in Chapter 15 when we analyze a proposed design to see if it would satisfy the requirements. Chapters 13 and 14 concern requirements and feasibility. We should have requirements before we evaluate feasibility so we know what we are evaluating the feasibility of. But in large projects we may define the requirements initially as part of the feasibility study; then if the project appears feasible, it is worth our while to develop the requirements more carefully and completely. Chapter 15 considers how to design the system; Chapter 16 considers how to build the system; and Chapter 17 concerns how to test and maintain the system. In reality the plans for testing are developed and some testing actually occurs while the design and implementation are being done. But, for convenience, we describe the testing as the last chapter, once we know something about what we are to test.

10

Documenting the Project

- Documents follow a uniform structure. Near the start should be a section explaining the document's purpose, audience, and author.
- The main premise behind document organization is that busy readers must be able to find the parts of the document relevant to them.
- Writers must begin with a clear sense of who their audience is so that their style and focus appropriately reflect the readers' interest and technical knowledge.
- Document style should be as clear and direct as possible. The communication model here is the short, active sentence.
- In the composition process writers may need to diverge slightly from the original outline in order to bridge the flow of text from one idea to the next.
- Two tools that can help writers to create outlines and revise text are the outline processor and the word processor.

Each phase of the project has as its output one or more documents. If these documents are to be worthy of the work that went into generating the information in them, they must be readable, understandable, and useful. In Chapters 12 through 17 we discuss each of the phases and present outlines of the major documents. The outlines show what is to be said in the documents. This chapter will discuss how to say it clearly.

Good writing is a skill that is important in all professions, and vital in software engineering. Moreover, effective writing is not limited to those with born talent; it is an art gained by practice. We can always improve on our own writing.

In Chapter 5 we discussed the importance of expectations. An organization can perform well only if those expectations are communicated well. Poor communication between users and creators of systems can waste time and effort building systems that do not meet expectations. It is cheaper to communicate carefully about what is wanted before the system is built than to build it over.

Poor communication can also result in poor coordination between those developing the system. Some team members may be building one system while others are building another. As software engineers, we may find that poor communication prevents us from getting the help we need from our user, our management, or those who have the information or resources we need. And it can cause someone to throw out our work and start over. No one can build on what we have done if he or she does not understand it.

Many people put off the documentation as an unpleasant chore to be done only after the "real" work is finished. It is often perceived as unpleasant because of its relative unimportance in the face of the pressure to do other work. This is a false understanding of the role of documentation.

Documentation properly done during the project is the glue that holds the work of the project together. As we said in Chapter 2, the requirements specification is needed to develop an agreement with clients as to what they want and can expect to get. It tells the designer what he or she is attempting to achieve. The design specifications tell how the system is to be built. The outlines of the project documents define the information needed to write the documents, and thus the work that needs to be done to obtain that information.

Most writers, including many very successful ones, find writing does not come easily. We should not expect it to. Writing, like creativity, comes to those who understand it is not merely a gift but a goal that requires hard work. As one writer put it, "I don't enjoy writing. But I do enjoy having written."

Document Structure

Each project should have a master document listing all the other documents to be written for that project. Organizations should have a **standards and procedures manual** that outlines and explains the format of each type of document commonly used, such as project proposal, requirements specifications, design specifications, and so on.

As documents are revised, each issue should have the version number and date on the first page. Often a cover sheet is used to show what changes were made. The changes may be marked in the document with bars to the right of the changed material. Then those who have read this document before do not have to read it all again.

If a document is longer than two or three pages, it should begin with a table of contents. When new terms or abbreviations are used, a glossary can define or identify them.

Each document has near the front a standard section that answers the following questions:

- What project does this document concern?
- What is the purpose of this document?
- Who is it written for?
- What questions will it answer or what will the reader be able to do as a consequence of reading this document?
- Who wrote it?
- Where can the reader get more information?

Every major section of the document should also begin with an explanation of the purpose of that section.

Writing Organization

Readers of a technical document want to get the information they need as clearly, accurately, and as quickly as possible, then get on with their work. Most project documents are not read word for word. Readers skip hither and yon looking only for what they want to know. The document must be structured with clear headings so anyone can find what he or she wants without unnecessary reading. Managers, for instance, may take just a few minutes to skip through our document before conferring judgment on months of our work. If they draw the wrong conclusion, our work is worthless. We must make sure they will find quickly what they need to draw the right conclusion.

Before starting to compose any text, we should write an outline. An outline is like a tree. It shows a hierarchy. We use the outline to make the document headings. We number the headings to show the structure of the outline—that is, what depends on what (see "Outline of the Project Management Plan" in Chapter 11). The number contains decimals to show subheadings. For example, the number 10.2.1 would refer to the tenth major heading or chapter, the second heading under that, and the first subheading under that. The headings may also be indented to show this structure. However, indenting should not be continued more than about three levels. Thereafter they should be numbered but not further indented. Later in this chapter we'll describe how outline processors can help us in this task.

Following the outline, we structure our communication hierarchically, like a tree. We write top-down, stating the *general* before the *detail*. We show the forest before the trees.

We state clearly *what* we are doing and *why* we are doing it before we get into *how* we are doing it. We should assume our readers know nothing about the background of what we are doing. We explain it to them. What is the system? How did this problem come about? We write short sections focusing on only one primary point. We write descriptive headings that tell how this section fits into the whole. If readers already know the information in a section and if the section has a good heading, readers can skip that passage.

Knowing Our Readers

We should always keep our readers in mind. We don't write just for ourselves. We cannot assume our readers already know what we know. We must tell them. If we cannot get our readers to understand a concept, that topic may best be omitted.

We must keep in mind exactly what we want our readers to know once they have read our document. We make sure we have written it so they will know that. We focus on it. We don't waste our shots on what we don't care whether they know.

Knowing who our readers are is not always easy. There may be many, and they may be all different. If we're addressing many readers with different backgrounds, we start each section with an introduction that the unfamiliar reader can read and the familiar reader can skip.

It is easy to forget that our audience is not as wrapped up in the project as we are and they may have to review documents on many different projects as they cross their desks. They may get confused and not know which project this document is about. So we must tell them before they get lost in it.

It is easy to write documents that overlook the obvious; an operating systems manual that fails to identify which computer the operating system runs on, or a systems document on a control system that does not bother to mention whether it is a control system for a flight to the moon or the Los Angeles sewer system. Some readers may not already know these things that the writer assumes.

We should not use a private language with words that have meaning only to us, or jargon that is understood only by us and our cohorts. We should avoid abbreviations. If we cannot avoid them, we should explain them clearly when they are first used. Also we should put them into a glossary. If readers have to stop to figure out unfamiliar abbreviations, they can't concentrate on their flow of thinking. We must make sure any new words or concepts are clearly explained when they are first used. If there is any shred of doubt, we should explain. Once we have used a term, we must consistently apply it in subsequent text. We don't use a new term that also must be explained.

Writing Style for Documents

Writing documents is not the same as writing fiction. Fiction writers often appeal to the readers' imagination. We enjoy reading a novel because the writer lets us fill in the details about the scenery, what people look like, what they may be thinking, and what's going on between the sheets. But when we write a document, we spell it all out. We don't leave anything to the imagination.

Documents need not be dull. But many of the tricks that writers of fiction use to make their writing interesting may interfere with clarity. We must use discretion. Only once we're sure we have achieved clarity should we consider introducing novelty.

If we can't explain an idea clearly, we should leave it out. We should not hint at something that's not explained. The suggestion only raises questions and slows readers down, causing them to lose their train of thought.

Imprecise qualifiers such as *often, frequently, many,* and *usually,* should be avoided. These words usually have no precise meaning. (Did the *usually* in the previous sentence add any meaning?) Our "many" may be five. Our reader's may be five hundred.

We should use good English grammar and punctuation. Remember theme, paragraph, and sentence construction. These rules were developed to improve understanding so we should use them.

Using the active "I did this" is clearer than the passive "This was done." Mixing active and passive voices creates more varied sentence structures and more interesting reading. But the active voice also leads to the greater use of "I," which some people find objectionable in documents. I happen to think that the greater clarity that comes from the active voice still argues in its favor.

Readers can digest no more than a limited amount of information at a time (Miller's Principle). We should not cram more information into that time than they can handle. A period gives readers time to digest. So by giving them more periods they can better digest what we have to say. That means short sentences.

Lincoln once wrote a letter in which he said something like "I apologize for writing you a long letter. I did not have time to write a shorter one." It is easy to go on and on. It is hard to write succinctly. We have to have our ideas well organized.

We must keep our sentences short. Shorter is clearer.

I have found myself writing long sentences because I fear that when I finally place that period I can be brought to trial and held accountable for whether everything has been completely stated and properly modified. I shouldn't.

We should write a short sentence, then write another sentence to modify it.

By varying the sentence length and structure, we can make the sentence pattern more interesting. We should use the same sentence structure for two or more sentences only when we want to describe parallel ideas: "This is this way. That is that way."

We should go (back) over our sentences and eliminate every word (that) we can without changing the meaning. *Then* and *that* are good candidates. (Is "going back over" really different than "going over"?) When we use a word processor, we can easily delete words.

This is Miller's Principle in action again. Putting more in a sentence to clarify a meaning may only clutter the sentence, making it harder to understand. We write a paragraph to say what we want. Then we go back and cut out words unmercifully. Depending on our initial writing style, we may find that, when we have cut out a third of the words, we will see a clearer paragraph emerge. (It's fun to do this on a word processor. Zap, zap, zap.)

I once thought that if I wrote something two or three ways, the reiteration would give readers more ways to see my point; each repetition would offer a different insight. Hopefully one version would hit the mark. More likely, though, the repetition gives readers more ways to be confused. They wonder why it is being repeated. Is there some new point each time that they are missing? The uncertainty slows them down and leaves them confused.

We shouldn't say too much. Anything we say beyond what is absolutely necessary can become a hook on which the readers may hang a misinterpretation. Saying too much to make a point more often buries the point. Length adds to confusion. If it is obvious, can be assumed, or is not necessary, leave it out.

When we don't have the information we need, we may have to make an assumption. That's okay. But we should label it as such. We must show which conclusions depend on that assumption and which do not. If the assumptions are clearly labeled, readers can decide for themselves whether they want to accept the assumption and what follows from it. It may be appropriate to offer an opinion. Again, readers can reject or accept the opinion, as they see fit, if we label it and show what depends on it.

Process of Writing

Writing begins with collecting our thoughts about who our readers are and what we want them to know or be able to do after they have read what we write. We develop an outline to structure both our thoughts and the document. To prepare the outline we can begin with an established standard outline and adapt it to the needs of the specific project. We then research information to fit the outline. This information may be collected and categorized in a data base to be organized and retrieved as needed to fit the outline. Finally we can write the text into the outline.

There is a danger, though, in being a slave to the outline. As we write the text, we must develop bridges to establish the flow from one idea to another. Some outlines just do not work when we try to write the text. So they have to be reworked. As we write the text and develop the details, concepts come to mind that did not appear, and may not have fit, in the original outline. After trying to write to the first outline, we may find it necessary to make a new outline and rewrite to it. Paragraphs written for the first outline can often be pulled into the new outline. But we may end up with two good paragraphs that say the same thing. It is hard to throw either of them out. But we have to make a firm decision. If it's got to go, it's got to go. Zap.

Tools that Help in Writing

Outline processors and word processors are two valuable tools in writing and maintaining documents. An **outline processor** is similar to the tree editor we discussed in Chapter 4. Such a tool can help us write either a program or a document top-down. With it we can organize our ideas in outline form, opening or closing the details below a heading to expose as little or as much detail as we wish. If we are trying to see the structure of the whole document, it is easier if only the main headings are brought together and are not separated by pages of intervening detail.

We may start with a standard outline for the type of document we are writing, then delete, add, and reorder to fit the needs of our particular project. Once we have a good outline, we can fill in the text to produce the report. The outline becomes the headings. If two or more people are working on the same document, it is often very useful to begin their collaboration by working together on the outline.

Over the course of a project, documents change. If, in the design phase, we find that a requirement cannot be satisfied, we must renegotiate the requirements with the customer and modify the requirements specification. A word processor is particularly useful for making such changes. A word processor can also be used to copy from document to document standard paragraphs such as the introductory paragraph describing the project.

Documents should be written so they can be changed. It is easier to change a document if there is no redundancy. Then each idea that is changed would require a change in only one place in the document. However, some redundancy is important if the document is to be structured so that various parts of it can be read by different people for different purposes. As a result, revising one idea requires checking every place in the document affected by that idea. The search feature of a word processor can be useful for finding the other parts of the document that must also be changed.

Using a word processor can be seductive. It gives the false impression that since it is so easy to make changes, one can afford to be sloppy on the first draft. However, we face the same problem in writing documents that we have in writing programs. Once they get beyond a certain complexity, they are too hard to change without introducing errors. It is better to use care in developing a good outline and get the text close to being correct the first time.

Several problems commonly arise when we try to make major rearrangements of a document.

1. When paragraphs are moved around, we have to be careful that we don't then have a paragraph that uses a word appearing before the paragraph that defines it.
2. When we rewrite we may have trouble being certain that we didn't omit material that was there before, or introduce the same thought more than once.

It is hard to recognize these problems in our own writing because we are too familiar with what we intended to write.

Summary

Good documentation is extremely important to the success of a project. It provides the coordination between the various people doing the work. It also organizes the work by focusing on what needs to be done to produce the information called for in the document. The documentation is an integral part of the coordination and planning that works for us during the project. It should not be left to the end of the project.

Documents follow a standard structure. They identify the version and the date on the first page. If they are more than a page or two, they have a table of contents. Near the start a section describes the document's purpose, audience, and author(s) and explains what questions it will answer.

The text itself should be organized to enable busy readers to find what they need to review as quickly and effortlessly as possible. Outlines help writers to organize their thoughts into a hierarchical order. Sections must be clearly focused and identified by descriptive headings.

When planning a document, writers also have to keep a clear picture in their minds of who their audience is. This awareness ensures that the concepts are explained appropriately for the intended audience.

The goal of document composition is to write simple, clear, unambiguous sentences. The prose need not be dull, but the first priority is clarity and precision. The active voice is preferred; sentences and paragraphs should be as short as possible.

Exercise

1. Find a piece of systems documentation. Critique how it is written. Develop an outline to show how you would rewrite it.

References

Flesch, Rudolf. *The ABC of Style: A Guide to Plain English*. New York: Harper & Row, 1980.

Flesch, Rudolf, and A. H. Lass. *A New Guide to Better Writing*. New York: Warner Books, 1982.

Rubin, Martin L. *Documentation Standards and Procedures for Online Systems*. New York: Van Nostrand Reinhold, 1979.

Strunk, William, Jr., and E. B. White. *The Elements of Style*. 3d ed. New York: Macmillan, 1979.

Van Duyn, Julia. *Practical Systems and Procedures Manual*. Reston, Va.: Reston Publishing Company, Inc., 1975.

11

Planning the Project

- The configuration management plan delineates how the current description of the product is to be maintained throughout the project.
- To guarantee that the product will meet certain expectations, the quality management plan describes the standards that are to be followed and the review procedure.
- The main process document is the project management plan, which spells out what is to be done, who is to do it, and how long it will take.
- The project management plan consists of an introduction and overview, list of deliverables, work breakdown, schedule, delineation of external and internal responsibilities, and assumptions.
- Company policies for work on all projects are defined in the standards and procedures guide.

In Chapters 6 and 7 we discussed methods of planning. In this chapter we will consider how these plans are documented in the **project management plan**. The project management plan is the primary process document for the project, just as the requirements specification is the primary product document. The project management plan shows what we intend to do and how we intend to do it to produce the product. It describes what specific tasks are to be done for this project, who is responsible, how long it will take, when it will be done, what resources will be needed, and what it will cost.

As we said in the last chapter, the organization producing the product should have a standards and procedures guide, which describes in a general way how one should proceed on any project. This guide contains procedures for keeping track of the latest version of the product (called configuration management) and guidelines for conducting the checks, reviews, and tests needed to ensure the delivery of a quality product (called quality assurance management). The project management plan describes how these generic standards and procedures are applied to a specific project.

Configuration Management Plan

The configuration is the current description of exactly what the product is planned to be—its parts, their assembly, and any assumptions about the product. The **configuration management plan** tells how this configuration is to be maintained throughout the project. It contains the procedures to consider the consequences of proposed changes, to decide whether the changes should be made, to notify those whose work would be affected by the change, and to make the changes in all the appropriate documents so as to record the change's effect on product configuration. Configuration management may also be called configuration control or change control.

These procedures usually call for proposed changes to be written on a form, sometimes called a **change proposal**, showing why the change is needed and suggesting its effect on the product, its cost, and the schedule. The proposal is submitted to a responsible body, often called a Change Control Board, which reviews the suggestion, makes an independent evaluation of its consequences and merits, and approves or rejects it. They may consult with those whom they believe could offer advice about the effects of the change. If the change is accepted, the board initiates the process of making the change. This may involve changing plans, updating documents, and notifying people whose work will be affected. Changes are often held until a set of changes can be issued all at the same time. This grouping allows changes to occur at defined times when they can be expected rather than occurring continually. Change control was discussed briefly in Chapter 7.

Quality Management Plan

The **quality management plan** describes the procedures used to guarantee that a quality product will be delivered. These guidelines specify how reviews are to be managed so that the product is developed according to defined standards of quality, points in the project where the reviews are to be made, what is to be reviewed, who has what responsibilities in the reviews, how the meetings are to be conducted, and how the recommendations are kept track of to see that they are followed. The standards and procedures guide shows the company policies for quality management. The specifics as they apply to a project are included in the project management plan, possibly by reference to a separate quality management plan.

Project Management Plan

The goal of project management is to plan and control the project so that we:

- Satisfy the customer's expectations.
- Deliver the product on schedule.
- Do it within budget.

To deliver the project on schedule and within budget, we must plan and control. **Planning** is deciding what is to be done, how, and when. **Controlling** is making sure that the work is done how and when the plan calls for, and if it isn't, to understand why and take corrective action. It may be necessary to change the plan or the way the work is being done or both to achieve the original goals. If the original goals cannot be achieved, new goals and a new plan must be renegotiated.

The project management plan describes to the participants what they are to do. It shows what work is to be done, who is to do it (or at least what skills they must have), and when and how they are to do it. The plan also coordinates the work so that the sum of all the people each doing his or her own work will produce the right final project at the right time for the right price.

The project management plan is a road map that management uses during the project to see where the project has been, where it is, and what yet needs to be done. Progress is measured by comparing work done to the plan.

If client and management decide to proceed, management selects a project manager, defines to whom he or she will report within the organization, and provides the manager with the authorities and responsibilities needed to do the job. Then this project manager will be responsible for staffing the project and developing the project management plan.

An important issue concerns whether the project management plan is shared with the client (that is, considered an external document), or is held privately within the organization (that is, considered an internal document). Some parts of the project management plan—such as the list of items to be delivered, their delivery dates, the description of what the client will supply, and the client's responsibilities to the project certainly must be shared with the client. Other aspects of the project management plan concern what goes on inside the project organization to deliver the product. Generally organizations consider these internal matters to be none of the client's business. Management usually prefers to provide clients with only those parts of the project management plan that relate to how they, as clients, interface with the project.

However, if the organization has not yet established a good track record, the clients could have some questions about how the work will be done. Since they may feel they are sharing some of the risk, they may insist on reviewing how the work is being planned. If such a review must take place, then it must. But normally management is inclined to resent this intervention by clients and tries to avoid it.

Contents of the Project Management Plan

Introduction and Overview

The project management plan begins with an overall description of the project. This section is characteristic of all the documents to be used by the project and can be copied from document to document using a word processor.

The overview also refers to the pertinent documents that might need to be consulted while reading the project management plan. Certain documents, called **controlling documents**, constrain how the project is managed. Among them are the standards and procedures guide, the project proposal, and the requirements specification.

Deliverables

To plan a project we must know what is to be done. To determine what must be done (and just as important, what we can avoid doing) we begin by asking: "What must we deliver to the client." These items are called **deliverables**. Then we can ask: "What must we do in order to deliver them?"

Deliverables include the software programs and any hardware that may be required. Usually they include documents such as the requirements specifications, user's manual, design specifications, and maintenance manual. They may also include services such as training courses, conversions of data bases to run with the new system, and consulting help in the operation and maintenance of the delivered product. Deliverables are everything we agreed to provide to the client.

Once the deliverables have been identified, we determine when we must provide them. Clients may want the deliverables delivered at specified times throughout the project instead of all at once at the end of the project. They may phase their work to coincide with ours so that we are really working out a joint schedule.

Dates upon which we make deliveries to the client are determined by a combination of scheduling and negotiation. We may use a critical path schedule to estimate when we can deliver. But clients may want us to deliver earlier. Then we must determine whether we should use more resources or proceed differently in order to meet the clients' date, or whether we can negotiate a new date. However, it is not uncommon for management to make a commitment to clients on a delivery date, and only then ask subordinates to work up a schedule to determine how or whether these dates can be met.

Work Breakdown Tree and Tasks

Once we know what we must deliver, we ask what tasks are required to produce those deliverables. The immediate tasks required to produce the deliverables are defined first, then the tasks required to produce these in turn, until we have developed a full work breakdown tree, as discussed in Chapter 6.

Also part of the tree may be nondeliverable products such as the programs we must write to develop test files, which are not part of what we deliver to the customer. These are sometimes called "throwaways," but we do include them in the work breakdown tree.

Two types of relations are established between the tasks:

1. Hierarchical relations with other tasks that make up this task. These are needed for developing the work breakdown tree.
2. Precedence relations between these tasks to show what tasks must be done before other tasks can be started. These are needed for developing the critical path schedule.

Schedule and Plan

Once we have established the tasks by the work breakdown tree, we must determine for *each task*:

- What other tasks must be finished before this task can be started? (These are the precedence relations between the tasks we use in computing the critical path schedule, as in Chapter 7.)
- Who will be responsible for the task?
- How will we define when the task has been done?
- How will we define whether the task was done correctly?
- How can we tell before the task is finished whether the task is going well or badly?
- What resources will the task require?
- What will be the task cost for:
  - Personnel?
  - Facilities?
  - Overhead?
  - Services?

We also define the equivalent information about the whole project:

- When should the project be completed and each deliverable delivered?
- How will we define when each deliverable and the project have been done?

- How will we define whether the project was done correctly?
- How can we tell before the project is finished whether the project is going well or badly?
- What resources will the project require and when will they be required?
- What will the project cost and when will the money be needed?

In Chapter 7, when we showed how to develop a critical path schedule, we worked backward from one required project completion date to compute the latest times for each task. But, as the preceding lists have shown, there may be a different required date for each deliverable. So we must change the way we compute latest times. For each deliverable we insert the required date for that deliverable as the latest finish time for the last task in the sequence that produces the deliverable. Then we compute backward from these times.

When we make up the critical path schedule, we may identify several key dates in the schedule as milestones. Some managements want to look periodically at the whole critical path schedule. Others may be interested only in following these few milestones to evaluate how the project is progressing.

Financial Analysis

The financial analysis will show the budget for each type of expenditure (personnel, overhead, subcontracts, and so on) and the cash flow as a function of time. We will discuss cash flow in Chapter 14.

External Responsibilities

Clients may just state what product they want, take delivery of it, and pay. On the other hand, they could be more actively involved. Our work may be part of a larger plan that requires the client to take responsibility for doing some part of the work. Clients are sometimes responsible for producing test data or for conducting tests. They may have to review documents in such time that they will not delay our work, or they may provide facilities that we require in our work. All these client responsibilities need to be stated.

Other external responsibilities associated with subcontractors, service providers, resource providers, interfaces to other projects, and vendors of supplies and equipment must be stated clearly.

Internal Organization and Responsibilities

The internal organization should be spelled out with the name of the project manager and to whom he or she reports, the organizational structure, the type of persons required to fill these positions, and where they might come from (internal or external to the organization). Chapter 5 discussed the various types of organization.

Assumptions, Contingencies, and Risks

As we make our plan, we also make many assumptions. For instance, we assume that computer time will be available when we need it. We expect, too, that clients will meet their obligations on time. Indeed we often make more assumptions than we are aware of. One key to making a good plan is to think through the assumptions and ask ourselves what is the chance that these assumptions may be violated. Then we should make some allowances in our plan to account for these hazards. For the major assumptions, if there is a good chance that they will be violated, we may want to make contingency plans.

Not everything that can go wrong will happen, but much will. We just don't know which things are actually going to go wrong. Looking at the hazards should give us an idea of whether we are dealing with a straightforward project where we don't have to make many allowances for hazards, or whether we are dealing with a high technology project requiring many things to be done that have not been done before, thus requiring more allowance for hazards in the cost and schedule.

We may often be faced with decisions between safe ways and ways that carry more risk. Usually it's best to play it safe if we can, but sometimes it may be necessary to take some risks to cut the schedule time and costs. If our gamble fails, it may become more costly and time consuming than if we had taken the safe approach. We need to think carefully about our approaches and their possible hazards.

Outline of the Project Management Plan

Objective:

- The primary process document—To define the work to be done and plan how it will be accomplished.

Outline

- Title page (standard)
  - Project name
  - Document type (project management plan)
  - Document identification number
  - Revision number and date
  - Authors / contributors
- Table of contents
- 1.0 Introduction (standard)
  - Overview description of this project (very brief)
    - Whole project
    - Part (or all) of project to which this document pertains
  - Description of this document
    - Purpose of this document and how it is to be used
      - Intended audience
    - Content and organization of the document
    - Authorship
      - Author
      - Date written

Procedure for changing this document
Controlling documents and references
Definition of minimum terms, acronyms, and abbreviations needed to read this document
2.0 Deliverable products and dates
(Objective: To define external view—what was negotiated to be delivered and when)
Software
Documentation—Content and level of detail
Services
Training, maintenance, operations support, etc.
Hardware
Acceptance criteria for all deliverables (may refer to requirements specifications)
3.0 Work breakdown
(Objective: Define tasks)
Nondeliverable products
Documentation tasks
Management/support/overhead tasks
Work breakdown tree
For each task: Durations, precedences, resources and skills, effort, costs, technical factors and risks
4.0 Management methods and procedures
(Objective: Define management practices)
(This may be handled by reference to the company standards and procedures guide)
Development phases used in this project
Development approaches (e.g., will prototyping be used)
Documentation structure and standards
Handling of reviews and walkthroughs
Testing methodology (top-down, bottom-up, etc.)
Configuration control procedures
Quality assurance procedures
Programming standards and conventions
Handling of customer interface
Automated aids
Other standards
5.0 Assumptions, constraints, and external project interfaces
(Objective: Define what ties our hands and other people's hands)
Customer obligations
Test and other data to be provided
Facilities and services to be provided
Responsibilities for reviews
Responsibilities for tests

- Other actions required
- Special contract provisions
- Subcontractor
  - Justification for the subcontracting
  - Obligations of the subcontractor
- Interfaces with other projects
- Supporting services and facilities required (noncustomer provided)
- Hardware and languages to be used
- Other assumptions
- Risks and hazards and what might be done about them

6.0 Organization

(Objective: Define who does what)

- Project head
- Person(s) to whom project head reports
- Members of executive review committee
- Members of task force
- Organization chart
- Responsibility for customer interface
- Responsibility for each task (see work breakdown tree)
- Responsibility for each position on organization chart

7.0 Plan

(Objective: Summarize plan for use in reviewing and controlling the project)

- Critical path schedule
- Bar chart
- Critical delivery and external interface dates
- Manpower and resource levels versus time
- Total planned costs and allocations
- Budgets
- Cash flow—Expenditure by month
- Risk analysis (probabilities and consequences) of not completing within schedules and costs

Glossary

Appendices (material that does not fit in its place above and supplemental material)

Standards and Procedures Guide

The standards and procedures guide defines company policies. It tells people how things are to be done. For example, part of the guide may describe the methods used to write programs and how to document them. Another part may state that systems are to be described in the requirements specifications using Two-Entity Data Flow Diagrams and that programs are to be developed and described using

Trees. The standards and procedures guide should list the documents to be produced for each project, an outline of their contents, the method by which these documents are numbered, whose responsibility it is to produce them, and where they are to be filed.

If the organization has a standards and procedures guide that will be used on this project, it should be referred to in the introductory section of the project management plan as a controlling document. If the organization does not already have such a guide, all this material must appear in the project management plan. Having a standards and procedures guide can save us a great deal of effort in making up a project management plan.

The standards and procedures guide may include standards and procedures that relate to configuration management and quality assurance, or these may be treated in separate documents. They are so interrelated that they may be combined as one document. In various organizations this one document may be called: standards and procedures guide or manual, configuration management guide or manual, or quality assurance guide or manual.

The outline in this section shows what might be in a combined standards and procedures guide including configuration management and quality assurance. The outline is not exhaustive, but it should be suggestive.

If we are evaluating an organization to do work for us, we might ask to see their standards and procedures guide. It can tell us a lot about the quality of the organization. If possible, we should also check how they use the guide and how well they train their people to use it. However, the company may consider their standards and procedures guide sufficiently important to giving them a competitive edge that they will consider it proprietary and won't let us see it.

Outline of the Standards and Procedures Guide

1.0 Description of this guide and how it is to be used
2.0 Project management procedures
  Method of handling of interfaces to the customer
    Person(s) authorized to negotiate with and make commitments to clients and person(s) not so authorized
3.0 Documentation (see Chapter 10)
  Standard introduction used in each document
  List of standard documents and their numbering conventions
  Reference outline for each document type
  Method for using documentation tools such as:
    Program to extract comments from code to form documents
  Document review procedures with checklist of who is to review and what is to be reviewed for each document type (see review procedures below)
4.0 Configuration/change control
  Composition of and procedures for configuration/change review board
  Procedures for submitting and handling proposed changes

Form for submitting proposed change showing why change is needed, consequences of not making the change, and its effect on product, cost, and schedule
Form for soliciting comments from knowledgeable and affected people about likely effects on product, tasks, documents, schedule, and costs
Form for change notification announcing decision to make change, to reject the change, or to renegotiate with the client before approving the change

5.0 Quality assurance and review procedures
- Aspects to be reviewed
- Person(s) to do reviewing
- Convening a review team
- Forms and check-off lists to be used in conducting review
- Procedures for sign-offs and approvals
- Procedures for follow-up

6.0 Data element dictionary
- Format for dictionary
- Procedures for entering and changing data elements
- Procedures for notifying people whose work is affected
  - (Changes to the data element dictionary may be considered the same as the configuration control of any other change, or may be handled in its own way)

7.0 Formats of programs
- Use of structured techniques
- Format (spacing and indentation) for each statement type
- Formats and information to be included in comments
  - Description of variables
  - Decision as to whether comments are to be extracted from code for use in other documents and, if so, how they are to be formatted
- Conventions for choosing names of variables
- Use of include files

8.0 Standards for machine-human interfaces (see Chapter 9)

9.0 Librarian procedures
- Person(s) making what type of computer runs
- Method of documenting and filing these runs
- List and description of useful utilities
- List and description of standard general-use modules

10.0 Development and testing procedures (see Chapter 17)
- Definition of development phases and review stages for this project
- Use of top-down development
- Use of top-down and bottom-up testing
- Procedures for generating test data and gaining test coverage
- Structure of test plan
- Structure of test report
- Procedures for setting up environments for tests

Procedures for module tests
Procedures for system integration tests
Procedures for acceptance tests
Retention of test data so that it can be rerun at any later time
Use of profilers for testing performance
11.0 Use of automation programs and data bases in systems development
Tree editor
Data structure matrix data base
Documentation aids
Means to extract comments from code into documents
Outline editors and word processors
Source code control system
12.0 Glossary of terms

Summary

The plans for a software engineering project are laid out in a set of documents that record the work, procedures, responsibilities, and lines of communication necessary to get the project done.

The configuration management plan contains guidelines for ensuring that the description of the product is maintained during the course of the project. It includes procedures and lines of authority for making changes. Recommended changes are presented in a change proposal. Another document, called the quality management plan, outlines the steps required to deliver a quality product.

The chief tool used by management for overall planning of a specific project is the project management plan. This document identifies the participants, the tasks, and the schedules. The plan may or may not be shared with clients, depending on the client-organization relationship.

Project management plans are made up of certain standard sections. They begin with an introduction and overview, which provide basic identifications, list the contents, and refer to controlling documents. A deliverable section lists what the project is to provide the client. A work breakdown describes the tasks that need to be done. The management plan also contains a schedule, financial analysis, a separation of external and internal responsibilities, and assumptions.

Organizations spell out their policies in a standards and procedures guide. Such a document tells employees the methods and rules governing all projects.

Exercises

1. Write a standards and procedures guide for your imaginary company that will do the CAPERT study in Appendix A.
2. Write a project management plan for the CAPERT case study in Appendix A.

References

Blanchard, Benjamin S., and Wolter J. Fabrycky. *Systems Engineering and Analysis*. Englewood Cliffs, N.J.: Prentice-Hall, 1981.

Biggs, Charles L., Evan G. Birks, and William Atkins. *Managing the System Development Process*. Englewood Cliffs, N.J.: Prentice-Hall, 1980.

Bruce, Phillip, and Sam M. Pederson. *The Software Development Project: Planning & Management*. New York: Wiley-Interscience, 1982.

Buckley, Fletcher. "A Standard for Quality Assurance Plans." *COMPUTER*, 12, 8, August 1979, 43–50.

Evans, Michael W., Pamela Piazza, and James B. Dolkas. *Principles of Productive Software Management*. New York: John Wiley & Sons, 1983.

Softky, Sheldon D. *The ABC's of Developing Software: A Primer on Essentials of Software Development*. Menlo Park, CA. The ABC Press of Silicon Valley, 1983.

12

Analyzing the Current System and Expectations

- A systems analyst is asked to conduct a preliminary study when someone recognizes that a system's behavior doesn't match the expectations of someone of influence.
- The preliminary study tries to determine the scope of the proposed system, its environment, and what expectations are not being met by the current system.
- An important part of the preliminary study is figuring out exactly what the problem is—defining the need and the means to satisfy that need.
- Before making a change, we must identify the change, simulate it, decide if we like it, and determine if it warrants the cost.
- During the fact-finding phase of the preliminary study, systems analysts use a number of methods to help them collect information. Logs and files, for instance, provide a way of storing and tracking the information.
- Among the documents to research are organization charts, standards and procedures manuals, preliminary study reports on the current system, journal articles, and vendor literature.
- Interviews allow the analyst to obtain information from and get the cooperation of the people involved in the system change. But the interviews must be carefully planned so that the answers are really relevant and that no one's time is wasted.
- Scenarios help the analyst think through how the new system would be used.
- Cause and effect graphs are models for organizing the information collected in a preliminary study. They show what actions or conditions cause what effects.
- The system analyst's findings are contained in a preliminary study report.
- A project request describes work that an organization needs done. A project proposal describes the work an organization will do to satisfy the request.

The term *analysis* refers to the study of a system to determine how it behaves, why it behaves that way, and what the consequences are of that behavior. We are concerned with analysis on two occasions during the development process. When we are considering making a change, we analyze the current system, the expectations people have for that system, and the changes that could be made to it. And in the design phase we analyze proposed designs to determine whether they would meet the requirements specifications. This chapter focuses on an analysis of the existing system and the needs for a new system or system change.

Considering Change

We analyze the current system or situation to decide whether we want to make a change. If that change would be easy to make and easy to undo if we don't like it, then we make the change, see if we like it, and undo it if we don't like it. That's easy enough.

But what if the change is not easy to make? And what if, having made the change, there is no turning back, or at least it is costly to undo? Then we would like to know whether we will like it before we do it. That's what the preliminary study is about.

First we have to decide whether we like things the way they are. If we are satisfied, there is no need to contemplate a change. Changes are often difficult and traumatic. We don't change unless we feel we have to. If we feel uncomfortable with things as they are, then we must ask ourselves why. What expectations do we have, and why are they not being met?

We want to analyze the existing situation to see how green the grass is here before we look at the grass over there. We make an analysis, therefore, so that we understand the existing system well, and so that we know what works and doesn't work and why. Perhaps some small fix is all that is needed.

How do we know whether we would like the change before we make it? An initial step obviously is to identify the change. Then we must simulate what it would be like once we have made the change. We can simulate it in our minds, with the help of paper and pencil, or perhaps with the help of a computer. Then having created it by whatever abstraction is appropriate, we ask, "Do we like it?"

If the change is attractive, we ask what it will take to make the change. If we think we would like it well enough to warrant the cost of making that change, we make it. If not, then having thought it through is a good lesson and may save us a great amount of grief.

How good our decision is depends on how well we are able to simulate the new situation that would exist after the change. If we overlook something, we might think the change looks good, go ahead and make the change, and then later regret it.

The purpose of the preliminary study is to look before we leap so that we make the right decision about whether to make a proposed change. In the process we might decide not to make the proposed change. Or we may propose another change.

Preliminary Study

When someone thinks there may be a problem or sees an opportunity to solve someone else's problem, a systems analyst is asked to conduct a **preliminary study** and write a **preliminary study report**. The preliminary study report will answer the questions discussed in this section and propose what action should be taken as a next step. The study may be done by a systems analyst who is concerned with just this study, or by the software engineer as part of his or her wider range of responsibility.

A preliminary study is typically initiated when someone recognizes that the behavior of a system does not match the expectations that someone of importance has for it.

This situation arises because there has been a change in the:

- behavior of the system
- expectations of how the system should behave, or,
- perceptions of how the system behaves and how that behavior matches expectations

This situation may be resolved by changing the:

- behavior
- expectations, or,
- perceptions

Before we undertake making these changes we must determine:

- What is the behavior of the system?
- What are the expectations for that behavior?
- What are the perceptions?

But we must also determine the following:

- What is the **scope** of the system being considered—that is, what is included in the system and what is not included?
- What is the **environment** of the system being changed—that is, what other systems interface with it, who or what type of people will operate the system, and under what conditions will it operate?
- Who recognized the failure of the behavior to match expectations?
- Who has these expectations, and why is it important that we care what he or she expects?

Note that the preceding questions can be derived from the italicized sentence at the beginning of this section.

We must also consider an additional set of questions:

- What caused the change in the behavior or expectations?
- What are the consequences of this change? What would be the cost of not doing anything about it?
- What alternative actions can we take to develop a new match between behavior and expectations?
- What are the consequences of these various actions?
- For each of these alternative actions, what is the cost to take the action and what are the costs of the consequences?
- What action do we recommend?

Our plan for this preliminary study and the outline for the preliminary study report arising from this study will closely follow these questions.

Often when we explore these questions, we find that the problem or possible opportunity is not based on valid perceptions. Our dissatisfaction with the current system's performance may be based on a false perception of how that system is really behaving, or on false assumptions about what practical options exist to change the system, whether the appropriate technology exists, and whether costs justify the benefits of making the change.

We must also ask if we care whether this person has a problem. If he or she is the boss or has some influence on us, we care because not to solve the problem would then present us with a problem of our own. But systems analysts must be careful about pursuing problems that they may feel are important but do not have management's endorsement.

The preliminary study develops information from which the client can decide whether:

1. A project should be initiated to develop a system to meet certain specified objectives, or,
2. The project should not be pursued further because either:
   a. The facts indicate that the perceived problem or opportunity does not really exist,
   b. It does not appear that there is a possible means for meeting the expectations, or,
   c. The estimated benefits do not justify the estimated costs, or,
3. A greater commitment of resources is required to make a more complete study before any decision can be made.

A project ends when either:

1. The parties involved recognize that there is no feasible means to match the need that will satisfy expectations, or when,
2. A feasible means has been built, matching the need to the client's satisfaction.

What Is the System?

If the existing system is not meeting our expectations, before we change it or build a new one, we want to analyze that system so we know what is wrong with it and why. We also want to know what is right with it so that, when we change it or build a new one, we can learn from the old one.

An additional step is to look at systems other people have that are similar to the one we want. What characteristics of their system are desirable or undesirable? How are their needs or environment different from ours and how does that affect our conclusions?

If possible, we want to have some working reference system to start from. In a complex system there are so many things to be considered that if we start from scratch, we usually overlook something. But if we can make a small change from a reference system, we can rely on the reference for those functions that work as we want, and we can focus on just the changes.

Sometimes we may be doing something for the first time. Then we have no existing system to work from, nor does anyone else have a system that we can look at before we begin. The Apollo project to put men on the moon was such a project. In such a case we must look at what exists that we can build on. In the Apollo project what existed was the current rocket technology, a national commitment, and the resources that come from that commitment.

What Is the Problem?

The systems analyst usually must do some detective work to determine exactly what the problem is. The problem is usually presented to the analyst as an idea—someone already has in mind what they conceive of as a need, and at least a vague notion of a possible means for a solution.

A project could also begin with the emphasis on a means instead of an emphasis on a need—for example, a new technology or product that might provide a possible market, or a new method that will save time and money.

Often the need as expressed by the users is not what they really need. The systems analyst may have to work with them to find out exactly what their real need is. Here the analyst works like a psychiatrist, who asks, "Let's figure out what is really bothering you." To work on satisfying the need as first stated may lead to an expensive project that won't be accepted when it is finished.

To think you know what the real need is and go ahead without the user's concurrence can be a disaster. This often happens when the user and the systems analyst do not communicate well, which is an important reason to have good methods of representation and communication.

An analyst was called in by a building manager and asked to find a way to increase the speed and capacity of the building's elevators. The building manager had received a number of complaints about long waits.

A study was made of the number of people waiting at each floor and which floor they were going to at each time of the day. It was thought that maybe a new elevator control discipline could be used to decrease the waiting time. A simulation showed that no significant decrease could be obtained. A study of adding another elevator shaft onto the building showed it would be too expensive.

An employee observing this asked to be allowed to try out his idea. Since nothing else seemed to work, he was given a chance. He made the observation that the waiting times were not that long compared to waits in other buildings; they just seemed long because people had nothing to occupy their thoughts while waiting, and so got bored. He set up mirrors so that people could entertain themselves by watching each other without being obvious. The complaints dropped sharply and the building manager was happy with the solution (Rudwick, 1969).

The real problem here was to eliminate the complaints. The request to the systems analyst was couched as a means instead of a need. A speed-up of the elevators was only one conceivable means. A great deal of time and money was saved by getting at the real problem hidden behind the request.

A systems analyst was called in by the president of the company and asked to set up a new order entry system. He had heard of many complaints from customers about the time required to get their orders filled. The analyst checked the records and found the order response time was excellent by industry standards. When he checked into the source of the complaints, he found that most of them came from only a few customers. When he called several of them, he found their complaints were due not to the order response time but to other problems, such as not being notified quickly whether their order could be met, thus losing time before they could seek the item elsewhere. Small changes were made to the existing system and the problems were straightened out to the customer's satisfaction. The situation was then explained to the president and he was satisfied.

Problems are often not what they seem. One of the systems analyst's first tasks is to look carefully at the origin of the problem, who raised it and why, what the problem really is, and whether it is valid and worth solving.

When studying the problem, the analyst may unconsciously make judgments, letting personal values creep into the problem in order to decide what he or she thinks the needs are. The analyst's mind is filled with all the exciting means that are available and that he or she would just love to have the chance to apply to a good problem. The analyst may then unintentionally misunderstand or distort the problem to suit these means. But when the system is finally finished, if it is not what the client expected, the system is a failure.

Often the neat solution is not the right solution. It is not so important that the solution be logically rational and mathematically neat as it is that people want to use it because it solves their problems. The analyst may make suggestions to the user about what the user might want, but the final judge has to be the user.

Anthropologists understand this problem well. They are taught that they cannot understand an unfamiliar culture if they judge it by their own sets of values. They must learn to understand the culture by *its own* set of values. This same methodology holds whether we are anthropologists dealing with the Tasaday peoples, or systems analysts dealing with a business executive (Weinberg, 1971).

A frequent error in defining the problem is to look at only one aspect, ignoring the broader picture.

> Someone noted that the computer paper was a large cost item and so recommended a cheaper grade of paper. Unfortunately the new paper didn't stack as well and this caused another problem. The computer operator was spending too much time straightening out the paper stack and not enough time tending to other operations. In the long run it would have been less expensive to use the more expensive paper.

The paper costs showed up on one ledger, while the wasted machine time caused by the delay in the computer operations showed up on a different ledger and was not attributed to changing the paper. This makes it harder to trace the real problem.

> A company making electrical components found that they had what appeared to be a seasonally varying business. This led them to lay off employees at one time of the year, rehiring them or training new people later when the demand picked up. This upset the lives of the employees and was expensive to both the employees and the company. There was no obvious answer like Christmas or summer fashions to explain why their business was seasonal.
>
> They hired a consultant, who did a simulation. He found an apparent seasonal variation had been introduced in the order cycle by the delays that occurred when the salesmen ordered from the district warehouses, the district warehouses ordered from the factory warehouse, and the factory warehouse orders were used to set the factory production rates. Changing the reordering procedures got rid of the yearly cycle and saved both the employees and the company much grief (Forrester, 1958).

Systems not only have many different parts but have many different people with different interests. Management wants to improve cash flow, marketing wants to reduce order cycle time, manufacturing would like to reduce inventory, and so on. The systems analysis often must integrate and arbitrate between these various points of view to find a common solution. Not until the systems analyst pulls together all the individual views to see the whole picture, or at least a larger part of it, can he or she determine whether the problem really exists, and if it does, how to solve it.

Fact-Finding Methods

Before a system change can be proposed, the systems analyst needs to collect as much information as possible. The main sources of information are documents, interviews, questionnaires, observations, and tests. The approaches to each of these sources must be carefully planned, and the results should be stored in an orderly and accessible fashion.

Logs and Files

It is important, particularly during the fact-finding process, to keep a log and a file. The log should be a diary of what we did and when. It should reference the documents we consulted, show who we interviewed and when, and summarize what we learned. Any documents we collect should be put in the file, and the log should reference them and show how they can be found in the file. The log should contain a summary of the valuable items of information digested from the material in the file.

Documents

We will want to know as much as we can from the available documents before we conduct interviews so that we will know what to ask. When we ask good questions that show we have done our homework, we will get more respect and cooperation from the interviewee. No one wants to have his or her time wasted by someone asking stupid questions that could easily be answered by reading the available documentation.

Several types of internal documents may be used to help understand the problem and the circumstances under which the problem arose:

- Organization charts and job descriptions aid in determining who may have the necessary information when it comes time to interview.
- Standards and procedures manuals describe to employees how various standard types of operations are done. Most mature organizations have them. We can consult these manuals to learn how the organization expects to deal with the situations we are studying.
- Systems operating manuals for the systems we are studying tell how the people who built the system intended it to be operated.
- Preliminary study reports on the current system originally analyzed the need and explained how this system was expected to meet that need. It can be illuminating to read those reports to understand the intentions the developers had in developing the system and to compare these with how the system is being used now. Sometimes the troubles with a system are due to people not using it as it was intended to be used.
- Reports written by people who have previously studied this problem, who have reported problems, or who have made proposals for changes may teach us something. But we must also ask ourselves, if this person has already studied the same problem, why are we being asked to do it again? We should see why the earlier report was not accepted and consider what we can do that will not repeat their mistake.
- Operations logs record the day-to-day operation of many systems. Going over the logs can be a tedious job, but it can reveal a great deal about what has been happening.

- Stockholder's reports give a broad idea of what the company does and its approach to doing it. If we are working in an unfamiliar company, as we might if we are an outside consultant, we may wish to consult these documents.

We may also wish to look at how other people have solved similar problems by seeking information from external documents:

- Articles in trade and business journals, newspapers, and newsmagazines often describe the new methods or products that have been produced to solve various problems and the experiences people have with them. Often, depending on the source, the information presented may be superficial, but there may be references to people or other sources of additional information. These journals may also tell who the vendors are who sell products to solve these types of problems.
- Vendor's literature and manuals may offer information in the general discipline as well as the specifications of their product. Since the vendors want to sell us their products or methods, they occasionally provide this broader information to educate us in the field. We may also find it useful to get from them the names of people who have used their products and who would be willing to talk to us. We must remember to consider, though, in what way our needs and environment are similar to or different from their's before we decide how their experience would hold true for our needs and environment.
- Reference services, available by subscription, often review and compare various vendors' products from a common perspective.

Interviews

Once we have gathered the best possible background about the organization and the problem by reading documents, we can plan what we need to know, who we will interview, and what questions to ask. As the series of interviews proceeds, we must be prepared to change our strategy if the information we get leads us into fruitful new areas. But we should not be distracted by side paths that don't lead where we want to go.

There are several reasons for interviews. One goal, of course, is to obtain information. Another, more subtle reason is to obtain that person's cooperation in supporting the system we are going to build. If the interviewee has information about how things are done now, he or she may also be involved in the new system. We will need his or her cooperation to make that system work. We are also seeking to get references to other sources of information and authorizations to interview other people or have access to certain information.

When we interview people, we are taking up their time. They must feel there is some reason the interview is worth their time. If they are under pressure from their management, as is often the case if we are looking into a problem, they will feel that they could be asked by their management why they were spending their time talking to us instead of doing their job or solving the problem themselves.

People we may interview might also feel threatened by the simple fact that we want to interview them. They may think we believe *they* are the problem, or that we may tell their management that they are the problem. They may also have become fed up with a long line of people interviewing them about this same problem.

Generally we will want to interview starting high in the organization and work down. We begin at the highest level of management that supports the study and get them to write a letter to the people we will interview showing support for what we are doing and asking for their time and cooperation.

We may want to interview certain people first to obtain information or authority that we will need to interview other people. We must plan who we will interview, in what order, and what information or authority we would like to get from each interview.

We plan our interviews so that we do not waste the time of the persons we are interviewing. The better we are prepared, the less the interviewee will feel we are wasting his or her time. We will obtain more respect and will get more cooperation if the interviewee senses we have done our homework.

Before the interview we should make notes about what we want to accomplish in the interview. What questions do we expect this person will be able to answer? What questions will we ask? To what sources of information can this person lead us? What authority to get information from other people can the interviewee give us?

We should write a letter to the persons we want to interview stating what we would like to talk about and why. Then our source can be prepared, bringing along any papers that would be of help. We state who has asked us to make the study and who may have authorized us to talk to him or her.

We arrange an appointment at a time the interviewee would find most convenient and in a place where we are not likely to be disturbed. If we meet in the person's office, we may find that the interview is continually interrupted by people walking in or the phone ringing. We will not be able to gain the continuity we need to discuss anything in depth, nor will we be able to develop a relaxed atmosphere in which the person is likely to unwind and reveal what we want to know.

Just before the appointed time, we should confirm the appointment and establish that it is still a good time. It is hard to interview someone who is distracted by another problem.

We keep the interview short—no longer than an hour at the most. It is hard to keep the attention of busy people for much longer than that. If we have more to discuss, we make an appointment for another interview later. It will give both of us a chance to pull together our thoughts to make the next interview more productive.

A tense person will fight us; a relaxed person will be more likely to help. We should take some time before the interview to plan a way to break the ice. Talk to people who know the interviewee to find out this person's interests—sports, entertainment, pastimes.

We start by introducing ourselves, explain what we are studying, who authorized us to make the study, and what we want to discuss. Then we go into our ice breaker. Five minutes to talk about sports or the weather may seem a waste of time. If the interviewee's reactions are nervous, we may want to cut this part short. But generally five minutes or so can pay off in letting the person feel more relaxed, making the whole interview more productive.

It is a good idea to take notes during an interview, not just so we'll remember the details, but so the interviewee will feel confident that what is being said is sufficiently important to make it worth writing down. However, sometimes it is a good idea to ask permission to take notes. It is always a good idea to ask permission if we intend to tape record the interview.

We should be prepared with specific questions to which we want specific answers. But we also want to be prepared to hear answers we didn't expect. If an answer gives us a lead that will help solve our problem, we ask questions to explore it further. But we don't let the line of questions and answers drift in directions that will not be productive to our investigation.

We may want to ask the interviewee for an opinion on what the problem is and how it should be solved. We must listen. We don't discuss how we think it should be solved. And we don't argue. If the response seems ridiculous, we let it stand and don't comment. If it makes sense, we may have gotten a valuable insight. If the interviewee gives us an idea that we can use, we should be prepared to give him or her credit. But be discreet. If the suggestion is to fire the boss, the interviewee may not want to be given credit. If we don't know whether to assign credit, we ask.

If the interviewee cannot answer our questions, often he or she can suggest how we can get an answer, who we might talk to, or what document will give that information. If our source recommends that we talk to an associate, ask for an authorization to be conveyed to the associate so that he or she will be prepared to talk to us. If possible, we get the interviewee to introduce the associate to us, or ask the interviewee to tell the associate that we will be contacting him or her.

We should beware of interviewees who cannot say they don't know. They are dangerous because they will make up answers to save themselves from having to admit they don't know. You cannot trust what they say. Some interviewers test interviewees by asking a question they know the respondent cannot answer.

Often the person interviewed will have documents or records that would be of value to the study. We can ask for a copy for our files. If we are looking for procedures that require forms, we can get a copy of a blank form, and if possible, a sample of the form as it has been filled out.

After the interview we will want to organize our notes, record in our log who we interviewed and when, what we learned, and what conclusions we drew.

We should write the interviewee a thank-you letter. We may want that person's help again. The letter is a courtesy. We summarize exactly what we understood from the interview and invite the interviewee to correct any mistakes in interpretation. However, we must keep in mind that some things may have been said in confidence and may upset our source if they appear in print. We must use our judgment. If we think something is being said in confidence, we should ask how we may or may not use that information.

Questionnaires

Questionnaires are used when there are many people from whom the same information is required. The questionnaire is used only after it has been determined exactly what information is required and how the responses can be processed and interpreted.

Trial interviews should be conducted first. In a trial interview one can get open responses and ask questions to explore where there may be possibilities for misinterpretations of the questions, what responses are likely, and what they may mean. A good questionnaire can take a great amount of hard work to develop.

Before the questionnaire is put out, we must consider what types of conclusions we want to be able to draw, how we will analyze the results, what statistical techniques we will use, and how we will enter the data and process it on the computer. All these concerns can affect the design of the questionnaire.

Observations and Tests

If existing documents and our interviews don't yield the information we need, we may have to use either observations or tests.

We should be very careful about any observations or tests that may interfere with the ongoing work. First, the interference may distort the data we want. Second, it may interfere with people getting their work done.

Authorization for the observations or tests must be obtained from those who are responsible for the people or equipment involved. We must also get the cooperation of the people who will participate in the test, and we must convince them that it is worth their while to allow whatever interruptions will occur to their work.

Models for Organizing Information

There is a difference between collecting facts and understanding what they mean. Fact collection is an iterative process. The early facts that we collect and understand will guide us in the collection of further facts.

To understand the facts we have gathered we need to develop a model that will organize this information so that we can make sense of it.

In Chapter 3 we studied several types of models and methods of description that will be valuable to us in organizing the data we collect about the system: data flow diagrams, data element dictionaries, and annotations.

As we collect this information we may wish to put it in a computer data base. The data flow diagram and data element dictionary can be stored as a data structure matrix. The annotations are entered with a word processor. We may enter and organize our notes using an outline processor (that is, a tree editor).

We will now introduce another type of model we did not consider in Chapter 3, the cause and effect graph.

Cause and Effect Graphs

Once we know what behaviors of the system we don't like and wish to change, we then seek to determine what causes these behaviors and which of these causes we are able to affect. Thus, the **cause and effect graph** helps us decide what things we might change (causes) that will produce the desired changes in the behavior (effects).

As we gather information, we find we have the descriptions of many conditions that exist. We want to put these conditions together to see what the relationships are between those things we can change (our controllables) and those things we want to change (the object behaviors).

As we collect information we will represent various conditions as nodes in a graph. We indicate that one condition causes another with an arrow from the cause to the effect. (See the discussion of graphs in Chapter 7.)

We can make use of the transitivity of cause and effect, by which we mean:

If A causes B $\qquad A \longrightarrow B$

and

If B causes C $\qquad B \longrightarrow C$

then we can imply that

A may cause C $\qquad A \longrightarrow C$

If we have a path in our graph from X to Y, we might assume that if we change X, then Y would change. But we must be careful not to imply too much. Simply because we find a path from X to Y in the graph does not always mean that we can change Y by changing X. And if we can change Y by changing X, it may be that the change we are able to make in X will not produce the change we want to make in Y. But if we do find a path, it suggests a possibility that we can then consider further. So we are looking for paths from conditions we *can* change to conditions we *want to* change.

As we saw in Chapter 7, we can represent the same information we see in a graph also as a matrix. This is convenient because the matrix can be stored in a computer data base.

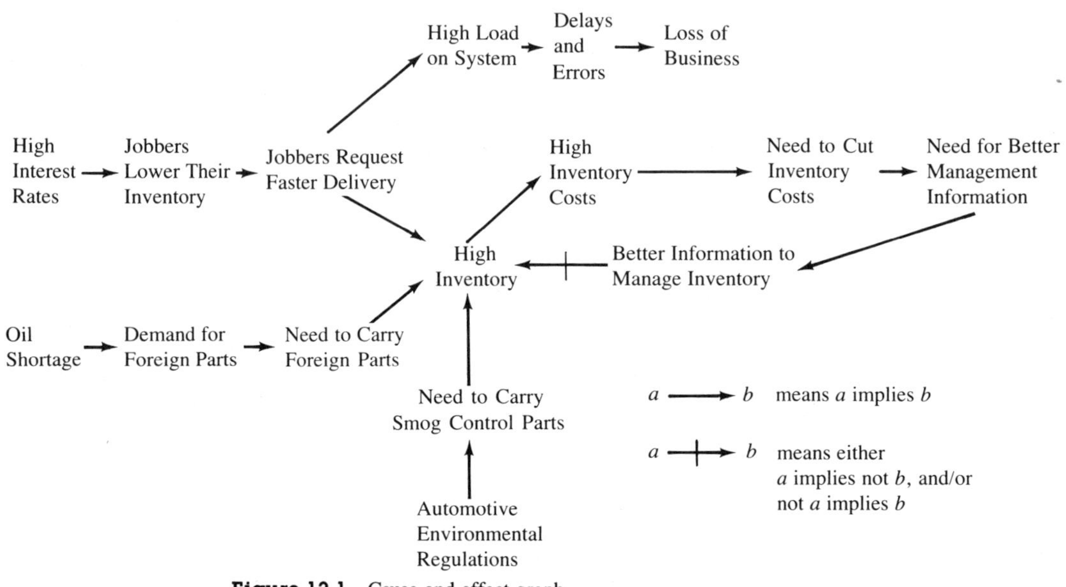

Figure 12.1 Cause and effect graph.

The best way to see how a cause and effect graph might be used is to apply it to an example such as the case study in Appendix A (see Figure 12.1).

We note that we can trace the delays and errors that have raised complaints back to the oil shortage. It was not within our power then to have solved either the oil shortage or the demand for foreign parts that resulted. But we are in control of the decision of whether we want to carry foreign parts. If we don't carry foreign parts, we may lose business because there is an accompanying lower demand for the domestic parts being carried.

We notice that high inventory plays a central role, and there are several conditions that cause a high inventory. Perhaps we can find ways to handle that high inventory more effectively by improving how we manage the inventory. That requires better information about the inventory. This reasoning causes us to add new conditions to the graph such as "Need to Cut Inventory Costs."

Generally, working with a cause and effect graph will help us organize what we know about the system so we can decide what we want to change.

Other diagrams and lists can also help us organize our collection of data.

Transaction Tracing

A transaction is a set of input data and its transformations as it moves through the system. A data flow diagram is a static description of a system; it shows each process and how it transforms data, and how these processes are connected together. But the data flow diagram does not give the dynamic perspective of what happens to a transaction as it moves through the system. This perspective can uncover many problems. It is a good idea to develop several typical transactions including transactions that represent likely input errors. We use the data flow diagram to trace what happens to these transactions as they flow through the system.

Scenarios

A **scenario** is a play-by-play description showing an example of how the system would be used. It is an excellent way of thinking through the use of the system to find problems that may have been overlooked, or of communicating with the client to gain a mutual understanding of what the system will be and how it will be used. A scenario is a verbal simulation. It can be written in the form of a narrative, or as a play script with the name of the actor in the margin next to what that person does.

Let us illustrate a scenario by using it to describe the operation of an Electronic Office Coordination System (EOCS) in the handling of insurance claims. (see Figure 12.2). EOCS is a way of using an electronic office to keep track of tasks. We may assign a task to be done. When the task is completed, we will be notified. If the task is not completed by an expected time, we will be so notified. EOCS saves a person from having to keep in mind all the various pending tasks that have to be monitored. It can also be used to keep track of tasks that we assign to ourselves.

You can see from this partial scenario that systems analysts can get a good idea of how the system works, what information has to be maintained by the system, what information should appear on the screens, how the users will use it, and so on. The more imagination we can use in coming up with both the usual and unusual cases, the better we can do at anticipating all the needs of the system.

Organization Charts

The **organization chart** is a graph, usually a tree, showing what position reports to what other position. The boxes in the chart can give the title and the name of the person holding that position. Sometimes it is useful to show the names of each person who has held that position and when they held it.

Events Lists or Chronologies

An **events list** shows the occurrences that are important to the study and when they occurred. It should be sorted by date.

Expectations Lists

As we interview people, certain expectations become apparent. An **expectations list** keeps track of what those expectations are and who has them.

Sally: 9:30 A.M. Sees a task message on her screen requesting her to handle a claim that has been received by claims entry and entered into the system.

"Task B65 / Please process this claim. See files B65.CLM and B65.POL / from Claims Entry."

She accepts the task and replies by entering:
Accept
Begun
Planned Completion: 10:30 A.M.

She reads the claim by pulling up on her screen file B65.CLM and looks over the policy by pulling up B65.POL. She finds there is a problem she does not know how to handle, so she addresses a task to Sam.

Task ID B65.1: to Sam / Reference Task B65. / Please look at this claim and advise on how to handle it. Notice that the claim is for a death. The insured died apparently of natural causes while a passenger in the car, but the car with the body was stolen before any death warrant could be issued. / Could you get back to me by 1:30 P.M. Thank you. / Requested Completion 1:30 P.M.

She also changes the Planned Completion time for her task, B65. She enters:

Reset Completion of B65 to 3:00 P.M.

Sam: 1:05 P.M. Sees and reads the task message for Task B65.1. He enters:

Task B65.1:
Accept Task
Reset Completion to 2:00 P.M.
Message: Didn't get your task until 1:05 P.M. Will get back to you by 2:00 P.M.

Sally: 1:37 P.M. Sees on her screen that the Planned Completion has expired. She checks on Task B65.1 and sees that Sam has reset his expected Completion to 2:00 P.M. She finds that acceptable.

Sam: 1:48 P.M. Responds with:

Task 65.1 / Message: You will need to make out a form C322 stating the circumstances of the death and the reasons a death certificate was not issued. You will need to have someone who is not a beneficiary to certify the circumstances of death. / from Sam.

Figure 12.2 Scenario.

Open Questions Lists

As some questions are asked, other questions occur. An **open questions list** just keeps track of those questions that have not yet been answered.

What to Look for

As we organize the information with our models and make our interpretations, we must keep in mind what we are looking for. Generally there are certain primary things to look for. There are other secondary things we should be prepared to recognize if they should arise. Before we begin we should list these primary and secondary objects of our analysis. We should be prepared to ignore information that does not pertain to our list; otherwise we'll waste our energies going in too many directions.

Primarily we are looking at the system to determine why it is not meeting its expectations. So we list those expectations that are not being met and seek to find out why.

If our expectations say that the system should process transactions faster, we look for what now limits how fast they are being processed. If our expectations say we should not be losing so many transactions, we look to see where transactions are being lost.

As we build our models and analyze the data with them we will recognize the need for additional information. Gathering and analyzing information is an iterative process. We may not know what information we need until we have done some analysis. The analysis tells us not only what information to gather but what not to gather so we don't waste time collecting more information than we need or can possibly analyze.

Each expert has a list of things to look for in analyzing certain types of systems. We should develop our own lists. Here are some suggestions.

For the product that is produced by the system:

- Does it meet its expectations?
  If it interfaces directly with the client's customer then:
    - Can the customer use it easily?
    - Does it give the customer what he or she wants?
- What properties of that product are essential?
  - How it looks?
  - How it behaves?
    - Speed, timing?
  - Ease of use?
    - Is it consistent and simple?
      (See discussion of software-people interfaces in Chapter 9.)
- What properties of that product are not essential—for example, need it look good, or must it only behave correctly?

For each process that makes up the system we might ask:

- What is the purpose of this process? How does it contribute to each of the essential properties of the product?

- Need this process:
    - Be done at all—perhaps it doesn't contribute to an essential property of the product?
    - Be done better—either to make a better product or to reduce the cost or time involved in another process?
    - Be not done as well—with a cost or time saving without compromising the end product?
    - Be not done for all transactions—for example, can only a few be checked instead of checking them all?
- Can some transactions be detected as requiring special consideration (like errors) and be sidetracked to keep them from causing a logjam on the main track?
    - Have all contingencies or exceptions been considered so we know what to sidetrack?
- Can something be saved now so that it needn't be done again?
    - Would it be better not to save something because it can easily be done again?
- Should we check or test:
    - More often so that errors are caught sooner?
    - Less often to save time and cost?
    - Are the tests made at the right place?
- Is there unnecessary duplication?
    - Should there be more duplication to check against error or protect against fraud?
- Can we combine it with another process so they can be done at the same time or with the same equipment?
    - Should we select out part of a process to be a separate process that can be done at the same time or on another machine?
- Could it be done more simply, say using common means such as rubber bands instead of special means like harnesses?
- Is the right equipment being used to do it properly, and is this equipment not unnecessarily expensive?
    - Are the right supplies being used? Can we use cheaper supplies, or would more money spent on better supplies make a better product or produce savings in other processes?
- Are things being prepared properly so they don't cause delays in later processes?
    - Will more time or care in this process save time in a later process?
- Should there be more maintenance to prevent downtime or deterioration of quality?
    - Could it be done with less maintenance with a consequent cost saving?
- What processes are on the critical path, and what is the bottleneck in the critical path?
    - Will more resources put here speed things up?
- Is the best technology being used?

    - Would a newer technology do it better?
    - Is the technology used too new, too unreliable, or too risky?
- Are the people who operate the process:
    - Properly trained?
    - Properly motivated?
    - Equipped with proper procedures and guidelines to turn to when there are questions?
    - Able to turn to supervisors when they need to resolve special cases?
- Would better machine-human interfaces:
    - Make the process more efficient?
    - Make the process more human, less fatiguing, more interesting?

Our job is frequently made easier by recognizing Pareto's law, which we discussed in Chapter 9. This law says that only a small fraction of the actors usually account for the major part of the effect we are concerned about. Pareto noticed that a small number of people in Italy at that time held most of the wealth. In our case, for example, when detecting input errors, we usually find that a small proportion of the transactions account for most of the errors. Or the total time to process transactions may be dominated by the time to process only a small proportion of all the transactions.

Pareto's law often allows us to reduce the size of the problem so that all we need do is isolate and analyze the small number in the population with which we really must be concerned.

When we find an error or some exceptional case, we may take a long time to handle it. During that same time we could be processing many normal transactions. Usually instead of tying up the processing of these normal cases, we shunt the abnormal one off to the side to give it special attention where it won't hold up the mainstream (Matthies, 1968).

The Preliminary Study Report

The systems analyst writes a preliminary study report on his or her findings. The outline of this report is a guide to the information to be obtained during the preliminary study.

Outline of Preliminary Study Report

1.0 Overview

Who is the client?

What is the pertinent system?

Did this begin as a:

Request to solve a problem beginning with a need, or a

Proposal to profit from an opportunity beginning with a means?
What is this problem or opportunity?
What event triggered this request or proposal?

2.0 Who's Who (Name and Organization)
Who is making the request or proposal?
Why is he or she making it?
Who will use the new system?
Who is the client who will pay for it?
Who recognized the problem or opportunity?
When did they first recognize it?
Who thinks it can be done?
Why does he or she think so?
Who will be affected:
a. By not doing it?
b. By doing it?
Who has control over the needed resources?
Who is to say how it will be done or constrain how it will not be done?
Who has to be convinced for this project to fly?
Who supports this system change?
Who will fight this system change?
(These last three items should be considered but might best be left out of the document.)

3.0 Scope and Environment
Which systems are included and which are excluded?
Which systems will be changed?
Which systems will not be changed?
Which systems will be affected and how?
What are the interfaces between these systems?
What conditions (environment) affect the operation of the system?

4.0 Existing System or Situation
Description of the system.
Performance of the system. (See Chapter 8 for performance variables.)
External description of the system—that is, data flow diagrams and data element dictionary.

5.0 The Old Expectations and How They Failed
What are the expectations?
Whose expectations are they?
How does the present system fail to meet these expectations?
What are the symptoms?
Are the symptoms valid?
What are the causes of these symptoms?
What are the costs and consequences of the problem or of not taking advantage of the opportunity?
Who would these consequences and costs affect?

What are the estimated benefits and detriments?
Who gets them?

6.0 Goals and Objectives—The New Expectations

What are the objectives—that is, what is to be achieved by the system change?
What should the new system do?
What should the new system not do?
Why should it do it?
How well should it do it?
How can you determine whether it does or does not do it?
Which features of the old system should be retained in the new system?
Which features of the old system should be changed in the new system?

7.0 Constraints and Capabilities

What can and cannot be changed? (controllables)
What resources can and cannot be used?
Who can and cannot be used in this project?
What organizations must we work within?
What are the policies and procedures of these organizations?
What are the ground rules about what one can do and cannot do in:
a. Building the system?
b. Using the system?
What laws, procedures, management mandates, and so on must be followed?
What previous experience does management have that will affect their predisposition toward this project?
What are the dependencies on other projects?
What external constraints exist—for example, environmental policies, government regulations, and so on.

8.0 Conceivable Approaches

For each approach:
Describe the approach
What experience or precedent exists for this approach?
What reference system do you have that represents this experience or precedent?
How does this experience or precedent relate to our situation?
How can we model this approach?
Is this approach feasible?
What would it cost to develop this approach?
What would it be worth to have used this approach?

9.0 Proposal

Which approach is recommended?
Why? (Show a comparison.)
What should be done and when? (proposed schedule)
Who should do it?
What estimated costs and resources will be required?

In the later phases (requirements and feasibility) we will be expected to improve on some of the estimates and schedules shown here. By that time we should have more detailed information and experience available on which to base these estimates.

Project Requests and Proposals

After the preliminary study report is reviewed by management, they may wish to initiate a project. A project may require a **project request** of another group to do work, or a **project proposal** to another group that we do work for them. Outlines that might be used for these two documents are as follows:

Project Request
- Identify:
  - Who is making the request.
  - The person of whom the request is being made.
  - The scope of the request—that is, what is to be included in the system or work to be done and what is to be excluded.
- Define the problem as you see it.
- Propose a likely solution if you have one in mind.
- Relate the solution to the capabilities of whom the request is being made.
- Define:
  - Proposed plan.
  - Limits on cost and schedule.
  - Proposed contract provisions.

Project Proposal
- Identify:
  - Who is making the proposal.
  - To whom the proposal is being made.
  - The scope of the proposal.
- Define the problem—that is, your understanding of the need the client has.
- Define the solution you propose.
- Relate the solution to your capabilities:
  - What similar projects have you done?
  - What indicates your organization can be relied upon to get the job done?
- Define:
  - Proposed plan.
  - Cost and schedule.
  - Proposed contract provisions.

The structure of the two documents is rather similar. In each case we want to identify those items needed to make a match between the party doing the work and the party for whom the work is done. In the project request the problem should be well defined, but the solution will only be a concept. The project proposal indicates our concept of what need the customer has and shows how our solution will satisfy it.

A common practice is to make a request in two stages: 1. make a request (called a **request for proposal** or **RFP**), 2. receive answers to the request in the form of proposals. The RFP may go out to a number of potential respondees and the best response is chosen from the returned proposals.

Summary

The first step in any system change is to analyze the current system to decide whether we really want to make the change. We look at what meets expectations and what doesn't. Part of this analysis includes examining what the change would mean. We define the change, simulate it, ask ourselves if we like it, and gauge whether it's worth the cost.

When someone recognizes that a system's behavior doesn't match the expectations of someone of importance, the organization asks a systems analyst to conduct a preliminary study and to write a preliminary study report. The preliminary study attempts to define the problem, to determine whether it exists, to establish the scope of the system, its environment, the expectations to be met, alternative actions that might be taken, and their consequences. The systems analyst also compares the desired system to similar systems that may exist at other organizations.

As they collect information in the fact-finding phase of the preliminary study, systems analysts maintain a log of research activities and a file of assembled documents. Internal and external documents may be consulted, including organizational charts, standards and procedures manuals, preliminary study reports on the current system, operations logs, journal articles, and vendor's literature.

Interviews offer opportunities for the systems analyst to obtain information and to get the cooperation of the interviewee in the planned system. The analyst must, however, do some homework before the session so that the respondent's time isn't wasted and so the dialogue addresses the important issues. Questionnaires, observations, and tests can be used for fact finding when the required information is fairly clearly defined.

A number of models, including data flow diagrams and data element dictionaries, can help to organize information once it has been collected. Cause and effect graphs, another model, show what events cause what effects. Analysts may also wish to organize their information with transaction tracings, organization charts, events lists, and expectations lists.

An important part of organizing information is knowing what to look for. Analysts can focus their attention by drawing up a list of primary and secondary objectives and then ignoring what lies outside these lists. Pareto's law, which says that a small fraction of things account for a major part of the effect, lets us narrow the field to the relevant matters.

The systems analyst reports his or her findings in a preliminary study report. This document covers who is involved in the system change, the scope and environment of the system, the current system, old and new expectations, constraints and capabilities, possible approaches, and a recommendation.

After management has reviewed the report, they may issue a project request to describe what they want done and how it's to be accomplished. Correspondingly, groups offering to do the work describe their proposed contribution in a project proposal.

Exercises

1. Make up each of the following for the CAPERT case study in Appendix A:
   a. Organization chart showing when each person held each position.
   b. Events list.
   c. Expectations list.
   d. List of objectives.
   e. Open questions list.
2. Make up a data flow diagram with annotations (see forms in Chapter 3) for the CAPERT case study. Where do you see problems that require more information? (Look for processes without adequate inputs, or outputs that are not used.)
3. Develop the cause and effect graph showing how the recession ultimately led to selling through chain stores and a higher load on the sales staff.
4. Make up several typical transactions and follow them through the system to see what happens to them.
5. Use a DFD from Chapter 3 or from the CAPERT case study, choose durations, and make a critical path analysis of how long it takes for a transaction to move through the system. Note that it will take different amounts of time for different conditions the transaction encounters.
6. Propose a specific project for CAPERT that seems reasonable in scope considering the resources and constraints present. Justify your proposal. (This project need not accomplish all the corporate goals, but it must provide substantial progress toward these goals within the time constraints given. Identify what corporate goals will and will not be met within that time.)
7. Propose the composition of a steering committee by listing individuals according to their position and potential contribution to the project. Indicate what type of people from which groups should be on the task force.
8. List, in order, the first seven people in CAPERT's organizational chart to be interviewed regarding the order entry and inventory control systems. Show what information you intend to be able to get from each of them.
9. Prepare a short (half page) memo to all employees of the inventory control

department outlining the purpose of the current investigation into the order entry and inventory control systems in preparation for interviews.

10. Interviews have been arranged with Curtis Billings, Director of Inventory Control for the Chicago Distribution Center, and with Justin Martin, Director of Data Processing Operations for the Trenton Distribution Center. For each interview prepare a list of at least seven questions. Prepare some sufficiently open-ended questions to provide the opportunity for the interviewee to express his or her feelings and opinions.
11. Write a preliminary study report for the CAPERT case study.
12. Make changes to and complete the scenario for the EOCS given in this chapter. Develop other scenarios for other ways of using EOCS showing how the system would work. Do you think that these scenarios would give you a good understanding of the system before you were to design it?

References

Cougar, J. Daniel, and Robert W. Knapp, eds. *Systems Analysis Techniques*. New York: John Wiley & Sons, 1974. (Part II: Techniques for Analyzing Systems)

Davis, William S. *Systems Analysis and Design: A Structured Approach*. Reading, Mass.: Addison-Wesley, 1983.

Forrester, Jay W. "Industrial Dynamics: A Major Breakthrough for Decision Makers." *Harvard Business Review* 36, 4, July–August 1958, 36–66.

Gause, Don C., and Gerald M. Weinberg. *Are Your Lights On?—How to Figure Out What the Problem Really Is*. Boston: Little, Brown, 1982.

Gore, Marvin, and John Stubbe. *Elements of Systems Analysis,* 3d ed. Dubuque, Iowa: Wm. C. Brown, 1983.

Kowalski, Robert. "AI and Software Engineering." *Datamation,* November 1984.

Matthies, Leslie H. *Basic Systems Techniques*. Colorado Springs, Colo.: Systemation, 1968.

Osborn, Alex F. *Applied Imagination: Principles and Procedures of Creative Problem-Solving,* 3d. rev. ed. New York: Charles Scribner's Sons, 1963.

Parkin, Andrew. *Systems Analysis*. Cambridge, Mass.: Winthrop, 1980.

Polya, George. *How to Solve It*. Garden City, N.Y.: Doubleday, 1957.

Robertshaw, Joseph E., Stephen J. Mecca, and Mark N. Rerick. *Problem Solving: A Systems Approach*. Princeton, N.J.: Petrocelli, 1978.

Rubinstein, Moshe F. *Tools for Thinking and Problem Solving*. Englewood Cliffs, N.J.: Prentice-Hall, 1986.

Rudwick, Bernard H. *Systems Analysis for Effective Planning: Principles and Cases*. New York: John Wiley & Sons, 1969, pp. 58–62.

Silver, Gerald A., and Joan B. Silver. *Introduction to Systems Analysis*. Englewood Cliffs, N.J.: Prentice-Hall, 1976.

Semprevivo, Philip C. *Systems Analysis: Definitions, Process, and Design,* 2d ed. Chicago: Science Research Associates, 1982.

Taggert, William M., Jr., and Marvin O. Tharp. "A Survey of Information Requirements Analysis Techniques." *ACM Surveys,* 9, 4, December 1977, 273–290.

Thierauf, Robert J., and George W. Reynolds. *Systems Analysis and Design,* 2d ed. Columbus, Ohio: Charles E. Merrill. 1985.

Weinberg, Gerald M. *The Psychology of Computer Programming*. New York: Van Nostrand Reinhold, 1971.

Weinberg, Gerald M. *Rethinking Systems Analysis and Design*. Boston: Little, Brown, 1982.

13

Specifying Requirements

- The requirements specification describes the system's environment, function, performance, logistics, and cost and schedule.
- Clients should participate in the development of the requirements specification so that they understand and are satisfied with the system they are to receive.
- Requirements specifications describe the product from the outside; design specifications describe the inside, or how it works to make the product behave as it should.
- The requirements specify not only the logical aspects but also the interface to physical equipment. They concentrate on the product, not the process.
- Effective requirements specifications are unambiguous, verifiable, traceable, modifiable, and consistent.
- User's manuals are written to explain to users how they can utilize the system. The user's manual can be included as part of the requirements specification.

During the analysis of the current system we defined the problem (or opportunity), determined whether it really existed, and estimated what it might be worth to solve it. Now in the requirements analysis phase we will define what the product must do if it is to solve that problem or realize that opportunity.

The requirements analysis phase ends with having produced the requirements specification, which defines the behavior of the product required to solve our problem. A preliminary user's manual may often be included at this time to show how the user will interact with the product.

The client may look at an early version of the requirements specification and say: "This is not what I want. But using this I can show you what I do want." Then we modify it. We may go through this process several times. Eventually the client looks at it and says: "Yes, that's exactly what I want. I don't care how you make it; as long as that's what it will do, I will be satisfied."

We can define the role of the requirements specification as follows:

> The Requirements Specification shall clearly and precisely describe the essential functions, performances, design constraints, attributes, and external interfaces. Each requirement shall be defined such that its achievement is capable of being objectively verified by a prescribed method, for example, inspection, demonstration, analysis, or test. (IEEE ANSI, 1981)

The requirements specification tells clients what they are buying. It's like showing them a picture in a catalog of what they will get if they place their order. It is the contract between clients and the systems developer that tells what the clients will get for their time, effort, and money. The premise of specifications is: If what you get is what you specified and you are unhappy, it is your fault.

We have referred to requirements and design as a recursive process involving a continuous development of new detail. To relate this to the more classical view of requirements as a customer-oriented activity that occurs before design, we will take the position that requirements are the depth into the recursive process that the client is willing to become involved.

The information in the requirements specification can be put into the following categories:

- *Environment*—what system is our system part of, what system does it interface with, who operates it?
- *Function*—what does it do, what inputs does it do it to, what outputs does it produce, under what conditions does it do it, and how do you control what it does?
- *Performance*—how fast and accurately does it do it, how much data and how big a problem does it handle?
- *Logistics*—how is it operated and maintained, how reliable is it, what controls does it have against bad data and misuse?
- *Cost and Schedule*—there is some difference of opinion whether cost and schedule are legitimately part of the requirements specification. Clients want to see cost and schedule as well as the other requirements to make a decision whether

they want to contract to get the system. Thus, cost and schedule might reasonably appear in the same document with the other requirements.

The first four of these (environment, function, performance, and logistics) are called external dimensions.

Client Involvement

The development of the product should not proceed until clients understand and sign the requirements specification, thereby acknowledging that it represents unambiguously and precisely what they want. There should be no question in the clients' minds that if that product is delivered and does what the requirements specification says, no matter how it happens to achieve it, they will be satisfied.

Clients tend to be loath to read complicated documents. Thus, they try to take the apparently easy route of substituting their trust in the systems developer for their own hard effort to understand the requirements specification. As the requirements specification is being written it is a form of communication. Once it is written, it is a contract between the clients and the systems developer. The clients' interest is not served if they say: "I can't be bothered. You just write me something that says what I think."

The systems developer's and client's interests are served only if the requirements specification is complete and the client understands it thoroughly. The best way to obtain the client's complete understanding is to get him or her to write it, or at the very least to get the client to participate actively in its writing. If the client is not prepared to make this effort, the systems developer should consider very carefully whether it is worthwhile working on this project. Projects in which the client gets involved have a better chance of successful completion and reflect better on the systems developer.

It is a common feeling among clients that by some magic the systems developers will know what the clients want, thus saving the clients from the hard effort required to get on paper exactly what they want. It is also a common feeling among systems developers that they know better than clients what the clients need. But if the clients do not participate in the development of the requirements specification, the results are usually disastrous.

To get the clients' participation it is necessary to have a set of methods that make it easy for them to represent and understand what the system will look like and what it will do for them when it is built. (We made a point of the importance of simple methods of representation in Chapters 3 and 4.) The systems developer may have to teach these methods to the client.

Clients are usually under a great deal of pressure at the time the requirements specification is written. The system is being built or changed because they have a problem; something doesn't work as it should. Clients want the new system to solve the problem and reduce that pressure. They are probably putting in overtime to work around the problem until the new system fixes it. Now is a very difficult time to ask them to learn a new language and discipline in order to define the system they need. Thus, the most important time to get their commitment is also the most difficult time to ask them for that commitment.

Clients may have the money to do the project. They may have the staff and the resources. But at this point the most precious commodity is the commitment of their time and their thinking. That is impossible to buy. It may not be possible for anyone else to do it for them. It is another item on an already crowded agenda of things to worry about. But making that commitment is the cost of developing a successful system.

The requirements specification should clearly state what the system will do and how we can tell whether it does it. It should also state the bounds on the cost and time to deliver the product. As a working document the specification is used to communicate between the systems analyst and the client. When completed, it represents a contract that states that the systems developer will produce the system for a given cost to be paid by the client.

Properties of the Requirements

A number of general properties characterize requirements specification. In this section we'll look at five areas in which the requirements are defined.

External versus Internal

We say that the requirements specification is an **external specification**. It describes what we see of the product from the outside—that is, how it behaves and what it will cost. To request something to solve our problem we need to know only what it does, what it will cost, and when it will be available. We don't need to look inside to see what makes it do that.

The design specification, which we consider in Chapter 15, describes what is inside that makes the product behave on the outside as specified by the requirements specification. Thus the design specification will be referred to as an **internal specification**.

The external-internal convention makes a clear, logical, and convenient distinction between the requirements specification and the design specification. The requirements specification is done by someone outside the box, and the design specification is done by someone inside the box. So the requirements specification is the interface between the person outside and the person inside. All the communication between them must be documented by that requirements specification.

Since external refers to what the client sees, any significant change to an external document must have the client's approval. Changes to internal documents do not need the client's approval, provided the changes have no external effects.

As we design and implement the system, we may discover possible changes that we believe would be inconsequential to the client but could save him or her money. Such suggestions really should be reviewed with clients. Sometimes we try to avoid bothering clients by thinking for them. But when we do that, we must recognize the risk.

If we propose a change and get it approved, we should get the approval in writing so there is no misunderstanding later. (This is the subject of change or configuration control.)

Not Overconstrained or Underconstrained

Those who set the requirements should not set any unnecessary constraints on those who do the design. The designers should be given the greatest possible freedom to satisfy the requirements in the best way they know how.

In Italy, if we ask for ice cream, we may be disappointed. But if we ask for something cool and refreshing, we leave open the possibility of getting spumoni, which we may find very satisfying.

When clients, with the aid of systems developers, make requirements specification, they should be prepared to accept anything that meets the specification. Designers should be allowed to use their best judgment about how to meet specifications with the least cost and greatest reliability.

If designers can give us something that they can show meets the requirements specification, but it does not satisfy us, then it is our fault for having written an incomplete requirements specification. Don't blame the designers.

To test for the completeness of our requirements, we should play the game of seeing whether we can think of anything that meets the requirements yet would not satisfy our client. If we can, then the requirements must be made more specific. Each requirement should be considered carefully for its intent and justification. This is particularly true of the performance specification—for example, why is it necessary that the response time be less than .5 seconds? Otherwise one might specify requirements that are more restrictive than are really needed, unnecessarily adding to the cost and schedule to develop the product.

Should Requirements Suggest the Design?

There may sometimes be advantages of suggesting a design that would satisfy the requirements. If you specify that you need so many square feet of living space provided as cheaply as possible, you might end up with something like a long extruded Quanset hut with no windows. Stating that you want a house would be a help by bringing to mind many requirements that you might otherwise have taken for granted or forgotten to specify—like having windows.

Thus, in the requirements specification, we may suggest a specific design to imply some of the requirements, but we should not require that it be designed that way. We must make it clear what is a suggestion and what is a requirement.

Some people include a high-level data flow diagram as part of the requirements specification. Technically we might say that a data flow diagram is properly part of the systems design, not the requirements. However, we must admit that sometimes it is appropriate for the design to intrude into the requirements to make them clearer, just as earlier we found it appropriate to specify "house" as a means of illustrating what we wanted.

When elements of design are included in the requirements, it should be to illuminate, not to restrict.

Logical versus Physical

Logical refers to the information; physical refers to the equipment or people. Since the purpose of an information system is to obtain useful information, the requirements on that system are primarily logical.

It would make it simpler if we could classify the requirements specification as strictly a logical document. However, the requirements pertaining to external interfaces must often be physical. If the specified system has to interface to a specific communications system or external piece of hardware, or the client wants to use a specific display terminal, the requirements for the system must specify the physical characteristics of those interfaces.

In Chapters 3 and 4 we discussed data flow diagrams and annotations for the processes, data flows, and files. These can be either logical or physical depending upon whether we include information about the physical equipment used.

Product versus Process

Another distinction that is important in the definition of the requirements specification is the product-process distinction. The requirements specification defines the product that will solve our problem. It does not specify the process by which that product is made. The project management plan, which we discussed in Chapter 11, describes the process.

We must clarify several distinctions between what is product and what is process. How about costs? Should the cost be specified in the requirements specification as part of the product, or in the project management plan as part of the process? We take the position that the cost and schedule are properly part of the contract with the client and should be included in the requirements specification. The cost of producing or obtaining the product is also a constraint placed upon the project management plan.

Thus the requirements specification is the external, largely logical specification of the product, the design specification is an internal specification of the product, and the project management plan is a specification of the process.

Properties of the Requirements Specification Document

Several characteristics that a good requirements specification should have are explained in this section with the key words italicized. Note that these are characteristics of the document, not of the product that the document describes.

A requirements specification should be *clear and unambiguous*. Every requirement should have only one possible interpretation. The interpretation that clients have when they write or approve the requirements specification should be the same as the designer has when he or she designs it, and the person who tests it has when he or she tests the product to verify whether it satisfies the requirements.

To see how hard it is to use the English language to write unambiguously, consider the following examples (Freedman and Weinberg, 1982):

"The control total is taken from the last record" may be interpreted as either:

1. The control total is taken from the record at the end of the file.
2. The control total is taken from the latest record.
3. The control total is taken from the previous record.

"All customers have the same control field" may be interpreted as either:

1. All customers have the same value in their control field.
2. All customer control fields have the same format.
3. One control field is used for all customers.

"All files are controlled by a file control block" may be interpreted as either:

1. One control block controls all files.
2. Each file is controlled by its own control block.

The requirements specification should be *complete*. If a product meets the specification but disappoints your client's expectations, the requirements were not complete. It is hard to ensure that the requirements are complete. The requirements tree, which we'll discuss later in this chapter, is a discipline to help develop a complete set of requirements.

The requirements specification should be *verifiable*. That is, it should be possible for a third party to read the requirements specification and test the product to determine whether or not that product satisfies the requirements. Could anyone be expected to determine objectively whether the response was "fast" or "fast enough"? The requirements would have to say something like "It responds in less than one second from the carriage return until the response begins printing on the screen." Thus a performance specification should be *measurable*. If there is any possible question about how to make the measurement, then this needs to be spelled out too.

The requirements specification should be *traceable*. One should be able to begin with a requirement and trace it forward all the way through the design and into the implementation to see that the product will satisfy that requirement. It should also be possible to start with the implementation or design specification and trace it backward to the requirement that made it necessary.

Clients can change their mind, and much to our consternation often do. Or once we get into the design, we may realize that some requirements cannot possibly be met, or it would cost too much to meet them. It is often possible to go back to the client to renegotiate a requirement to reflect a practical way the product can be made, yet still be satisfactory to the client. These changes must be documented. Thus a requirements specification must be *modifiable*. Putting documents on a word processor makes them easier to change. Marking just the changed lines on the printed document makes it possible to review the change without having to reread the whole document.

The requirements specification should be *consistent*. It should not say one thing in one place and another thing in another place. If it says in one place the color is to be blue, then it shouldn't say in another place that the color is to match that of the something that turns out to be pink.

It is difficult to make a document both consistent and modifiable if it contains any redundancy. When a change is made, it must be made every place that is affected. Otherwise the change can introduce an inconsistency.

It is often useful to include some redundancy to make the document clear. But there should be cross references so that changes can be made in all the relevant places. These cross references can appear within the document itself such as "see Section 6.3.1." Or they can be in a separate document in the form of a table that shows that Section 5.2.6 and Section 6.3.1 contain redundant information.

External Dimensions of the Product

A product may be a whole system or only a module of that system. Each product has certain external descriptors that we use to specify the requirements for that product. We call these the **external dimensions** of the product. The external dimensions we use for information systems are:

Environment
- Environment—what is external to the system that can affect it?
  - Where and in what organization does the process run?
  - What are the conditions under which it runs that affect it?
- Interfaces, which we further define as interfaces with:
  - User
  - Communications systems
  - External hardware
  - External software
- Hardware required to run it

Software required to run it:
Operating systems
Calling procedures
Called procedures
Function
Function—what does the product do and why?
Output
Where is the output used and for what purpose?
Data elements that appear on that output
Format and layout of those data elements on the output
Input
Source from which the input comes
Data elements in the input
Format and layout of the data elements as entered into the system
Files and data bases used by the process that are external to the process
Process—the algorithms by which the outputs are defined in terms of the inputs
Control
By an interactive user
By batch input data
By operations
Handling of conditions that are either:
Normal
In error
Unusual
Performance
Performance requirements (that is, speed, size of problem, accuracy, and so on), how they are measured, and the criteria for their acceptance
Logistics
Security—how well must it be protected?
Reliability and maintainability—how easy must it be to maintain it and how infrequently?

Hierarchical Specification of Requirements

There are two approaches to developing the requirements of a complex system through the use of a hierarchy. One, we can use a requirements tree, as described in this section, to represent the functions of the system hierarchically. Or two, if we use data flow diagrams in a requirements specification, we can make use of the hierarchy in the data flow diagrams (see Chapter 3) to break the system into modules. We'll discuss each of these approaches in turn.

Functional Requirements Tree

Part of the requirements of a system is a statement of the functions it must satisfy. These functions can be developed in the form of a tree. At the root of the tree the function of the whole system is stated. The branches off each function show the more detailed functions needed to perform this function. Developing the tree is a discipline that helps to develop a more complete set of functional specifications.

Sometimes we can treat several functions of the system independently as though at one time the system performs one function, and at another time it performs another function. The system may be used to make a request from the data base at this time; at another time it may be used to make a change to the data. Then the tree begins with a root, which names the whole system, and a limb showing a choice of these independent functions.

We will illustrate with an example that defines the requirements for a system whose function is to warn ships of icebergs. We will show it first in tree form (see Figure 13.1).

We can also show this same information in outline form (see Figure 13.2). In the outline we will expand the detail of one area so we can illustrate a point. We can maintain these requirement outlines on the computer with conventional outline processors.

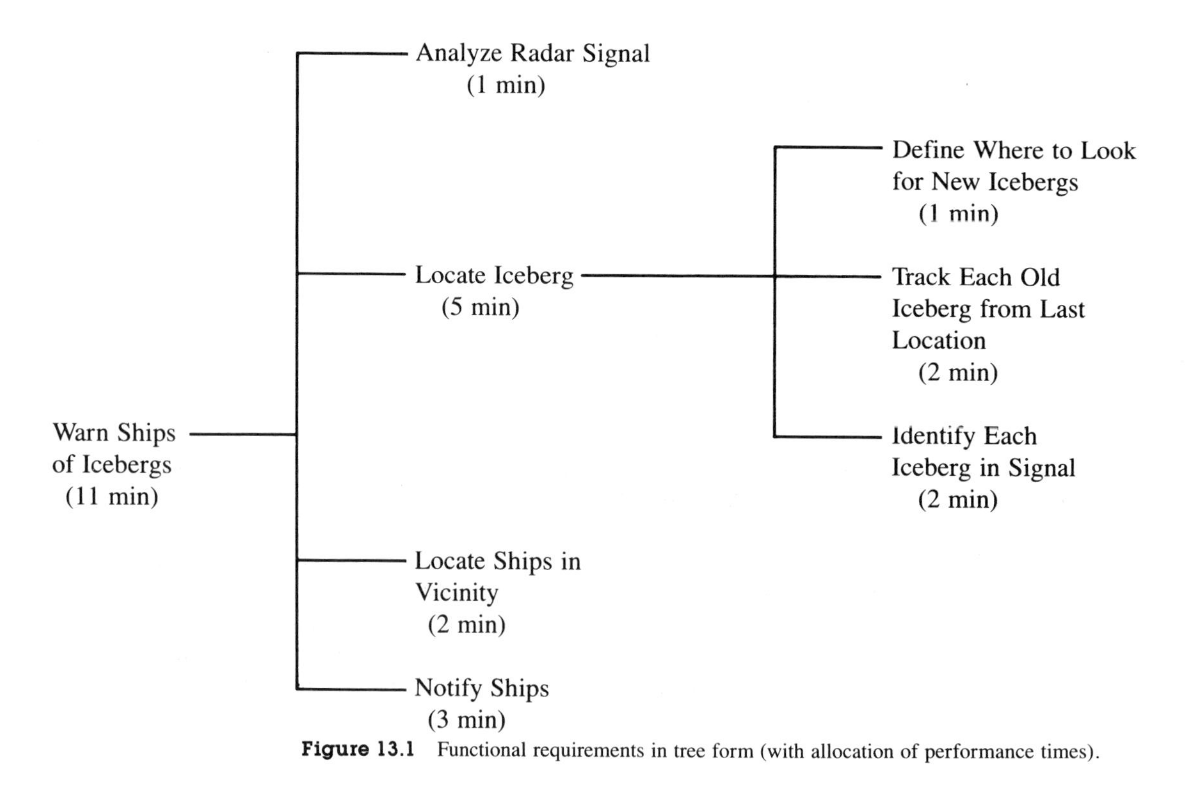

Figure 13.1 Functional requirements in tree form (with allocation of performance times).

0. Warn ships of icebergs (11 min)
1. Analyze radar signal (1 min)
2. Locate iceberg (5 min)
2.1 Define where to look for new icebergs (1 min)
2.2 Track each old iceberg from last location (2 min)
2.3 Identify each iceberg in signal (2 min)—for 75 icebergs:
2.3.1 Compute RMS match between received signal and stored template (.001 accuracy) (1600 msec)
2.3.1.1 Compute inner product between two vectors (.0005 accuracy) (800 msec)
2.3.1.2 Compare RMS with acceptance limit (500 msec)
2.3.1.3 If accepted, add coordinates to list (300 msec)
3. Locate ships in vicinity (2 min)
4. Notify ships (3 min)

Figure 13.2 Functional requirements in outline form.

Allocation and Trade-off

The detailed function on the leaves of the tree may specify the **allocation** of resources used by the product, such as execution time, memory space, and accuracy. Figure 13.2 shows the allocation of execution time and accuracy to the individual detailed functions. If one takes longer, another must be done faster. During the design it may be appropriate to trade off these resources between different leaves—for example, one function specifying less accuracy and less time means another function must have more accuracy but can be allowed more time (see Figure 13.3).

The assembly *A* is allocated 1600 milliseconds to do its function. It is assumed that the sum of the times taken by each of its components should add up to no more than 1600 milliseconds. Initially an allocation is made of the time allowed to each component, and an attempt is made to design each of the components to function within its allocated time. During the design and analysis of the system's performance one of the components may be judged too costly or otherwise unable

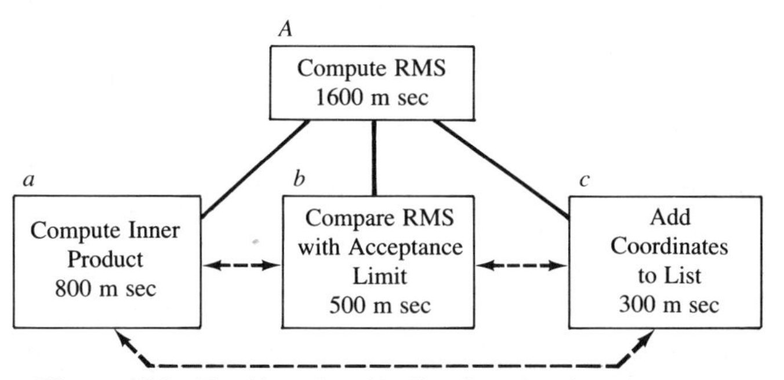

Figure 13.3 Tree hierarchy with allocation of performance.

to meet this allocation. But it may be possible to get the extra needed time from other components that run in less than their allocated time. This is called making a **trade-off**.

Let us add arcs between the leaves to represent the trade-offs. If two leaves have the same resources that can be traded between them, we draw an arc between them. The graph we get from these arcs is not a tree but a graph with circuits. The functions on the leaves can be assigned to design tasks where it is resolved how to make the trade-offs. The arcs provide some of the precedence relations between the design tasks that are used in the design structure system. The design structure system can be used to schedule these design tasks as we discussed in Chapter 7.

Data Flow Diagrams as Requirements

Let us assume that we subscribe to the idea that the requirements specification does suggest a high-level systems design, which can be shown as data flow diagrams. Now we can talk about how the requirements of the whole system and the requirements for each of the modules can be used to represent a hierarchical system of requirements specification.

In Chapter 3 we saw how to develop data flow diagrams hierarchically. We made a data flow diagram that represented the whole system. The processes in this diagram represented modules for which we can also make data flow diagrams. This gave us a hierarchy of data flow diagrams.

The requirements specification can be set up as a hierarchy in the same way. We specify the requirements for the whole system, then with a data flow diagram show how the system is made up of modules. Then we specify the requirements for each module.

The environment for a module can be interfaces with other modules within the system, or interfaces to things outside the system that interact directly with this module.

In Figure 13.4 we can see that *D*'s environment includes interfaces *d* and *e* as inputs from other modules within the system, *g* as an input from outside the system, and *f* as an output to outside the system.

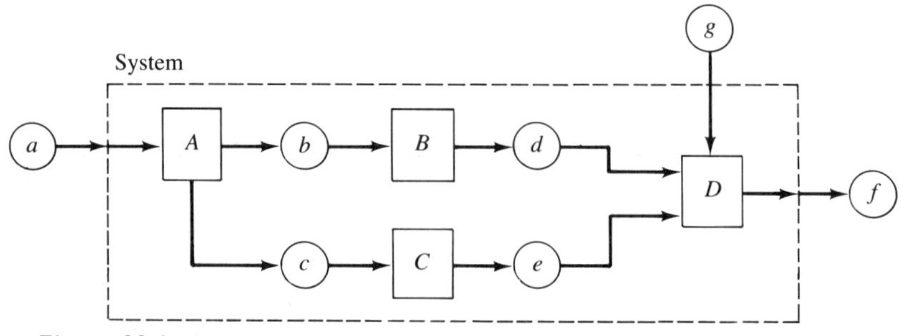

Figure 13.4 System level data flow diagram.

Outline of the Requirements Specification Document

Title page (standard)
- Project name
- Document type (requirements specification)
- Document identification number
- Revision number and date
- Authors / contributors

Table of contents

1.0 Introduction (standard)
- Overview description of this project (very brief)
  - Whole project
  - Part (or all) of project to which this document pertains
- Description of this document
  - Purpose of this document and how it is to be used
    - Intended audience
  - Content and organization of the document
  - Authorship
    - Author
    - Date written
    - Procedure for changing this document?
- Controlling documents and references
- Definition of minimum terms, acronyms and abbreviations needed to read this document

2.0 System description
- 2.1 Environment
  - Who is the user, what are his or her characteristics, and how does he or she interact with the system
  - Interfaces to communications, external hardware, and existing software
  - Software required to run the system including operating systems, calling procedures, and called procedures
  - Hardware required to run this system
  - Operations and site facilities to be used
- 2.2 System functions
  - Descriptive overview
  - For each specification wherever it is not obvious show how it is to be verified including:
      - What measures are to be used, and
      - What is the criteria for acceptance

For each function of the system describe the:
Output
Input
Process
Algorithms
Handling of normal, error, and unusual conditions
Files and data bases used
Control by interactive user, batch input, or operations.
Show what the screens would look like.
2.3 Performance (see Chapter 8)
Speed, response time, number of transactions per unit of time
Maximum problem size
Maximum allowable rate of lost data during communications
2.4 Logistics
Security needs
Maintainability needs
3.0 Hierarchical requirements by module
For each module of system x:
(use same outline as for the system in 2.0)
(see previous section in this chapter titled
"Hierarchical Specifications of Requirements")
3.x.1 Environment
3.x.2 Function
3.x.3 Performance
3.x.4 Logistics

4.0 Reference design—optional

5.0 Design and implementation constraints
What technologies are to be used
What standards are to be followed
What hardware is to be used
What software is to be used

6.0 Constraints on time, cost, and resources
Cost of operation
Cost allowed to produce or obtain system
Time allowed to produce system

Appendices—detailed tables and charts that are too bulky to be included in the body of the report such as:
Functional requirements tree
Data flow diagrams
Glossary
Index

User's Manual

The **user's manual** is written explicitly to explain to users how they can utilize the system. Thus the user's manual states precisely what the system will look like to users. If users can read the manual before the system is developed and imagine themselves running the program, then they can make intelligent criticisms as to whether that is what they want. Thus it fulfills a function similar to the functional requirements specification, which is part of the requirements specification.

A user's manual written only after the product is completed most often shows users how to deal with the idiosyncrasies of the system as the programmer happens to write it. A user's manual written before the design and implementation reflects what users want, not what they must learn to live with.

For a small system the user's manual alone can almost fulfill the role of the functional requirements specification. What the user's manual does not include, which needs to be specified in the requirements specification, are the performance and logistics. How fast does the system respond, how many transactions per day can it process, how seldom is it likely to fail, and how easily can it be maintained, and so on?

A good program is useless if users can't learn how to use it. As computers are becoming more widely used by people who have less training on how to use them, the need for good, clear, simple user's manuals becomes more important. On-line help, which we discussed in Chapter 9, can supplement the user's manual. We will consider just the user's manual here.

As important as they are, many user's manuals are poor. The type of talent required to write a program is apparently not well suited to writing manuals. Some companies have hired people who specialize just in writing manuals.

A user's manual can have many audiences. It might be looked at by the person who is considering buying the program. Or a user's manual might be read by someone who wants to understand what the program will do and how easy it will be to use before he or she pays someone to write the program for him or her. The manual will be read by users before first using the program. And it will be referred to by users from time to time as they want to do things they have not done before, they have problems they have not dealt with regularly, or they come across errors with which they are unfamiliar. And the user's manual will be read by the person who must install the program before it can be run on a particular machine.

One approach to manual writing is to write an introductory section that describes the program and is read by everyone at least once. First-time users will want also to read the second section, which will introduce them to all the general principles involved, and the third section, which will take them step by step through an example. When they have questions they will want to refer to the reference section. The installation is done only once, so rather than place it in the body of the manual for everyone to stumble over, it should be placed in the appendix. A good table of contents tells readers where they will find what they want to know.

Outline of User's Manual

1. Introduction—What does this system do?
    - What functions does it perform?
    - Who is the system intended for?
    - Who developed the system?
    - Who can answer questions?
    - What system and configuration does it run on?
    - Briefly, how does it do it?
    - What are its special features that may distinguish it from other related systems?
    - What are the limitations in its use?
        - Size limits
        - Assumptions

2. Principles and procedures
    - General:
        - What are the outputs?
            - How are they interpreted?
        - What are the inputs?
            - Where are they obtained?
        - How is its operation controlled?
            - Menu driven, command line driven, and so on
            - Menu transition diagram or general command syntax
        - Conventions
            - How inputs are represented, outputs interpreted, or controls implemented, and so on
        - General procedures
            - Start-up / sign-on
            - Backup
            - Shutdown
            - Formatting disks
    - By function:
        - How to get the system to perform the function
        - Screens seen as you execute the function

3. Tutorial: Step-by-step walkthrough of example
    - Functions to be illustrated in this example
    - Narrative overview of example
    - Step-through example showing screens, input, and output

4. Reference
    - For each function (listed alphabetically):
        - Describe the function

- How and when do you use it?
- Structure of command or screen
- What happens
- How to use
- What errors can occur and what you do about them
- Examples

For each error:

- How do you recognize it?
- What does it mean?
- What do you do about it?

Appendix:

A. How to install the system before first use

Summary

The requirements specification, which is created during the requirements analysis phase, defines the behavior of the product, and tells clients what they are getting. This specification includes descriptions of the system's environment, function, performance, and logistics and may include cost and schedule.

Clients may be very busy during the writing of the requirements specification and they may not want to get involved with what they regard as a complicated document. It is crucial, however, that they do help to develop the specifications. Only with clients' participation can we be assured that they will be satisfied with the final product.

The requirements specification is delineated by a number of general properties. For instance, it is an external specification, describing what's on the outside, not what's inside making the product run. The specification should not be too constrained but should allow designers the greatest possible freedom. It is primarily logical—that is, concerned with the information—but it may also describe the interface with the physical equipment. Finally, it focuses on the product, not the process.

The document holding the requirements specification should strive to be unambiguous, verifiable, measurable, traceable, modifiable, and consistent.

To develop the requirements we can use a hierarchy in the form of a requirements tree or data flow diagrams. With a tree the branches hold the functions the system must satisfy. Trees may also show the allocation of resources and the trade-offs that sometimes must be made between components.

User's manuals are designed to help users learn how they can operate the system. They can also be used to convey some of the requirements of the system. Such manuals should be planned before design and implementation so they guide the development of a well thought out system rather than just explain the idiosyncrasies of a poorly thought out system.

Exercises

1. For the critical-path-scheduling-program case study in Appendix B:
   a. Develop a functional requirements tree.
   b. Write a user's manual.
   c. Write a requirements specification.
2. For the CAPERT case study in Appendix A:
   a. Develop a functional requirements tree.
   b. Write a user's manual for the order entry system.
   c. Write a requirements specification.

References

Freedman, Daniel, and Gerald M. Weinberg. *Handbook of Walkthroughs, Inspections, and Technical Reviews,* 3d ed. New York: Little, Brown, 1982.

Gehani, Narain, and Andrew D. McGettrick, eds. *Software Specification Techniques*. Reading, MA.: Addison-Wesley, 1986.

IEEE ANSI/IEEE Std 730-1981, IEEE Standard for Software Quality Assurance Plans.

IEEE Transactions on Software Engineering, Special Collection on Requirements Analysis, SE-3, 1, January 1977, 2–84.

The Software Engineering Standards Subcommittee of the Technical Committee on Software Engineering of the IEEE Computer Society, *A Guide to Software Requirements Specifications,* 1 August 1983.

Thurber, Kenneth. *Computer Systems Requirements*. Los Alamitos, CA.: IEEE Computer Society Press, 1980.

14

Appraising Feasibility and Cost

- Constraints define what we can and cannot do. Choices that satisfy all the constraints are called feasible.
- Alternative solutions may be sought by hiring consultants, searching the relevant literature, and speaking with vendors.
- Prototyping can help to assess feasibility.
- When comparing various options, we can use net present value to accommodate the differences in when the costs and benefits occur.
- The net present value adjusts for when the costs and benefits occur by making a comparison to a fixed interest investment.
- The payback period is the time required before the cumulative cash flow crosses over to being positive.
- The final decision regarding a system must fairly weigh the dollar value along with the intangible factors.

In this chapter we'll consider whether it is reasonable to expect to get the system we want. The requirements specification stated what we want the system to do. Now we must consider the various options we have as to how we build a system to do it, establish what options if any are feasible, evaluate them, and choose the best.

We make a **feasibility study** to estimate whether it is likely that for the system we propose producing it will be:

- *possible* to build it and make it work
- *affordable* considering the estimated costs to develop it and the expenses and revenue from operating it
- *acceptable* to the client who will pay for it and the user who will use it

There is usually not just one way of doing something. We'll have a number of different options as to how we do it. Among the options that meet these feasibility conditions we decide which one looks best to pursue.

The feasibility study seeks to determine the cost and value of the product so that we can decide whether to go ahead and develop it. We want to know if it's worth making the product before we make it. The sooner we know, the sooner we can drop the project if it is not worth doing.

But this raises a question. How can we determine the value of something if we don't know what it is? It seems obvious that we can't do a feasibility study until after we know the requirements that tell us what the product is to do, and know the design that tells us what it is.

This is a dilemma. We want to do the feasibility study before we decide whether to do all the work involved with the requirements analysis and design. But we need to know the requirements and design before we have the information to do the feasibility study. The solution is to take the standard way out of this type of dilemma. We make guesses, we start with a very broad, high-level view by ignoring details, and we iterate.

Some books show the feasibility study coming before the requirements analysis, implying that if the project proves to be infeasible, we can be saved from the extensive work involved in an elaborate requirements study. But when the process is shown this way, a primitive requirements study is really hidden within the feasibility study. We have shown the feasibility study after the requirements analysis. But it is all really an iterative process involving requirements, design, and feasibility. We either do a preliminary feasibility as part of the requirements analysis, or a preliminary requirements analysis as part of the feasibility analysis. In either case during the feasibility analysis we may make a rough design in order to estimate costs.

In Chapter 2 we showed the feasibility study with a set of parentheses. At the beginning we guess at the net value the product will have (left parenthesis), and after we have finished we compute the net value the product does have (right parenthesis). Then we can determine how good our guess was.

In this chapter we'll define some of the terms used in appraising feasibility. With an extended example we'll see how to judge a system's costs. Finally, the last section explores how to put together all the factors in order to make a decision.

Constraints and Feasibility: A Linear Programming Analogy

Before we discuss generating, evaluating, and choosing options we need to define constraint, feasibility, optimizing, and satisficing. A feasible design must be created subject to certain "constraints." And the client may ask whether the design is "optimal." So we'll need to make these concepts clear. They can be made so in the context of a simple linear programming example. By using just a simple diagram we won't have to get into all the mathematics, which we'll leave to another course.

Let's assume we have a plant and equipment in which we can make either cars or trucks with the following restrictions (which we call constraints):

1. Machine 1 has a capacity that would allow us to make at most an average of five cars or three trucks each day, or any linear combination provided we stay below line 1 in Figure 14.1.
2. Machine 2 has a capacity that would allow us to make at most an average of two cars or five trucks each day, or any linear combination provided we stay below line 2.
3. Both machines are required whether we make a car or a truck.
4. The average number of cars and trucks per day cannot be negative. These constraints are implied by staying to the right of the vertical axis and above the horizontal axis.

The shading in Figure 14.1 shows the outer side of each boundary. On the other side the constraint is satisfied. For this problem there is an area where all the constraints are satisfied. It is possible with the machines we have to make any combination of cars and trucks defined by a point within this area, say point *a*, and impossible to make a combination represented by a point outside this area,

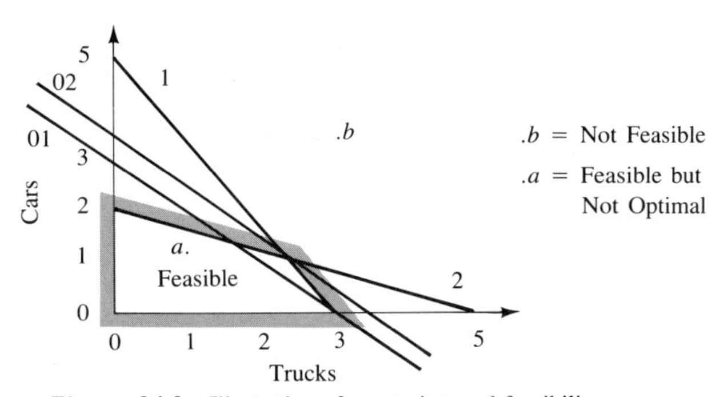

Figure 14.1 Illustration of constraints and feasibility.

say point *b*. We will assume we can make a fractional number of cars or trucks during a day because we can make the rest of them the next day.

Constraints tell us what we can and cannot do. In this example they still allow us to make any of a whole family of acceptable combinations. Any choice that satisfies all the constraints—that is, is inside the area—is called **feasible**.

When we lay down our constraints, we may find there is no feasible region, no possibility that satisfies all the constraints. Then we must either:

1. Reexamine the constraints to see whether they may be more restrictive than they have to be, in which case we may still be able to find a feasible region, or else,
2. Conclude there is no feasible solution and terminate the project.

If there is no feasible solution, we want to know as soon as possible. There is no point in spending money and time looking for a feasible solution when one does not exist. Thus, as early in the project as possible, we want to make a study that will tell us if the project is not feasible.

The feasibility study requires a quick first pass at the design to determine whether it is likely there exists a solution. The feasibility study does not have to produce the best solution. But if it can produce at least one satisfactory solution, just to establish that one exists, then we can go on to making a more careful study to obtain a better solution.

Often a very simple analysis with very primitive assumptions is sufficient to determine that a solution does not exist. Let's assume we wanted to build a solar electric system to store energy in a battery to run the lights in our home. We can quickly estimate the total sun energy that falls on an area the size of our roof under the best conceivable conditions. We can also determine the highest-efficiency solar cell that may be available. If this high estimate of the sun energy times this high estimate for the cell efficiency does not provide enough energy to light our house, we can reasonably conclude that it is not feasible. We need not refine our calculations any further. However, if we find that these high estimates would satisfy our lighting needs, we would have to make more careful calculations before rejecting or accepting this project. This is called "placing bounds on our problem."

If our estimate indicates that there might be enough energy, we would go to the next step. We would refine the constraints and add other constraints that apply, such as the orientation of the roof at each time of day and year, the maximum efficiency of the cells and battery constrained by what we can afford and so on.

If at the end of a step it is still feasible, then we add more constraints and continue until we have either shown it is not feasible, or have added all the valid constraints and still find it is feasible.

A good systems analyst familiar with the problem area will have developed a skill in making such a search, knowing which constraints to add and in what order to determine most quickly whether a feasible solution exists, and if so, to find acceptable solutions.

Returning to our simple linear programming problem of the cars and trucks, we have a family of feasible possibilities—any point in the unshaded area. We can now ask what would be the best choice from among these. For this we need more information: how much is the payoff from producing a car as compared to a truck? Once we know that, we can ask for the solution that has the greatest payoff among all those that are feasible.

An **optimum** is a solution such that no feasible solution has a better payoff. Point *b* is not feasible. Point *a* is feasible but not optimal.

Let's assume we are told that either a truck or a car generates the same profit. We can draw a line (01) such that a solution anywhere along this line will produce the same payoff. This is called a payoff line. Other parallel lines represent combinations of trucks and cars that produce different payoffs. Higher lines here produce more trucks or cars and thus higher payoffs. If we take a line parallel to 01 and move it up (increased payoff) as far as we can before it no longer intersects the feasible area, we see that the greatest payoff that is still feasible is where our line is just ready to leave the area, which is where the two constraint lines meet. We can determine exactly where this point is by solving for where these constraint lines meet. Solving these equations we get 20/19 cars and 45/19 trucks. This is the optimal feasible solution.

If the payoff line had a different slope, it might leave the feasible area at a different point. For example, what would be the optimum feasible solution if the payoff line showed that five cars generate the same payoff as two trucks?

You can see this by taking a right triangle and putting one side of the right angle along the payoff line and the other side on a ruler. Now if you slide the triangle along the ruler, the other side moves parallel to the payoff line. Note when this side last touches a point in the feasible area. That is the optimal feasible solution.

There is a story told about a systems analyst who wanted to demonstrate how linear programming could be used to find the most profitable mix of chicken feed that still contained an adequate amount of each required nutrient. By asking questions of the old hands who for years have been planning the mix of chicken feed, he was able to get for each nutrient the constraint line that showed the combination of ingredients that gave the minimum amount of that nutrient. He came back with a mix that cost only half as much as the present mix. But instead of cheering, they laughed. "But that mix contains so much molasses the chickens would have diarrhea." So he added another constraint to ensure that the chickens didn't get too much molasses and found a mix that cost 80 percent as much as the current mix. But again they laughed. "We can't sell that mix because farmers don't believe that chickens will eat feed with that color, and unfortunately the farmers buy it, not the chickens." His next try came up with a feed that was just a bit better than the old hands had produced, but by that time the cost of feed had changed.

Unfortunately, not many systems problems are so simple that they can be stated in the form of a linear programming problem. We are seldom able to obtain an optimum solution. Usually we are not even able to define how we would recognize an optimum if we found it. If we can recognize it, we are seldom able to afford the effort to find it. It would cost us more to find an optimal solution than we could save if we found it.

Usually we will work until we get an answer that is satisfactory and no further work for a better answer appears worth the effort. This is called **satisficing**. We also refer to such an answer as a sloppy optimum, or simply as a "sloptimum."

We have introduced linear programming only as a model to define the concepts of constraint, feasible, and optimum. If we were to study the mathematics of linear programming further, we would learn that the optimum feasible solution, if it exists at all, will be on the boundary. By using the simplex method, we can learn to compute the solution of complex problems.

Finding Alternatives

If we can show that there cannot be any feasible solution, there is no point in carrying our project any further. But if it appears a feasible solution does exist, we want to find one that satisfices—one good enough so that it is not worth the time or cost to search for a better one.

In the linear programming example just presented we could consider any number of solutions represented by a point in the feasible region. It is more usual to be faced with only a small, discrete set of approaches that we can consider. For example, in our CAPERT case study in Appendix A we might consider as one option a system in which all functions are handled by one centralized computer and data base accessed by terminals through a communication system. Another option might be a distributed system in which the centralized computer is a file server handling only those functions that cannot be handled by local workstations connected in a network. Each of these is a specific alternative to be considered.

Where do we find these optional solutions to evaluate? Generally we look at how other people solve the same or similar problems. We learn of this at meetings or search the literature for articles that discuss systems that might be applicable. We can hire consultants who have solved similar problems. We can also talk to vendors, who may provide equipment to solve such problems. Vendors can act as consultants too, and are less expensive, but we must be aware of their vested interests.

Once we can identify candidate systems that might solve our problem, we obtain as much information as possible that will help us evaluate how that solution might work on our problem in our environment. We might search the literature for more information, visit other people who have such systems in operation, or get more information from the vendors.

When our only initial source of information is the vendor, we might ask the vendor whether we could talk to customers who have used the equipment under circumstances similar to ours. If these organizations or companies are competitors in our business, they may not be willing to talk to us. But very often, although these organizations or companies perform similar functions, they are not competitors and would be willing to take the time to talk to us as a matter of goodwill or for what they can learn by seeing their system through someone else's eyes.

Once we have identified several candidate systems, we can start to look at costs and benefits.

Prototyping

Prototyping is a very useful technique for assessing feasibility. If there is some particular aspect of the system whose feasibility we have some question about, we can sometimes arrange a way to try just that aspect to answer our question without having to develop the whole system. In Chapter 9 we focused on prototyping to analyze the user interface. Prototyping can be used to analyze other potential problems too. For example, we may be concerned about whether an algorithm will run fast enough to make our system practical. Often we can write a primitive system to try just the algorithm to see how fast it runs without having to wait until the whole system has been written.

There is a potential difficulty in running a small problem to see if a larger problem will run fast enough. If careful mathematical analysis of the algorithm is not done, the results can be misleading. The time may increase, not linearly with the size of the problem, but perhaps with the square or cube of the problem size, or even faster. A mathematical analysis of the algorithm may show how fast its running time grows with the size of the problem, and then a few test points may be sufficient to extrapolate to the size of problems we want our system to run. (Horowitz and Sahni, 1978).

Accounting for Costs

In the planning of a system one of the important considerations, especially for the client, is the system's cost. How do we go about determining whether or not the money will be well spent on a prospective system? This section takes an example and describes some of the formulas that can help to figure costs.

Comparing the Value of Money at Different Times

We wish to compare the costs and benefits of producing a product by the various options to see if any option is feasible, and if so, what option is best.

Usually we must invest money now so that we have the product to use later. Once the product is in use, its operation will cost us expenses and will produce revenues. We might be inclined to assume that if the investment plus the expenses are less than the revenues, then the system would be worth building. But we must also consider when the investment, revenues, and expenses occur. The value of a dollar invested in our system now is not the same as the value of a dollar we will receive from operating the system in the future. This is because instead of investing the dollar in our system, we could have invested it somewhere else that would have paid us interest.

Two principles can be used here. One is that we can compare the value of different kinds of things by comparing the value of each to some common equivalent. Then we can make comparisons in the common equivalent. For example, we can compare the value of a horse to the value of a number of bushels of apples by knowing what each is worth in dollars. This method is the principle of using a common denominator.

The other principle is that of opportunity costs. When we use resources to do one thing, we have lost the opportunity to use those same resources to do something else. The loss of opportunity to do something else is called the **opportunity cost**. If we invest money in our new information system, we have lost the opportunity to invest the same money in the bank and receive interest.

We use these two principles together to solve the problem of how to compare money spent and received at different times. We compare each to a common denominator, the opportunity to invest the same money to earn compound interest. Let's look at an example. Consider that we make an investment of $100,000 today in our new system. Assume that the system is completed and put into operation at the end of the first year. At the end of each year, beginning with the second, it will cost us $10,000 in expenses to operate and will return $30,000 in revenues. At the end of the sixth year we sell the system for $15,000. This $15,000 is called the salvage value. Should we make this investment?

If we make no allowance for the difference in the value of a dollar at different times, we might argue as follows. We would get a net return of $20,000 per year for five years, which would cover our investment of $100,000, and then we could sell the system for $15,000 net profit. That's simple enough. But if we had the opportunity to invest this same money at 10 percent per year, would it be better to invest in our new system or invest in the bank?

A dollar invested in the bank this year at 10 percent interest is equivalent to $1.10 next year. And that $1.10 reinvested is equivalent to $1.10 × 1.10 the following year, and so on. Thus we have the following formula for the value in year y given a value of V_0 in year 0:

$$V_y = V_0 * (1 + I)^y \qquad (1)$$

Using this formula, we can then compare the amount of money we would have at the end of the last year if we had (1) invested the original money in the bank, or (2) invested the original money in our new system. To make the comparison fair we should take any money we get as a return in each year from our system and invest it in the bank for the remaining number of years. Investing this way, we get the following results:

1. For the simple investment in the bank:

$$V_{B6} = \$100{,}000 * (1 + .10)^6$$
$$= \$177{,}156$$

2. And for the investment in our proposed system:

$$\begin{array}{llll} & & & \text{End of year} \\ V_{S6} = & \$20{,}000 * (1 + .10)^4 & (= \$29{,}282) & 2 \\ & + \$20{,}000 * (1 + .10)^3 & (= \$26{,}620) & 3 \\ & + \$20{,}000 * (1 + .10)^2 & (= \$24{,}200) & 4 \\ & + \$20{,}000 * (1 + .10)^1 & (= \$22{,}000) & 5 \\ & + \$20{,}000 & (= \$20{,}000) & 6 \\ & + \$15{,}000 & (= \$15{,}000) & 6 \text{ salvage} \\ = \$137{,}102 & & & \end{array}$$

Clearly we would be well advised in this case to put our money in the bank. Not only will we get more money, but there will be less risk. If the investment in our system returned more than the investment in the bank, we may then consider whether the difference is worth the added risk.

If we look at the net value of our investment at the end of year y as the difference between the value of investing in our system and investing in the bank we get:

$$NV_y = NR_0 * (1 + I)^y + NR_1 * (1 + I)^{y-1} + NR_2 * (1 + I)^{y-2} + \ldots$$

where NV_i = net value in year i

NR_i = income − expense − investment in year i

= net return in year i

Instead of comparing the values at the end of the time period, the usual way to make the comparison is to reduce all future dollars to their value at the beginning of the time period. This is called the present value. To get the present value we solve equation (1) for V_0 in terms of V_y as follows:

$$V_0 = \frac{V_y}{(1 + I)^y}$$

So to get the net present value at time 0, we divide the equation for net value at time y by $(1 + I)^y$. This gives:

$$NV_0 = NR_0 + \frac{NR_1}{(1 + I)^1} + \frac{NR_2}{(1 + I)^2} + \frac{NR_3}{(1 + I)^3} + \text{etc.}$$

where:

NV_0 = net value at time 0 (that is, net present value)

Thus the net present value can be used as a measure of whether to make the investment in the system. We might be inclined to think that we should invest in the system if the net present value is positive, and that we should not invest if it is negative. However, there are other considerations.

Investing in our system may involve risks. Unforeseen problems may arise; the cost may be higher than we estimated or the benefits less, or the project may take longer than planned. Investing in the bank does not have these same risks. Thus, before we would invest in the system, we would want the net present value to be more than sufficient to compensate us for this risk.

A little thought will tell us that we will get a different result if we use a different rate of interest. A smaller interest rate may make the investment in the system look better, and a larger interest rate would make the investment in the bank look better. Thus, we want to use an interest rate that would represent a realistic opportunity to invest at that interest.

Another approach is to compute these investments at different interest rates to find the rate that makes both investments worth the same. We do this by changing the value of the interest to make the net present value given in the equation earlier equal zero. This interest rate is called the internal rate of return. Then we can compare this rate with the interest rate we would get at the bank.

Consider the example shown in Table 14.1. The first column shows the year. The second column shows the investment, the third column the income, and the fourth column the expense. The fifth column shows the income minus the investment, minus the expense, which we call the net return or cash flow. We assume that wc arc comparing this investment to a bank investment at an interest rate of I, where I for this example is 15 percent. Column 6 shows $(1 + I)$ raised to the y power, which is the worth in that year of $1.00 invested at an interest of I beginning in year zero. Dividing the net return in column 5 by column 6 gives the present value of that net return, which is shown in column 7. Column 8 is

Table 14.1 Present value and cash flow.

	1	2	3	4	5	6	7	8	9
	y	**Investment**	**Income**	**Expense**	**Net Return, i.e., Cash Flow**	$(1+I)^y$	**Pres. Val.**	**Cumulative Pres. Val.**	**Cumulative Cash Flow**
	0	100,000	0	0	−100,000	1.0000	−100,000	−100,000	−100,000
	1	0	0	20,000	− 20,000	1.1000	− 18,182	−118,182	−120,000
	2	0	45,000	5,000	+ 40,000	1.2100	+ 33,058	− 85,124	− 80,000
	3	0	82,000	2,000	+ 80,000	1.3310	+ 60,105	− 25,019	0
	4	0	80,000	10,000	+ 70,000	1.4641	+ 47,811	+ 22,792	+ 70,000
	5	0	115,000	5,000	+110,000	1.6105	+ 68,302	+ 91,094	+180,000
salvage	5	0	200,000	0	+200,000	1.6105	+124,185	+215,279	+380,000
	Given			Computed					

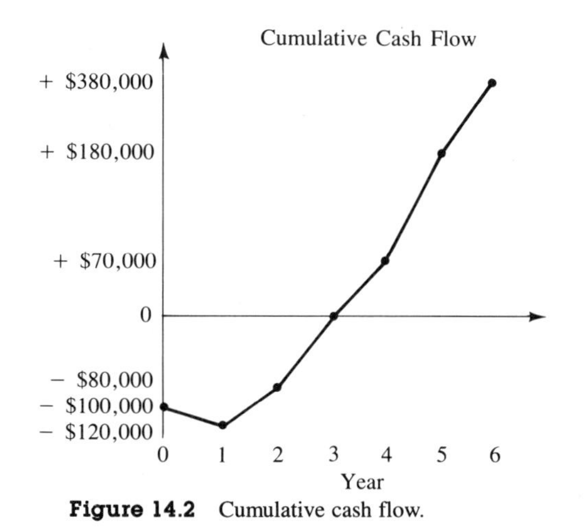

Figure 14.2 Cumulative cash flow.

the cumulative present value, which totals the present values through year *y*. Column 9 shows the cumulative cash flow, which is also plotted in Figure 14.2.

The data in Table 14.1 indicate that we made an investment of $100,000 in year zero. In the first year there was a large expense and no income. The system was put into operation in year two and started to produce an income. In the third year the expenses were decreasing and the income was increasing. But by the fourth year income leveled off and maintenance and improvement costs raised the expenses. Once the maintenance and improvements were completed, in the fifth year the income increased and the expenses came down again, which restored the net benefits, making the books look good so that the system could be sold at the end of year five for $200,000. If we use the present value calculation, the worth of this investment is $215,279. If we did not use present value, the worth would be $380,000.

Cash Flow

Using the cumulative cash flow in Table 14.1, we can plot the total of the amount of money that has flowed in or out, up to each year.

When making our plans, we must consider what the cash flow will be. It is not sufficient that once the product is operational that its benefits are worth its cost. If, at some time along the way, we have a negative cash flow that is more than the money we have or can obtain, we will go broke and go out of business before we can ever get to the point when the system would have shown a profit. When gamblers face this problem, it is called gambler's ruin. It can also mean businessperson's ruin.

How much negative cash flow we can handle will depend upon how much money we have or can borrow and how confident we are that we can replace that money. If our lender does not share that confidence, he or she may not be willing to lend us the money.

Payback Period and Internal Rate of Return

The **payback period** is the time it takes before the cumulative cash flow crosses over to being positive. We see that in our example at the end of the third year the cumulative cash flow was 0. Thus the payback period for this example is three years. (The figures here were set up to come out zero at exactly three years.)

The **internal rate of return** is the interest rate to which we compare our investment so that the cumulative present value at the end comes out zero—that is, our investment and the comparison would give the same return. The internal rate of return is based on a specified investment period and must include any salvage value in the calculation of the value of the investment. We can see in this example that since the cumulative present value at the end of the five-year investment period is positive that the internal rate of return must be greater than the 10 percent we used for the comparison. To get the correct internal rate of return would require that we use several values for the comparison interest rate until the cumulative present value comes out zero.

Net Present Value with Taxes and Borrowing

Now we can make the net present value analysis with the added consideration of taxes and borrowed money. Assume we are considering buying a computer and accessories for \$6000 and using it in a small business in which we hope to pay for the computer and make some extra money. The tax laws and specific percentages are subject to change, but this example will illustrate many of the principles.

Assume we don't have the \$6000 initially. We will borrow the money at 10 percent interest. We will assume that we pay the loan yearly at a fixed amount each year so that the principal we owe is reduced by the amount of our payment minus the interest we own that year. As the principal is reduced so is the interest, and thus the principal gets paid off faster as time proceeds.

The formula for the constant payment that reduces the principal to zero at the end of n payments with an interest of I is:

$$\text{payment} = \text{initial principal} \times I \,/\, (1 - (1 + I)^{-n})$$
$$\text{payment} = \$6000 \times .10 \,/\, (1 - 1.1^{-5}) = \$1582.78$$

Table 14.2 shows the value of the principal and the interest each year.

Table 14.2 Principal and interest.

Yr	Principal	Interest	Payment
1	\$6000.00	\$600.00	\$1582.78
2	\$5017.22	\$501.72	\$1582.78
3	\$3936.15	\$393.62	\$1582.78
4	\$2746.98	\$274.70	\$1582.78
5	\$1438.90	\$143.88	\$1582.78
6	\$ 0	\$ 0	\$ 0
			paid off at end of fifth year

Table 14.3 Depreciation.

yr	Value	Depreciation
0	6000	1200
1	4800	1200
2	3600	1200
3	2400	1200
4	1200	1200
5	0	0

The government does not allow us to deduct the amount we pay on the loan, but it does allow us to deduct the interest we pay and to deduct the amount by which their formula shows that the value of our asset (the computer) decreases. This is called depreciation. Given a five-year life allowed on this type of asset (a computer), the government allows us to use a straight-line depreciation, meaning the value is depreciated by one fifth in each year (see Table 14.3).

Table 14.4 shows the contributions to the cash flow for each year. The cash flows are then reduced to their present values assuming a comparison to a 10 percent interest rate. We assume for this example a tax rate t of 34 percent. The income and expense are multiplied by $(1 - t)$ as they appear as real income and expenses after taxes are paid. The depreciation and interest costs are multiplied by t to get the value in our pocket because they reduce the amount on which we pay taxes. Above each column we have shown the quantity that column is to be multiplied by to take care of the effect of taxes. Below each column we have shown the sum of the present values for that column, and this value times the appropriate tax multiplier so we can see the contributions of each column to the net cumulative present value of $1902.29.

If we sell our computer and accessories at the end of the five years, it is said to have salvage value. Assume that we sell it for $1000. The depreciated value is zero, because at the end of the five years it has been fully depreciated. Thus we have a capital gain of $1000 − 0 = $1000, which is taxed at the regular

Table 14.4 Present values and cash flows with taxes and borrowing.

Cash Flow = (Income − Expense) · (1 − t) − Loan payment + (Depreciation + Interest) · t

Yr	$+(1-t)$ Income	$-(1-t)$ Expense	-1 Loan payment	$+t$ Deprec.	$+t$ Interest	Cash flow	$(1+I)^Y$	Present value	Cumulative present value
1	400	750	1582.78	1200	600.00	−1201.78	1.1000	−1092.53	−1092.53
2	2500	450	1582.78	1200	501.72	+ 348.80	1.2100	+ 288.26	− 804.27
3	3500	400	1582.78	1200	393.62	+1005.05	1.3310	+ 755.11	− 49.16
4	4000	300	1582.78	1200	274.70	+1360.62	1.4641	+ 929.32	+ 880.16
5	4500	300	1582.78	1200	143.89	+1646.14	1.6105	+1022.13	+1902.29
	10585.57	1745.43	6000	4548.95	1532.80	+1902.29			
	Present value sums of each column								
	6986.48	−1151.98	−6000	1546.64	521.15	(+1902.29 is sum of this row)			
	Present value sums times tax multiplier								

rate. Thus, with a 34 percent tax rate we have left after the sale: $(1 - t) \times \$1000 = \$660$. We divide this salvage value by 1.6105 to get its present value and add it to the \$1902.29 to get \$2312.10 as the total present net value for this investment.

Choosing the Best Alternative

The use of the present value can have an effect on what option looks best. For example, consider Table 14.5. If we just sum the net returns without considering present values, Option 2 looks better with a total return of \$25,000 compared to \$20,000. But if we consider present values using a 15 percent interest rate, then Option 1 looks best. Option 2 does not even do as well as the comparison investment at 15 percent. The reason is that Option 1 returns more of its money in the earlier years, when the present value calculation makes each dollar worth more.

It is important when comparing alternatives to assume that each alternative is to be done in the best way possible so we are not comparing a good way of doing one alternative with a poor way of doing another. It is easy to reject good alternatives by thinking of poor ways of implementing them.

There may be intangible costs and benefits associated with building and operating our system—that is, costs and benefits for which it is difficult or impossible to assign a dollar value. Examples are customer satisfaction, or the reputation the company gains from having a system that serves its customers better. We may decide to invest in a system even with a negative cumulative net present value if we feel that this cost is worth the intangible benefits. On the other hand, we may not be willing to invest in a system with a positive cumulative net present value if we feel that this positive value is not enough to justify the risk.

Many features important to our decision cannot be assigned a dollar value. Among these are availability when the system is needed, ability to provide quick and reliable service to customers of the business, or freedom from worry.

Comparison Matrices

Now we must find a way to put together things we can reduce to common dollar values and intangibles we can't reduce to dollars in such a way as we can come to some decision.

We cannot do this in a completely objective way. The best we can do is to come to a decision with which we feel comfortable. An important aspect of our feeling comfortable with this decision is that we can make other people, such as management, feel comfortable with it. We wish to make a decision that we can easily rationalize.

We make up a matrix with the rows showing the factors that we'll consider in evaluating the options (see Table 14.6). The columns show the options. Within the matrix we show numbers representing relative values we assign to these factors for each option. We pick the values we use so that they are normalized to a

Table 14.5 Comparison of options using present value.

	Option 1				Option 2		
y	Net Return	$(1 + I)^y$	Pres. Val.	y	Net Return	$(1 + I)^y$	Pres. Val.
0	−50,000	1.0000	−50,000	0	−50,000	1.0000	−50,000
1	30,000	1.1500	26,087	1	5,000	1.1500	4,348
2	20,000	1.3225	15,123	2	15,000	1.3225	11,342
3	10,000	1.5209	6,575	3	20,000	1.5209	13,150
4	10,000	1.7490	5,718	4	35,000	1.7490	20,011
Totals	$20,000		$3,503		$25,000		$−1,149

roughly common scale. We wouldn't want to use dollars in the thousands for cumulative net value and a scale from 1 to 10 for ease of customer use. Obviously the cost would swamp the ease of use.

We also want to pick weights to represent how important each factor is. When we add the values for the factors, we multiply each by the weight for that factor. Then we choose the option with the highest value.

You may argue that the method of choosing these values and weights is very arbitrary. We won't disagree. It is just a means of coming to a decision with which people feel comfortable. If you regret that these numbers make Option 2 look the best, you don't have to change the numbers much to be more comfortable with Option 1. Show it to the boss. If he or she doesn't like these numbers, you can change them a little to make him or her feel comfortable. You generally are more comfortable when the boss is more comfortable, which makes his or her numbers satisfactory to you.

This method is not very scientific. But it does force people to think about all the important factors, which is not a trivial consideration.

When making decisions, we often have a tendency to ignore those factors that cannot easily be compared and place the burden on factors that can be distinguished. Sometimes these easily distinguished factors on which the decision is based are almost frivolous. A good salesman will find out what these factors are and make the most of them.

The author was involved in a decision made by a committee to choose between several vendors' computers to replace an existing outmoded computer. The deci-

Table 14.6 Matrix showing values of factors for each option.

	Option 1	Option 2	Option 3	Weight
Cumulative net present value	10.0	8.5	8.0	10
Expandability	8	7	6.5	3
Availability	9.5	7	5	7
Advertising appeal of novelty	3	9	2	6
Ease of use by customers	6	8	3	8
	256.5	273	170.5	

sion involved several millions of dollars and many factors that were hard to evaluate. The decision was finally justified on the basis that one machine had a 48 bit word while the others had 32 or 36 bit words. This was not a very important factor, but it was one that could be distinguished very clearly. The people on the committee felt comfortable, probably each for a personal reason that had nothing to do with the word length, and the decision was made. Without that factor on which to hang their hat, the committee would find it hard to reach a decision. Can anyone say the decision was wrong?

Justification

The matrix should be reviewed informally with those people who have to feel comfortable with it. Then a final justification is written stating in a convincing way why one option was chosen over the others. Remember that this is primarily a rationalization so that people can live with the decision. This justification should emphasize what has made the decision easy and deemphasize those factors that have made the decision difficult.

The purpose of the feasibility study is to decide which option to pursue through the design and implementation phases. Sometimes any option is as good as any other, or the choice may come down to two with no way to decide between them. But other times it may be very important that the right option is chosen, yet the information available after the feasibility study does not clearly justify one option over the others. Then, if the extra cost can be afforded, it may be appropriate to pursue more than one option through the additional phases until a good decision can be made.

Outline of the Feasibility Study Report

Title page (standard)
- Project name
- Document type (feasibility study report)
- Document identification number
- Revision number and date
- Authors / contributors

Table of contents

1.0 Introduction (standard)
- Overview description of this project (very brief)
  - Whole project
  - Part (or all) of project to which this document pertains
- Description of this document

Purpose of this document and how it is to be used
Intended audience
Content and organization of the document
Authorship
Author
Date written
Procedure for changing this document
Controlling documents and references
Definition of minimum terms, acronyms, and abbreviations needed to read this document

(2.0 Preliminary requirements specification if the feasibility study is done before the formal requirements specification. If this option is used, the remaining section numbers are increased by one. Otherwise the requirements specification should be referenced as a controlling document in Section 1.0)

2.0 Descriptions of options
For each option:
Description of the option
What precedent exists for the feasibility of this option?
Who is doing something similar?
What is their experience?
How does our environment or problem differ from theirs?
What are the implications to be drawn from their experience about the application to our environment or problem?
Basis for assuming that it is:
Possible
Affordable
Acceptable
Very preliminary estimates of:
Cost to develop
Expenses to operate
Revenues to be obtained
Intangible costs and benefits

3.0 Descriptions and bases for factors used to evaluate options
For each factor:
Describe the factor
How is the factor to be evaluated for the options?
How important is that factor to the overall decision?

4.0 Comparison matrix

5.0 Recommendation and Summary of Justification

Summary

Before we commit ourselves to large expenditures of resources we would like to have some confidence that we will be able to achieve what we set out to do within the constraints with which we must work. The feasibility and cost study is our method for getting an early evaluation of whether we can do it before we do it.

The feasibility and requirements go hand in hand. We need to know what the requirements will be before we can determine whether those requirements are feasible. Often in large projects we will spend a considerable amount of time and effort in developing the requirements. Before we do that we may wish to have some preliminary notion as to whether our project is feasible and thus whether doing the detailed requirements study is worthwhile. So a feasibility study including a brief requirements specification may be shown in the schedule as occurring before the requirements. Actually we can say that we continue to reevaluate feasibility throughout the project. If at anytime we conclude that the project is not feasible, we stop and spend no further resources, except possibly to record what we have done as a lesson for the future.

Constraints outline what is possible and what is not possible. Choices that satisfy all constraints are feasible. A quick analysis can sometimes determine if any solution exists. An optimum is the best feasible solution—that is, it has the best payoff. Optimums, though, often cannot be found. In that case we find a satisfactory answer such that no further work is worth the effort, which is called satisficing.

Sources of alternative solutions include professional meetings, literature searches, consultants, or vendors. We may speak with organizations that currently use a similar system. Prototyping allows us to analyze certain troublesome or questionable aspects.

Part of any estimate of a system's feasibility is judging the cost involved. Two important principles of cost comparison are the principle of using a common denominator and the principle of opportunity costs. To determine whether to invest in a system we use the net present value. We must also assess the cash flow. The payback period is the time necessary before the cumulative cash flow crosses over to being positive. The internal rate of return is the interest rate that gives the same return as our investment.

In addition to the dollar cost involved there may be intangible costs and benefits. Many features, such as availability and freedom from worry, cannot be assigned a dollar value. The decision to choose the best alternative should carefully weigh all factors together.

Exercises

1. Consider what options you would choose for the CAPERT project (see Appendix A). Would you agree with limiting the options to those brought up by

management in the discussions and interviews? Are there other options that management might not be familiar with because they have not kept up with information management technology? How about salesmen using portable PCs to enter orders and phone them in using modems?
2. Use these options to write a feasibility report.
3. Compute the internal rates of return using the figures in Table 14.5. You will need to try several interest rates.
4. What interest rate used in Table 14.4 would give a zero cumulative present value at the end of the period? You will need to try several interest rates.
5. Calculate the cumulative present values for the following net returns compared to an investment at 10 percent. Note that without using present value the return of each would be $30,000, making both options equally good. Which option would you choose if you use present values?

Yr	Problem *a* Net Return	Problem *b* Net Return
0	– 100,000	– 100,000
1	10,000	20,000
2	20,000	50,000
3	50,000	50,000
4	50,000	10,000

References

Cougar, J. Daniel, and Robert W. Knapp, eds. *Systems Analysis Techniques*. New York: Wiley, 1974. (Part III: Cost/Effectiveness Analysis)

Horowitz, Ellis, and Sartaj Sahni. *Fundamentals of Computer Algorithms*. Patomic, MD.: Computer Science Press, 1978.

15

Designing the System

- Design does not involve a completely systematic way of working backward from the requirements. It requires iteration between synthesis and analysis.
- The design process involves the two main steps of synthesis and analysis. Within synthesis we constrain the boundaries of the problem, search for alternatives, and construct a system on paper. In analysis we simulate and evaluate that system.
- To design algorithms we must break down the problem into smaller problems, decide how to represent the input and output, do the problem by hand, develop a variable dictionary, and transform the descriptions into computer language.
- Strategies for designing information systems include backward search, top-down search, and a method that creates program structures from the structures of the inputs and outputs.
- Structured design requires organizing programs so they'll be easy to maintain. This approach allows other people to make changes later without causing unexpected problems.
- Classical structured design involves five types of coupling, or preserving the integrity of the data flows, and seven types of cohesion, or keeping the integrity of the module.
- Once the logical design is done, we can begin the physical design, where we allocate the transformations that are to be made to the hardware, software, and people.
- The amount of memory for the data can be estimated from the data structures, and the amount of input-output processing can be estimated from the data structure matrix worksheets. The program memory size and running time are harder to estimate.

The system requirements specification is an external view of the system. It states what the system should look like as seen from the outside. The **design specification**, on the other hand, is an internal specification. It tells how the system is to be built inside so it will look like what the requirements call for on the outside. And, as we discussed in Chapter 2, requirements and design are recursively related. Design involves putting together components that either we have or whose requirements must be stated so they in turn can be designed. One person's design is the next person's requirements (see Figure 15.1).

Design is a search for the right components; it includes how to put them together and how to operate them to form a system that will meet the requirements specification. The result of the design is the design specification, which tells how to build and test the system.

The Nature of Design

Design involves synthesis and analysis. As we said in Chapter 2, synthesis is choosing and putting together components and how they are operated to form a system. Analysis is the process of determining the behavior of that system to establish whether it would satisfy the requirements specification.

We design our system by putting together components with pencil and paper rather than by putting together the real components themselves. This is because it is usually cheaper to change pencil and paper than it is to change the real thing. We may represent our design in the computer rather than with pencil and paper, but in either case during design what we are working with is abstract, not the real thing. To make our analysis we must determine how this abstract thing would behave before we are really ready to build it. Our principal means of doing this is simulation—logically working out step by step exactly how the system would behave.

Design has two aspects: technical and creative. The technical aspect concerns knowing what components are available and how they work. We discussed some technical considerations in Chapter 8. The creative aspect concerns the challenge of putting together components in the hope that they will work in a particular

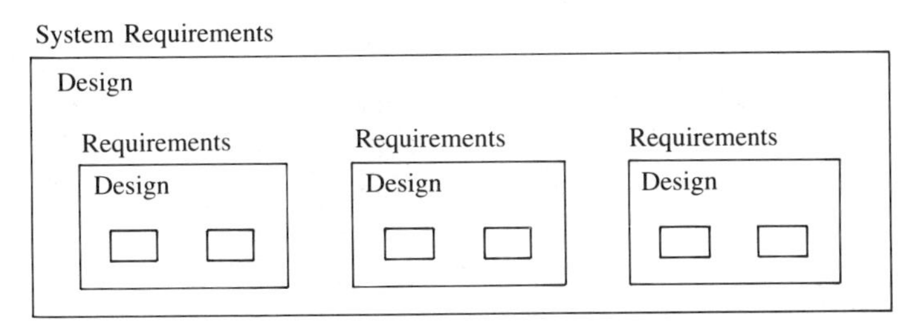

Figure 15.1 (same as Figure 2.3) Recursive requirements and design.

way, just as an artist puts together a creation with the hope that the result will affect the viewer in a particular way.

Some people feel that the creative aspect of design is an ability, not a skill, and thus cannot be taught. We prefer the point of view that it probably can be taught, but in the past we have just not known how. We will try to teach it here. Part of the problem in learning design is to recognize that it is hard work, it requires a lot of cutting and trying, and it can be frustrating. A design does not just come to us because we have some knack for it. We have to work at it.

Unfortunately, there is not a systematic way we can work backward from the requirements to guarantee we will get a design that will work. If the system is complex, when we start putting together components, we can only guess, but cannot know for sure, what system behavior we are going to get. We trudge along, and when we finish our synthesis, we analyze it to see what behavior it has. If our analysis establishes that what we synthesized does not meet the requirements specification, we make another synthesis. Design is an iterative process of synthesis and analysis.

Because design is not straightforward and systematic, as are most processes with which we are familiar, it can be frustrating. At first we tend to think design is easy, and when it isn't, we give up instead of trying again. But once we learn to expect to be frustrated, it can start to be fun. If we don't get frustrated, we're probably not doing it right.

Design is a creative and skilled process. If we make good guesses about what to put together and how, we won't have to make so many iterations. Good designers have the experience and skill to guess well. They must know the technology thoroughly so that they know what they might expect when they put components together. Designers must know what is likely to be easy to build and what may be hard, and what might work and what might not. Most other computer courses are concerned with learning the technologies. Here we are concerned with the art of using the technologies to put together systems.

Design is problem solving—solving the problem of how to put together a system that will satisfy the goal of meeting the requirements. Thus design, like problem solving, involves search. Interestingly, the search strategies we use in design are similar to the search techniques we study for artificial intelligence.

Good design involves making the simplest system that will satisfy the requirements. It is frequently easier to make a design complex than to make something simple. Each time a problem arises with what has been designed there is a temptation to add something to take care of the problem rather than to reconsider what has already been done. Sometimes what we conceive of as a simple idea turns into a Rube Goldberg design when given to others to work out. We must try to preserve simplicity. Generally simpler systems work better and are easier to change.

The effort of design results in a design specification, which shows the builder how to build it, and shows the tester how to test it to determine whether it was built as called for in the specifications. Thus the design specification shows what components are to be used, how they are to be put together, how the system is to be operated, and how it is to be tested.

We may think of design as the planning for how the system is to be implemented. What first comes to mind is planning for how it is to be programmed. But there are many other things that have to be planned also in parallel with planning the program. We may need to purchase hardware or software for which we have to write purchase specifications. Purchase specifications look very much like requirements specifications, except that the requirements are being placed on someone else, the supplier. Files that were used in the old system may have to be converted for use in the new system. There are manual procedures to develop, manuals to write, and people to train. Design involves planning many things that have to work together to make the system work. The "Outline of Design Specifications" at the end of this chapter will give some idea of all the various things that must be considered in designing a system.

Designers should specify their design carefully enough in the design specifications so that someone else can build and test the system and thereby enable it to work correctly. Thus, good design, which involves iteration, means that we can hope to have no iterations while building the product. A good, carefully developed design can save a great deal of time, effort, and cost during the implementation phase.

As we pointed out in Chapter 2, the burden of ensuring that programs do what they are supposed to do must be borne largely by careful design. If we have a poor design, we may never be able to get our system to work correctly by testing and fixing it.

If we can create our system from simple modules that fit together with a small number of simple interfaces, it will be easier to make a change in one module without upsetting other modules. We are more likely to have a system that works, or can be fixed without introducing additional new errors. The system will also be easier to change if we want it to do something later we did not plan originally.

Two Models of Design

If the design is of a simple small system, it may be done by one person. We don't need to fully understand how that person organizes his or her thought processes. But if the design is of a larger more complex system, it must be done by more than one person. Then we have to understand how to organize the work and coordinate the thought processes of these several people so that their joint effort produces the design of the whole system.

In this section we'll discuss two approaches to breaking down the design of large complex systems into design tasks that can be done by different people. In practice, though, we often use a combination of the two approaches.

Top-Down Design: System Then Module

Following the **top-down design** approach, we first do a system design to show how we will build the system out of modules. We specify the requirements for each of those modules. Then we design each module. This process can be repeated, designing the modules by building them out of smaller modules for which we specify the requirements, until we reach modules that are already available. We begin the system design with the data flow diagram we developed when we did the requirements specification. This is a basic pattern we have used throughout the book.

In the module design phase we design each of the modules so that it will satisfy the module requirements specification established for it in the system design. The system design determines what each module should do so that each module can be designed independently as a task by itself. Once the systems design is done, different people can be designing different modules all at the same time.

If for some module we cannot satisfy the requirements established for it by the system design, we must go back to the system design and define the modules differently (see Figure 15.2). We may need to make performance trade-offs between modules, requiring one module to do more because another cannot do as much. (In Chapter 13 we talked about such trade-offs between requirements.)

We can think of this system design and module design approach as an example of the "divide and conquer" technique used in problem solving. We solve a big problem by breaking it into smaller problems, assuming we can solve the smaller problems. If we can't solve the smaller problems, we go back to the larger problem and break it into a different set of smaller problems that hopefully we can solve. Similarly, we solve the problem of designing a system by designing it with modules, assuming we can design the modules; if we can't, we go back and redo the system design.

We generally use data flow diagrams to do the system design. The data flow diagrams show how the system is made up of modules and the interfaces between

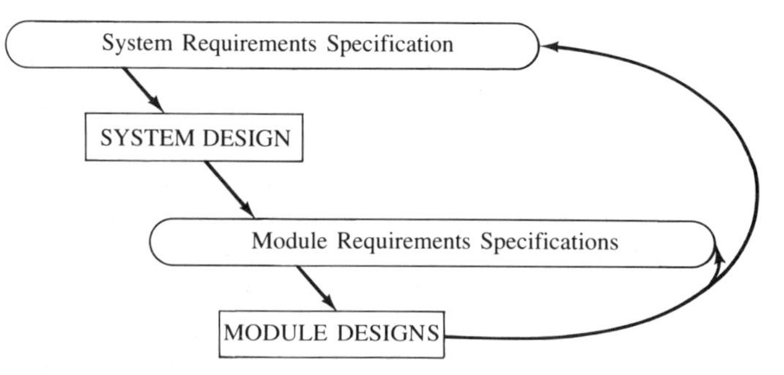

Figure 15.2 Hierarchical design.

the modules. We then use Trees to design the program modules. The data flow diagrams can be converted into a Tree so that we may thereafter maintain the whole system in one common method of representation.

In Chapter 4 we discussed how to use Trees to break the system into modules using the techniques of structured design. We defined the modules to minimize the coupling between them and maximize the cohesion within them.

The Design Structure System

It is not always possible to come up with a system design that defines modules and interfaces between them so that each of the modules can be designed knowing nothing about the others except their interfaces. We often cannot guess that well during the system design what these interfaces should be. There may have to be trade-offs of performance parameters between modules, as we discussed in Chapter 13. So during the design of a module we may have to reach beyond the interfaces of its neighboring modules to look into their design. Similarly the design of those other modules may have to look back into the design of this module. Now the modules are no longer black boxes. We can see into them.

But this means that the tasks to design these modules will depend on each other. They will have predecessors between them that we can represent by a graph. This graph may have circuits. Then we must decide where to tear some of these predecessors by making guesses so we can start the design. We iterate and make reviews to see that everything is consistent. These are the techniques of the **design structure system**, which were considered in Chapter 7.

The Design Process

Earlier we noted that design involves both synthesis and analysis. Now we will note that synthesis in turn is composed of constraint, search, and construction; and that analysis in turn is composed of simulation and evaluation. We can describe this in a tree as in Figure 15.3.

Both approaches—the design of the system or a module in top-down design, or the design of a module in the design structure system—involve an iteration through the processes of constraint, search and construct, and simulate and evaluate as shown in Figure 15.4. During the design there may be several iterations through these steps.

Synthesis: Constrain, Search, and Construct

Something is constrained by establishing a boundary between what is acceptable and what is not. This was discussed in Chapter 14 in the context of linear programming and feasibility. Here the initial constraints are established by the

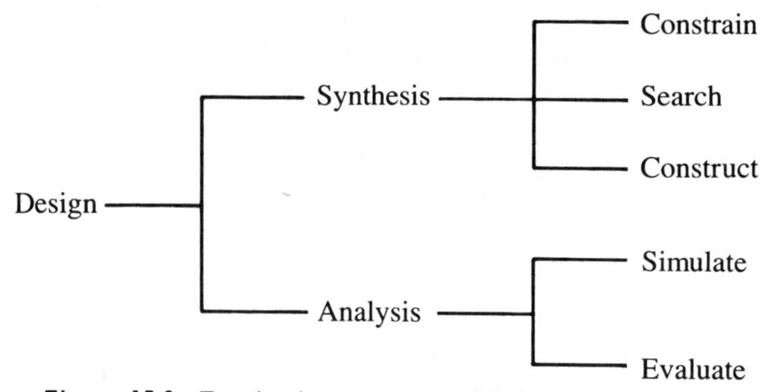

Figure 15.3 Tree showing components of design.

requirements specification, which shows the boundary between what will and what will not be accepted by the client.

We constrain to separate what we are looking for in the search from what we will not bother to consider. Are we looking for components that use the latest technology and may show very high performance possibilities, or are we looking for components that have proven reliability but perhaps have less performance? Are we looking for software or hardware to perform this function? Are we looking only for software that will run on our present computer, or can we also consider the possibility of purchasing another computer?

There are two general approaches to constraint. One is what we might call the "Ideal Approach." We start by looking for the best design with a minimum of constraints. Our imagination is allowed to go free. Initially we are not unduly

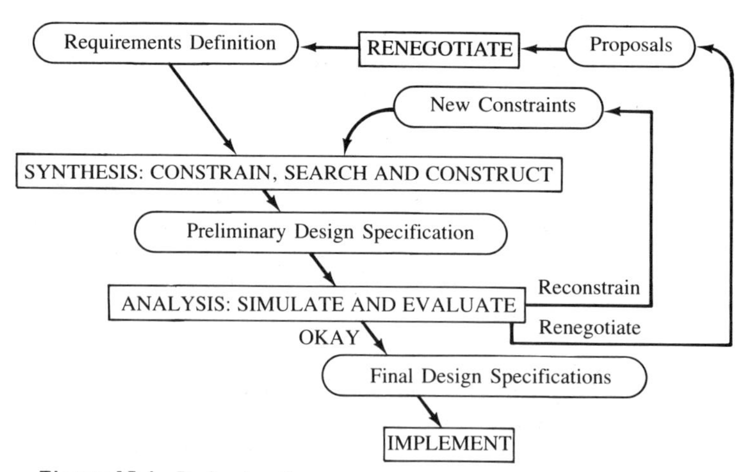

Figure 15.4 Design iteration.

concerned by what is within our means. Once we have an ideal design, we can start to add realistic constraints, pulling back from the ideal to what we are capable of doing. This approach gives us a good opportunity to come up with really innovative new approaches. Sometimes we will come up with an idea that may not meet the original requirements but will offer such advantages that the client may be open to renegotiating the requirements.

The other approach might be called the "Mundane Approach." All the constraints are put in our initial search. We never look at anything we cannot do or cannot afford. The only systems that are considered are those that use proven hardware and software, or that use the hardware we already have. Some people think this is the more practical approach. What we come up with will usually be easier to sell because it is not likely to be very novel in concept.

When developing constraints, we must be careful to be no more restrictive than we absolutely have to be. Unnecessary restrictions may eliminate good solutions.

Often the requirements specification will suggest a conceivable approach to the design. We should not limit the design to one proposed in the requirements specification. But we should be careful to preserve the requirements that this design might suggest (see Chapter 13). If there is a question, we may have to renegotiate with the client.

Now we look for components and how to put them together in a way that we believe will satisfy the requirements specification. The constraint limits our search so we don't waste time looking at everything under the sun.

The search may take the form of looking at catalogs and getting specifications of possible components from vendors or manufacturers. It may involve looking into the technical literature to see what has been done that we can use. We might visit other companies to see how they have solved similar problems. This involves data collection, which we talked about in Chapter 12. This data collection will suggest ideas that can be put together to construct our design.

We may need to make calcualtions or investigations to determine what components to choose and what their parameters must be—for example, we may compute that to satisfy the response requirement the terminal must run at 9600 baud. The search may involve calculations to determine what disc access time is required to satisfy a specified response time, then a search to find a disc that meets these specifications.

We construct a design by putting together components, not the real components but their representations on paper. We have to consider not only how these components will be put together, but also how the resulting system will be operated.

In the design search we must make decisions about what qualities are most important. Often some qualities must be traded off against other qualities. For example, we may have to lose some efficiency to obtain a program that is versatile and maintainable. Some of the qualities that may be involved in trade-offs are those mentioned in the discussion of metrics in Chapter 2.

Analysis: Simulate and Evaluate

When we started our project, we analyzed the existing system or situation to understand its strengths and weaknesses and to see how well it did or did not satisfy expectations placed on it. Similarly, we now analyze the design, which we are proposing as a replacement for the existing system, to see how well it satisfies the expectations placed on it by the requirements specification.

Before we can evaluate how well the design satisfies the requirements specification, we need to compute or simulate how we can expect it to perform. We must also consider the costs of making the system we have designed and the costs, benefits, and consequences of operating it.

The analysis is done by simulating the specified design and evaluating the results of that simulation. The simulation could be done by using a simulator on a computer. Or the simulation could be just the process of thinking through in our head or with paper and pencil all the consequences of using the system as we have designed it.

By this point in the iterative process we may have generated several designs. We want to evaluate which of them may best satisfy the requirements, and whether we think it is worth our while to develop yet another design. If we go through the cycle again, we may come up with a better design, but the advantages of this better design might not be worth the cost of searching for it. It is usually not possible to find an optimum design. We are usually concerned with finding a design that is satisfactory or that will be accepted; further effort to find a better one is not worth the time and effort. We have called this satisficing.

One very difficult decision for technical people to make is "When is enough enough?" If technical people can't make this decision, business people who may not be technically qualified will make it for them.

Renegotiate

During the design it may become clear that some requirements were not completely specified in the requirements specification. If these missing requirements have a significant impact on the external view of the product, they should be taken back to the client for renegotiation.

When we developed the requirements specification, either for the system or for a module, we assumed we could probably come up with a design that would satisfy these requirements. That assumption is not always justified. We may have to go back to the source of those requirements and ask for a renegotiation. If we are now in the modular design, we go back to the system design. If we are now in the system design, we go back to the client.

We can change the design without renegotiating provided there is no effect on the systems requirements. But if the change affects the systems requirements, it must be renegotiated with the client.

A renegotiation would never be necessary if it were always possible for clients to specify completely and accurately in the original requirements specification exactly what they want, and if it were always possible for us to provide what they want for what they are prepared to pay and within the time they are prepared to wait. Then we could proceed directly to a design search constrained only by the requirements specification. Unfortunately, this simple process works only with small, simple systems. In any complex, real-world problem clients are usually not able to state completely what they do or do not want. And we are not always able to produce what they want, or what we thought we could have produced and promised to them. So we may have to renegotiate.

Developing an actual design and showing a representation of it to the client may reveal details that clients, even with the help of good methods of needs analysis and requirements representation, would not have thought of. Such details often get fleshed out only when clients see the design, and sometimes not until they see the final product.

The object of the new structured techniques of representation is to help clients visualize the final system better and sooner so as to reduce the costs associated with renegotiations and redesign. But even though we try very hard to improve clients' visualization early in the project, we still must acknowledge that some of their visualization comes only after the design, or even after the final product has been built.

If we cannot renegotiate, the design may be very much harder to achieve, and the product will probably be more expensive and take longer to develop. We saw this in the COCOMO model for cost and time estimation in Chapter 6, where the organic project was more open to renegotiation and took less time and man-months to develop than the embedded project.

Design of Algorithms

The design of algorithms is a different animal than what we confront in most other computing courses. It is creative, not straightforward; it is likely to be somewhat frustrating and potentially very rewarding—and a lot of fun. Algorithm design uses the basic techniques of problem solving. The best way we can address this subject is by presenting first an example and then a discussion of the principles we use.

Assume we want to write a program to take an indented outline and print it out in the form of a tree. How do we develop an algorithm to do this?

As we develop our algorithm, we represent it as a Tree. We can always start the Tree by writing the root, as in Figure 15.5.

Next we make up a simple problem, which we do by hand, as in Figure 15.6.

As with many algorithms, we can make up a problem and solve it by hand quite easily. The trick is how to tell a computer to do it. Humans have the ability

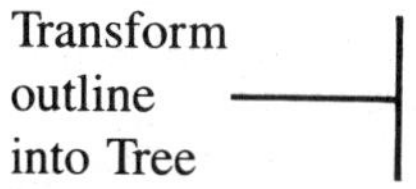

Transform
outline
into Tree

Figure 15.5 Root of Tree.

to see things as a whole—a trait called gestalt. The computer is nearsighted. It sees only one thing at a time. So we have to make a transformation from a gestalt process to a sequential process. Although we may solve the problem quite naturally, like riding a bicycle, we have to think about how we do it in order to tell the computer how it should do it. This is very much like teaching a skill to another person. We have to analyze what we do in order to tell them how they can do it.

Now we must consider how we represent our problem—that is, how we represent the input before the transformation and the output after the transformation.

Indented Outline	Line	Indentation Level
A	1	1
B	2	1
*B*1	3	2
*B*2	4	2
*B*3	5	2
*B*3*a*	6	3
*B*3*b*	7	3
C	8	1

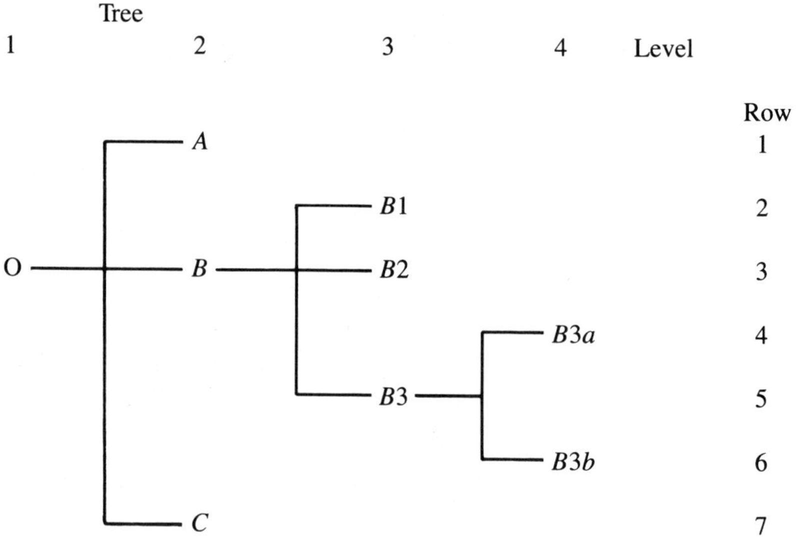

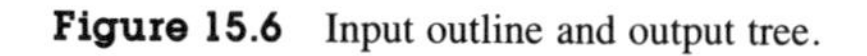

Figure 15.6 Input outline and output tree.

In this case we represent the input as lines with line numbers and a number for the level of indentation. We represent the output by the lines, the level of indentation, and the row number on which the line will be printed. It is a guess that if we use this representation it will help us in the algorithm. We plod on to see whether we are right.

We can see quickly that if we number each line of our outline by the level number and row, we can sort these lines by the row to print out our Tree in row sequence. Thus we can solve our problem provided we have the solution to these three other problems: number, sort, and print.

This technique is a common ploy; we conclude that we can solve one problem if we can just solve certain other problems. Then we proceed to try to solve these other problems, which hopefully will be easier to solve. This process of breaking down problems into other problems can be represented as a tree, in this case a solution tree. If we come to a problem we cannot solve, we retreat and try something else. Once we get to the situation where all the leaves represent problems to which we have a solution, we have the problem resolved. We solve the problems at the leaves and work back up the tree until we get to the root, giving us the solution to the original problem. Our tree so far looks like Figure 15.7. The Sort and Print are problems for which we already have solutions, so we won't consider them further here. Now we'll focus on how we'll do the numbering.

Here we can introduce a new concept in developing solution trees. We use a single limb (AND) to represent when all the subproblems must be solved, and a double limb (OR) to indicate when solving any one of the subproblems will solve the original problem. For example, Figure 15.8 says we can solve problem A if we can solve problems $A1$ and $A2$, and we can solve problem $A1$ if we can solve either problem $A1a$ or $A1b$.

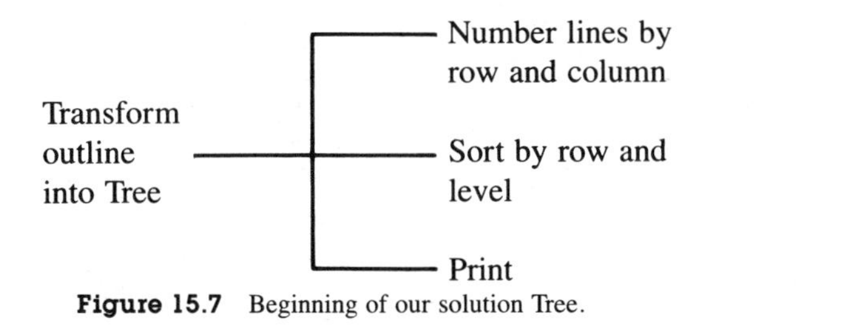

Figure 15.7 Beginning of our solution Tree.

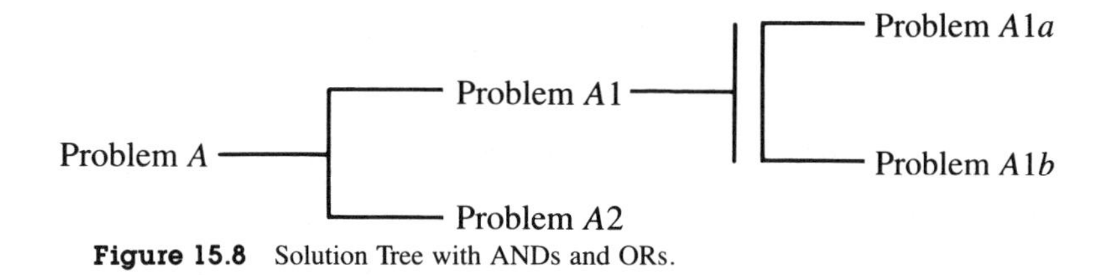

Figure 15.8 Solution Tree with ANDs and ORs.

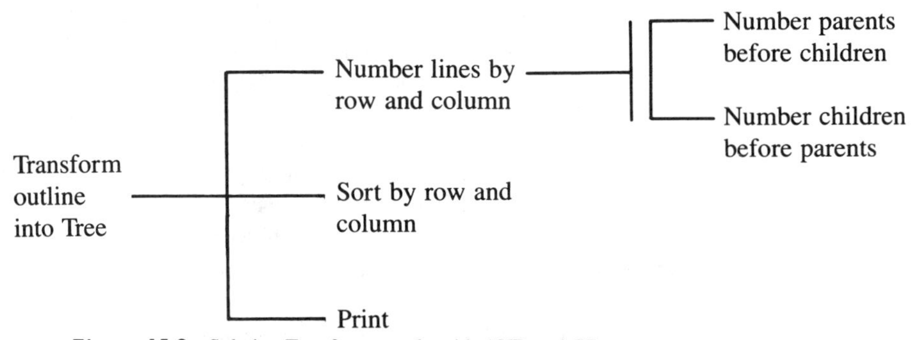

Figure 15.9 Solution Tree for example with AND and OR.

Applied to our current problem, we may guess that we can number the lines if either we can number the parents before the children, or number the children before the parents. Now our solution tree looks like Figure 15.9. We can try both numbering parents first and numbering children first in our hand problem to see which, if either, will work.

When we test these two hypotheses by hand, we see that we cannot assign a row number to a line until we have assigned row numbers to all its descendents. Thus we stick with the "Number children before parents" and reject "Number parents before children." Thus we can eliminate all but one of the OR alternatives in our solution tree.

Numbering children before parents suggests that we might need a stack to keep track of the branches as we pass through them on our way to the leaves, assign row numbers to the leaves, and then work backward assigning row numbers to the branches leading to those leaves.

Next we consider the decisions that need to be made. As we go down the lines of the outline looking at the next line, we must face several circumstances. The next line could be a child—that is, have a higher level number. Then we would add the current line to the stack to defer its numbering until after its children have been numbered. The next line could be a sibling—that is, have the same level number. Since the current line then has no children, we could assign it the next row number. Or the next line could be neither a child nor a sibling—that is, have a lower level number or be at the end. Then we know this line is the last sibling, so we can assign it the next row number, pop the stack, and number its parent. We can try this by hand with our sample problem to see if our guess works. If so, we continue in this way until we get a complete, successful solution. We have made a start in Figure 15.10.

Now we have to put our decisions in terms of computer variables rather than conditions that we recognize only by eye. For example, we have to set up an array and index to point to the current line and the next line and make these decisions in terms of these variables. We have to have a variable to keep track of the row count. The parent will be printed on a row that is the average of the

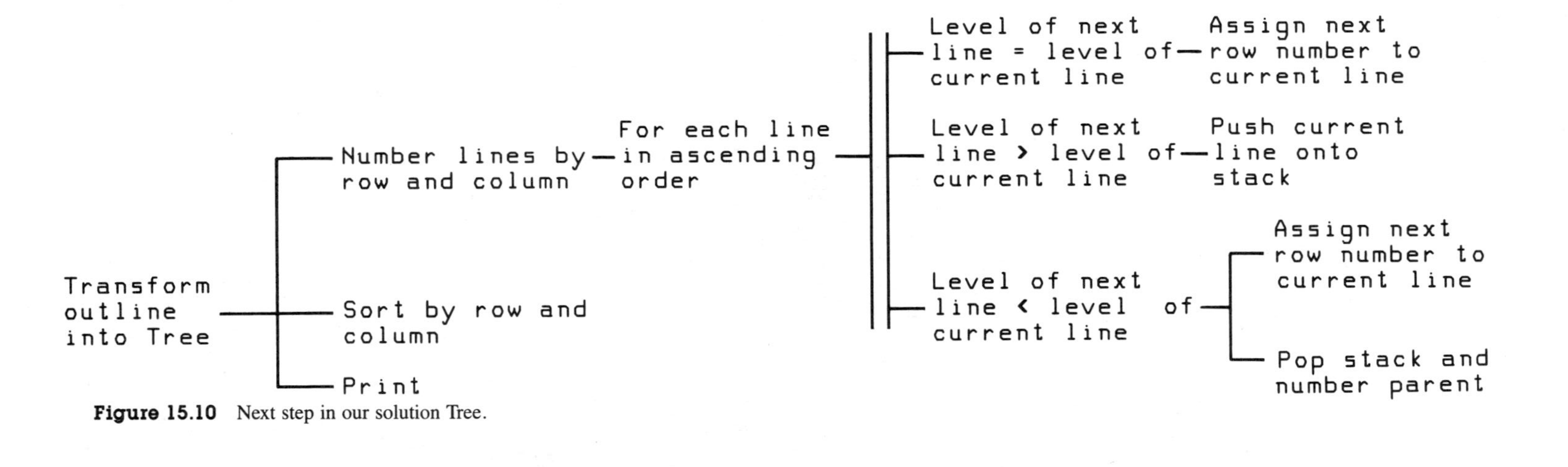

Figure 15.10 Next step in our solution Tree.

N	=	number of rows in outline
J	=	index of current line
LEVEL[J]	=	level of line J
L	=	level number
STACK[L]	=	stack
ROWCT	=	row count
ROW[J]	=	row number assigned to line J
ROWF[L]	=	row of first child
ROWL[L]	=	row of last child

Figure 15.11 Variable dictionary.

rows of its first and last child, so we have to keep track of where we print the first and last child. We guess at this time that this may have to be kept in a stack. In this way we start to build up a variable dictionary, as in Figure 15.11. We will also start adding the calculations of these variables to the tree, as in Figure 15.12. Note that as we change the control, add the calculation of variables, and add decision statements—all in computer language—the solution tree starts to become our program represented as a Tree.

We continue in this way, building our tree by defining problems in terms of other subproblems, adding the required decisions, and defining the variables we need to make these decisions. At each step we simulate our current program by hand, using an example.

Once we have an algorithm that we can show by simulation solves our simple example, we try it on more difficult examples. If we have any question about whether some aspect of the algorithm might work in certain cases, we make up an example that will test these cases. In this way we work with more and more difficult examples until we are convinced that the algorithm could handle any case that could be thrown at it. Since we are hand-simulating these cases, we try to make them so they require as little hand calculation as possible, yet reflect all the logical conditions we want the algorithm to be able to handle.

It is suggested that the reader finish this algorithm as an exercise.

Let us generalize what we have done here.

1. Our general strategy is to break down our problem of developing an algorithm into smaller problems such that:
   a. The solution to the smaller problems will result in the solution of the original problem.
   b. We believe that we have a good chance of solving these smaller problems.

   A solution tree can show how we are breaking these problems into smaller problems using ANDs and ORs. When we reach a problem we can't solve, we retreat and look for a different problem we believe we can solve, which will lead to the original solution.
2. We look for states—that is, places in the algorithm—where certain situations should exist. For example, in our example of a payroll program in Chapter 4 (Figure 4.28) "Get Match" is a state where all the information needed to determine an employee's pay and deductions has been assembled. We consider what must be done to achieve this state. This is like attaining subgoals along

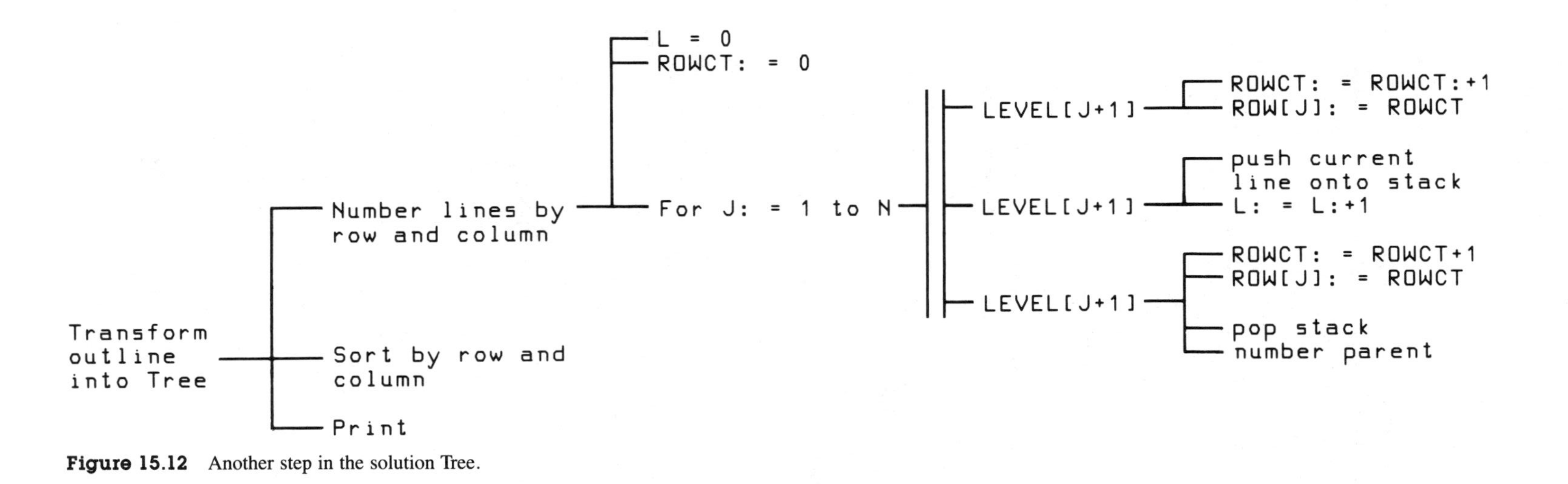

Figure 15.12 Another step in the solution Tree.

the way, which can help us break the problem of developing the whole algorithm into the problem of developing smaller algorithms.

3. We propose a way of representing the input state and output state.
4. We do the problem by hand and observe:
   a. What decisions we make
   b. What information we need to make them
   c. When we have and need that information
   d. How we get it
5. We determine what information we need to keep track of and develop a variable dictionary.
6. We transform the descriptions in the solution tree into computer language. Thus we transform the solution tree into a program Tree.

Design Strategies for Information Systems

When we make a design search, it helps to have a strategy that tells us where to look first to make the search most efficient. Let's consider some strategies that are useful in designing information systems.

Backward Search

Backward search concerns working backward from the outputs we want to the inputs we must have to get those outputs. If these inputs are available, the search is successful and we say we have obtained closure. But if we find such inputs are not available, then the backward search is not closed. We may have to try a new design or even renegotiate with the client to change the outputs the system will produce.

Generally to increase the likelihood of closure we start our search with what is most constrained and work toward what is less constrained. For example, our primary constraint is the outputs we need. We are less constrained by what inputs are or are not available. This is in part because our decisions about what outputs we want are tempered by our insights as to what outputs we can reasonably expect to get from available inputs. So working from outputs to inputs fits the pattern of working from the more constrained to the less constrained. Backward search is represented in Figure 15.13.

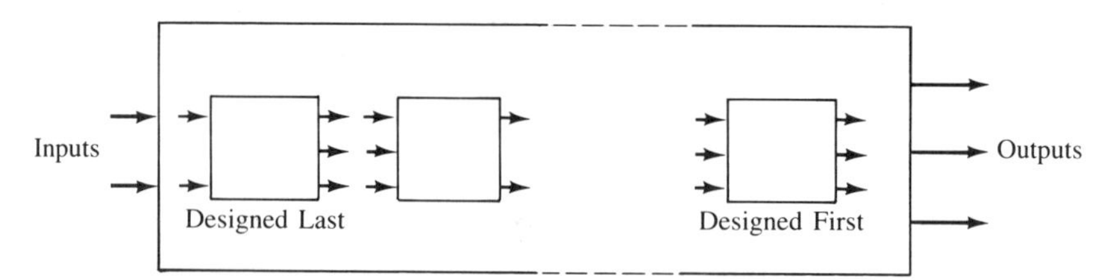

Figure 15.13 Backward search—from outputs to inputs.

Top-Down Search

Another type of search concerns the top-down approach to design, which we discussed earlier in this chapter and in Chapter 2. Most of our discussions of design up to this point have assumed we are using **top-down search**. We may design our system using processes that do not yet exist but that we believe we could build. These are called **abstract processes**. Once we have a design using abstract processes, we analyze it to show whether it satisfies the requirements. If it does, then we go on to design the abstract processes. If this recursion ends with our system described in terms of available processes, then we say the top-down search has reached closure. If we are unable to make top-down closure, we may have to return to our top level design and use different abstract processes. Top-down search is represented in Figure 15.14.

Data Structure First—Warnier-Orr and Jackson Methodologies

Yet another design approach says that the hierarchical structure of the program should be derived from the hierarchical structures of the outputs and inputs. The Warnier-Orr and Jackson methodologies take this approach. We will illustrate this procedure using as an example the billing procedure that appeared in the order processing system discussed in Chapter 3.

First, we make a tree describing the structures of the output and inputs. As we consider each part of the data structure, we add a corresponding process to the program structure. By this means we can move from the data structure trees in Figure 15.15 to the Tree for the program in Figure 15.16.

First the heading on the billing ledger is written out and certain initializations are done. Then there is an iteration as each customer's bill is computed and printed and the aged balances printed on the ledger. Then the total aged balances are printed on the ledger. We add the details to the Tree showing how these high-level actions are to be carried out. This "how to" now adds detail that goes

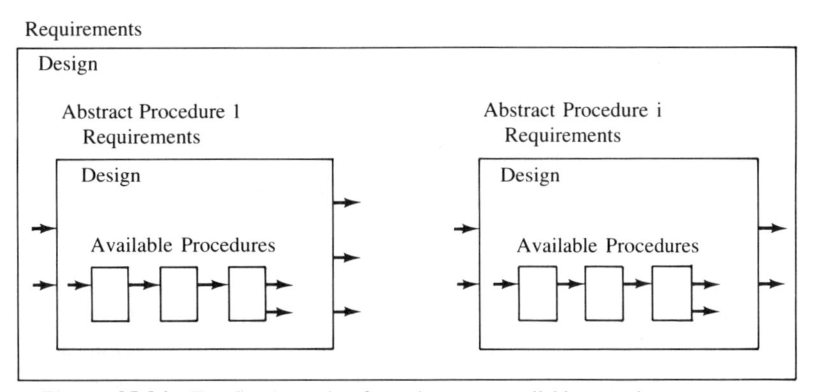

Figure 15.14 Top-down search—from abstract to available procedures.

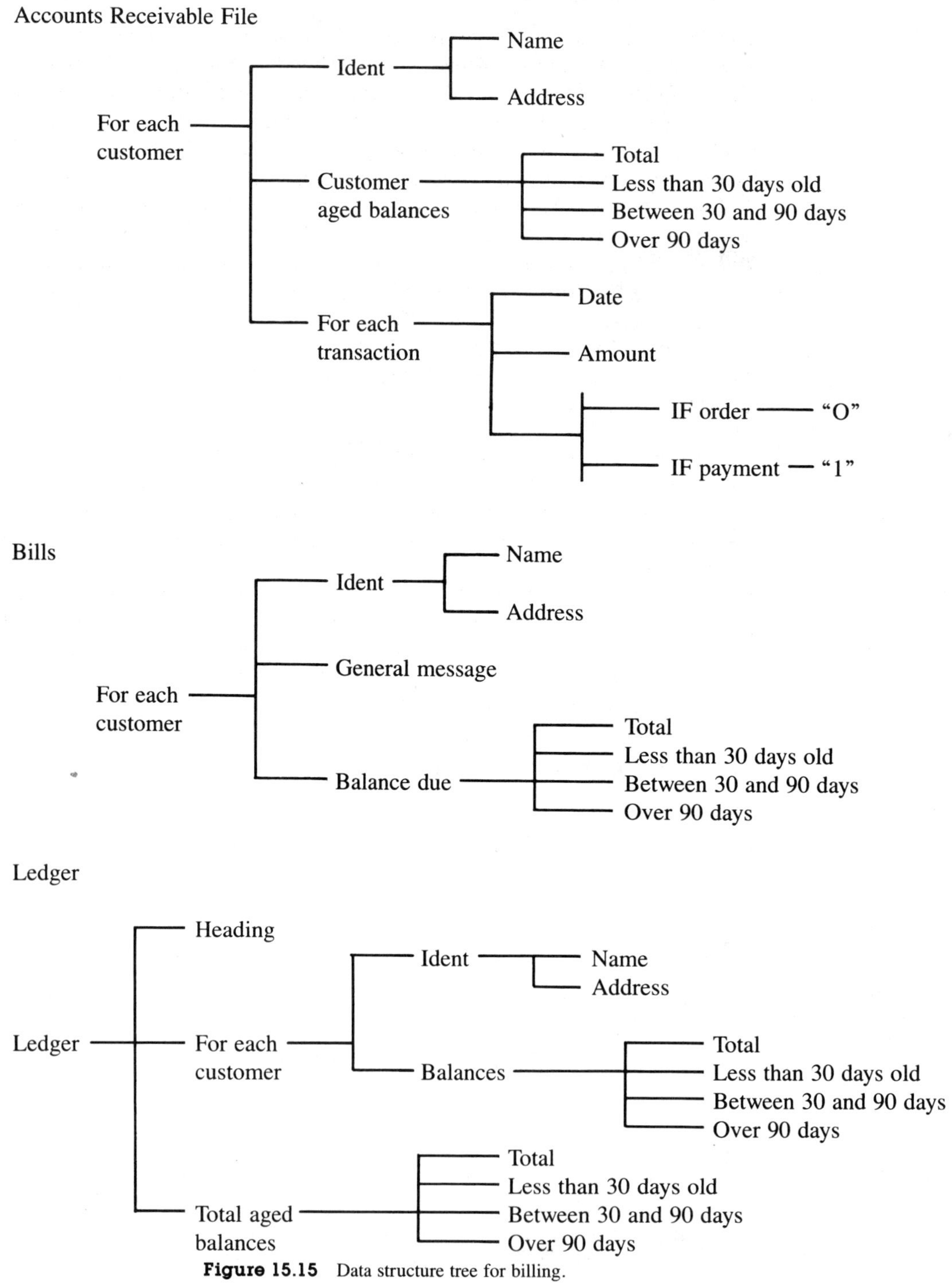

Figure 15.15 Data structure tree for billing.

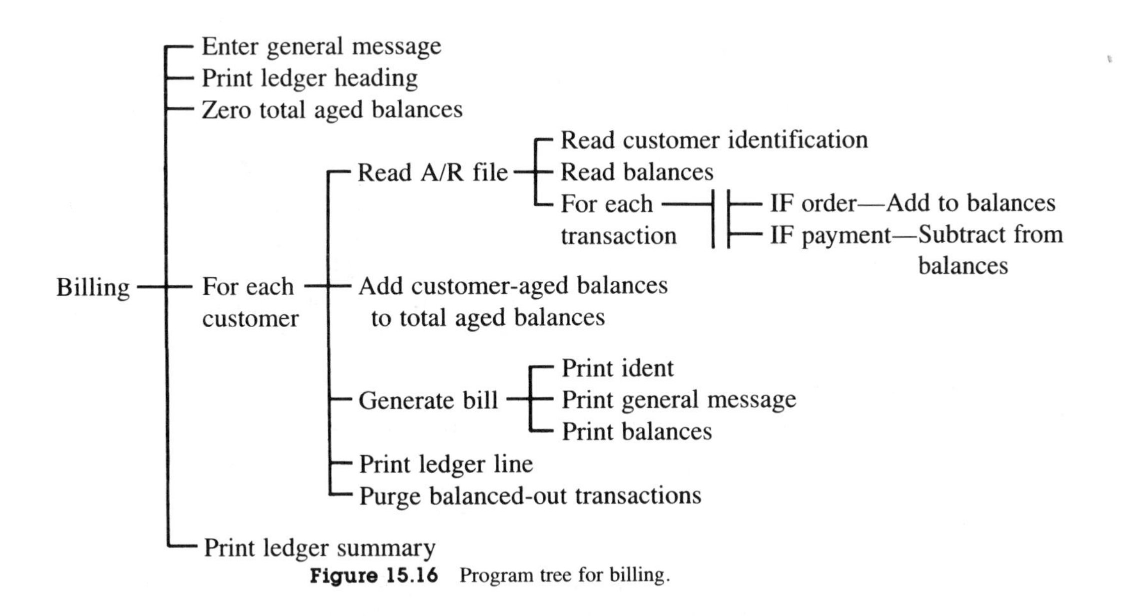

Figure 15.16 Program tree for billing.

beyond just a reflection of the structure seen in the data diagram. In our example, at this next level we detail how we handle reading the transactions and adding to the customer balances.

Note that this program structure minimizes the amount of information the program has to hold between input and output because we are able to print the output in the order in which it is computed. But suppose we were required to have the total balances printed on the ledger before the customers' balances. Now we have to hold all the results of computing the customer's balances until after we have all the information needed to compute and print the totals of these balances. Only then could we print out the customers' balances that we were holding. This is an example of what Jackson calls structure clash between the data structure and how the program structure must be developed.

Thus we cannot always follow the data structure exactly to get the program structure. If we have the freedom to define the input and output data structures, such as printing totals or summaries at the end, then we can usually design typical data processing programs by this method. If we do not have the freedom to control the structures of the input and output, we may still be able to use this technique if we introduce an intermediate file. The intermediate file is written as the input is processed, then reordered so it can be read in the order the output is generated.

Structured Design: Coupling and Cohesion

Structured design concerns how to organize programs so they will be easy to maintain. The basic idea is to construct the program so that later when someone goes into it to make a change, the change will be unlikely to have some effect that causes an unexpected problem. This approach implies that the module should have some integrity, meaning that the functions performed within the module should be so related that someone else can easily see what the module does and doesn't do. It also implies that the data flows between modules should have integrity, meaning that it should be easy for someone changing the program to see how changes in one module will or will not have effects propagated to other modules. The integrity of the module is called **cohesion**. The integrity of the data flows is related to **coupling;** high integrity yields low coupling.

Structured design has two aspects: (1) how to design the hierarchy, and (2) how to pass data. The former we considered in Chapter 4 as part of our discussion of Trees. The latter is properly a topic in programming techniques but is often included in software engineering.

In the classical discussion of structured design we are generally given five types of coupling between modules and seven ways of establishing cohesion within a module. These types of coupling or cohesion are ordered from good to bad. We should constantly look at the design of our modules to establish that the methods of coupling and cohesion used are as good as circumstances can allow. The following lists contain the five rules of coupling and the seven rules of cohesion. We will then give another representation of these concepts that leads to essentially the same results.

Coupling

Methods for coupling, from best to worst, are as follows:

1. *Data Coupling*—Just the data needed are passed between the calling and called modules through calling parameters. This is the best way.
2. *Stamp Coupling*—Data are passed between the calling and called modules through calling parameters. However, the parameters include data structures where only part of the data in these structures is needed in the call—that is, more is passed than is needed. This circumstance can create a problem because a change in one of the data structures to accommodate a change in either the calling or called module can have effects on other modules as well.
3. *Control Coupling*—One module passes data called control flags whose purpose is to control the actions within another module. Such other modules are frequently written to perform somewhat different functions for different calling modules. This arrangement can be avoided by having each calling module call a module specifically designed to perform just the needed function.
4. *Common Coupling*—Data are passed between modules by leaving the data in some mutually agreed upon location in a global data area (for example, the COMMON feature in FORTRAN). A change in one module may then require changes in other modules sharing the same data area.
5. *Content Coupling*—One module reaches into the internals of another module

to grab or deposit data or control its function. When a change is made to either module, it may require a thorough analysis of the internals of both modules to determine how to deal with the consequence of the change.

Although not ranked on this list, another type of coupling to be avoided is called "tramp coupling." In tramp coupling data flows through many intermediate modules to get from where they are produced to where they are used. When possible, we should try to have the module requiring the data closer to the module that produces them.

Cohesion

Methods for cohesion, from best to worst, are as follows:

1. *Functional Cohesion*—Everything in the module contributes to the same easily defined function. Thus, when we are maintaining that module, we know what to expect will go on inside it. This is the best way.
2. *Sequential Cohesion*—The various functions in the module are executed in sequence with the output of one becoming the input to the next.
3. *Communicational Cohesion*—The various functions in the module use inputs from the same source or produce outputs for the same other modules, but they do not represent a sequence where the output of one is the input to the next.
4. *Procedural Cohesion*—Control within the module flows in sequence from one module to another.
5. *Temporal Cohesion*—Functions within the module are not related by any of the preceding criteria. They just happen to be done at about the same time in the program, such as doing all initializations for all modules within an initialization module. It is better to initialize variables in the modules in which they are to be used.
6. *Logical Cohesion*—Functions within a module are related by the type of processing they do, such as printing various outputs. They are not related by any of the preceding types of cohesion. It is usually better to have separate modules to perform just the specific function needed.
7. *Coincidental Cohesion*—There is no clear reason why the things done within a module are done within the same module.

We can't always use the first, or best, type of coupling or cohesion on the list. For example, we might find it convenient to have a module that handles printing error messages. This method would require control coupling and logical cohesion, which are way down on the list. That's not necessarily bad, but we have to ask ourselves why we didn't choose a method of coupling further up on the list. The further down on the list we go, the more carefully we should consider why we are doing it that way and what effect it will have on the ease of maintaining the program.

Another Approach to Coupling and Cohesion

Another approach to the same end as using these rules for coupling and cohesion is the application of the following concepts:

1. Assign functions to modules (that is, subtrees) so that when people have to change the module, it will be easy for them to grasp exactly what the module will and will not do. Avoid having modules that will do many different types of things for different calling modules depending on the calling parameters. In Tree development a test of good functional cohesion is whether it is easy to write on the branch leading to the subtree a succinct statement correctly describing what the subtree does.
2. Minimize the number of data elements that flow on the limbs closest to the root (as we discussed in Chapter 4). The data flows between modules should have an integrity so that we can easily understand when changes are made exactly what effects of the change will be transmitted to other modules, requiring changes also to be made in them. If we are using a Tree editor, the editor should evaluate a coupling metric so we can compare various structures to see which is best.
3. Implement the method by which data are passed (for example, passing parameters in calls, leaving data in common data areas, and so on) so that at most the two modules involved in the passing of data will have to be changed if one of them is changed. Preferably, when changing a module, we should have to know only about the module being changed and the data flows into and out of it, not the modules on the other side of those data flows. Data should not be passed using structures that are also used to pass data between other modules.

Tools of Design

During the design process we invent and connect processes, data flows, and data stores to build a design. To keep track of what we are doing and to describe what we have done, we maintain either data flow diagrams or Trees. We describe the data we are working with in data element dictionaries, and show the structure of the data in data structure trees. We discussed these methods in Chapters 3 and 4.

In data flow diagramming the backward design is shown by designing the processes that produce the final output, then the processes that produce the inputs needed by these processes, and so on until we work back to available inputs. The top-down design process is shown by the development of a hierarchy of data flow diagrams.

In a Tree the backward design is shown by developing the Tree from the outputs that occur toward the bottom of the page, advancing up the page toward the top where the inputs occur. Top-down design is shown by working from the root at the left out toward the leaves at the right. Data structure first design means that we develop a tree for the hierarchical data structure, then derive from it the Tree for the program.

Physical Design

The preceding discussion has focused on logical design, which is where the design of information systems begins. Logical design is concerned only with the information and what transformations are made to it. Once we know what transformations are to be made, we can consider the physical design.

In the physical design we allocate these transformations to the hardware, the software, or the people who will perform them. If the transformations are allocated to hardware or software, specifications must be written describing what hardware or software is to be bought or how it is to be written. If the transformation is allocated to people, we must design the procedures they will follow.

The logical design emphasizes satisfying the functional part of the requirements specification. The physical design emphasizes the allocation of functions to hardware, software, and people, then satisfying the performance part of the requirements specification.

When we write requirements for computer hardware, we must consider such matters as whether the hardware will provide for adequate primary and peripheral storage for problems of the size specified, and whether it will be fast enough to satisfy the performance provisions of the requirements specification. If we write requirements for the purchase of software, we must specify such matters as the capacity of that software—that is, the size of problems it can handle, as well as how fast it can run these problems, and how easy it will be to train the people who will be using it.

Technical Considerations in Design

The technical aspects of design concern the properties we want in our system and the properties of the components we put together to get that system. These technical properties, which are of concern in both analysis and the design, are usually the subjects of other courses. We discussed some of these topics in Chapter 8 and will mention them briefly again here just to bring them into the context of design.

Hardware

The decision of what hardware to use becomes a constraint on writing the software. We may use off-the-shelf hardware we already possess or which we can purchase, or we can design the hardware as well as the software to run on it. The logical design of the software can occur even before the hardware is specified. The software, however, cannot be compiled until the hardware is specified, and cannot be tested until either the hardware is built or we have a hardware simulator running on an available machine.

Communications and the Distribution of Functions

A current trend in computer systems is to distribute work to personal work stations tied together by communications. The designer must consider the locations of the people using the system, where the inputs come from and the outputs go to, what data is stored and where, and where the computing occurs. Initially this can be shown on the data flow diagram. The design of distributed systems is complicated by such problems as how to maintain the integrity of the data when it is distributed in several places and operated on by various users.

Controls, Backup, and Security

Before we design the system, we must think of how it will operate so we can attempt to conceive of everything that can possibly go wrong. We consider problems arising from, among other causes,

- errors in input,
- difficulties occurring in the operation of the system,
- errors arising in communications,
- equipment malfunctions and downtime,
- previously undetected software errors,
- loss of files and data bases,
- physical or logical access to the system by someone who may destroy equipment or data, either accidentally or intentionally, and
- access to information by someone to whom that information should not be known.

Once we conceive of as many problems as possible, we must design the system to detect these problems, if and when they arise, and to take corrective action. How well we design the system to handle all possible problems will be a major factor in the success or failure of the project.

Design Estimation

The logical design should tell us the data structures that will be used. From these data structures we should be able to estimate the amount of primary and peripheral memory to be used by the data. The amount of input-output processing can be estimated with the help of the calculations we make on the data structure matrices and worksheet, as we discussed in Chapter 3.

Estimates of memory size required for the code and the speed are difficult to make. We may compare each of the program modules with similar modules that have been written earlier, making allowances for the differences in the size of the data structures. What constitutes "similar" is currently left up to experience. With sufficient data we might some day be able to make these estimates with equations that use parameters describing the amount of input-output, the level of checking required and the reliability of the sources of input data, the amount of calculations to be done, and whether it is a batch, on-line, or real time program, and so on. What usually saves us in these estimates is the freedom to allow plenty of memory or time. If the requirements squeeze us on memory or time, we can be in for difficulties.

Often it is worth our while to write prototypes of various critical parts of the programs such as key algorithms to help us estimate running time and memory. But we must make sure that the prototype properly represents the dimensions we wish to estimate. If the prototype compromises on the input checking, as we might do for a prototype to evaluate the human interaction as discussed in Chapter 9, we should not expect to get from this prototype a good estimate of the program memory size or running time. A prototype may give us the time to run a small problem. A mathematical analysis of the algorithm will be needed to see how this time scales with the size of the problem—that is, does the time increase linearly with size, with the square, or faster? When using prototyping, we must be very concerned with what we are trying to estimate and what is being compromised to be able to develop the prototype more quickly than the final program.

Often we do not know the memory or timing implied by the logical design until after we have gone through the physical design and preliminary implementation of the system. Then if we find that the implemented system is not satisfactory, we may analyze it to determine which are the critical items that must be redesigned and reimplemented. If the timing is not satisfactory, we may instrument the program by adding statements that will allow us to analyze how much time is being spent in each part of the program. We might redesign just the unsatisfactory parts. Problems in timing or memory may require that we change algorithms, particularly algorithms that affect how we use primary and peripheral memory.

Outline of Design Specification

The following outline of the design specification will give some insight into the various things that must be specified during design. The outline can be tailored to the needs of an individual design and used as a guide for documentation and in planning the design work. Some of the topics shown in the outline refer to technical considerations that should be covered in other courses, such as normalization of data bases, and so on.

- Title page (standard)
  - Project name
  - Document type (design specification)
  - Document identification number
  - Revision number and date
  - Authors / contributors
- Table of contents
- 1.0 Introduction (standard)
  - Overview description of this project (very brief)
    - Whole project
    - Part (or all) of project to which this document pertains
  - Description of this document
    - Purpose of this document and how it is to be used
      - Intended audience
    - Content and organization of the document
    - Authorship
      - Author
      - Date written
      - Procedure for changing this document
  - Controlling documents and references
  - Definition of minimum terms, acronyms, and abbreviations needed to read this document
- 2.0 Requirements overview (optional)
  - Purpose of system
  - Environment and interfaces
  - Functions
  - Inputs, outputs, and files
- 3.0 Purchase specifications (may be separate documents overviewed here)
  - (like requirements specification, but for purchased items—that is, environment, function, performance, cost, and so on)
  - Hardware:
    - Computers: mainframes, minis, micros, PCs, and so on
    - Peripherals: disc drives, printers, and so on
    - Terminals
    - Communications
  - Software: data base management systems, operating systems, utilities, applications, and so on
  - Facilities: space, power, air conditioning, lighting, communication lines, and so on
- 4.0 System design
  - Processing type: Batch, on-line, real-time, time share, personal computer, and so on
  - For each type of transaction:
    - How is the input information generated?
    - How is it processed?

- Selection of hardware: processors, storage devices, peripherals, communications, and terminals
- Selection of software: operating system, data base management system, utilities, and applications
- Developed software: functions, programming language
- Distribution/location of: users, functions, processing, data storage, and communications
- Controls and protections against: error, wrong access, mischief, and fraud
- Backup and recovery procedures
- Data flow diagrams with annotations of: processes, data flows, files
- Data element dictionary
  - For each data element:
    - Identifier
    - Name
    - Aliases—other names and where they are used
    - Description
    - Format:
      - Length
      - Type: Alpha, alphanumeric, integer, decimal (with decimal position and units)
    - Structure: part of (parent), made up of (children)
    - Original source
    - Verification criteria: coding or range
    - Where used: inputs, reports, screens, data flows, processes, files
    - Access—Who can: read, add, change, delete, or reorganize
    - Volume: minimum, maximum, average, expected growth
- Standards: programming, user interface, and so on

5.0 For each process/module:

- Interfaces:
  - Called by (parent), with what parameters
  - Calls to (children), with what parameters
  - Data flows: in and out to other modules, in and out to files
  - External I/O and controls: input, screens, reports
- Functions and options
  - How selected/controlled
- Algorithms and data structures
  - Narrative description
  - Hand-computed examples showing sequence of steps and data values
- Exceptions and how they are handled
- Controls/protections against: error, wrong access, mischief, and fraud
- Input/output: blocking factors, buffer sizes, and so on
- Data structure trees
- Program trees

6.0 For each record or data base:

- Type: input, report, screen, data flow, file, data base

Logical design:
  Data elements
  Structure: Relations between data elements
  Source and where used
  Keys and sorts
  Organization: indexed random, indexed sequential, and so on
  Linkage between files / Access diagram
  Normalization (1st, 2nd, 3rd, Boyce-Codd, 4th, 5th normal form)
  Volumes: minimum, maximum, average, growth
Physical design:
  Device or media: disc file, paper, screen, and printer
  Fields: contains what data element, size, position
Storage location
Backup and restore procedures
Retention period
Access/security—Who can: read, add, change, delete, reorganize
Controls against bad data: coding, range, and redundancy
Test data: values to be used in testing and for Mortar First—Bricks Later implementation (see Chapter 16)
For screen records:
  Control/selection method: menu, command, and function key
  Layout of: titles, data, and columns
  Use of: windows and icons
For input records:
  Source
  Coding/interpretation
  Method of verification: check against file, built-in redundancy, double input and compare, and so on
For output records:
  Type: detail, summary, exception
  Layout of: titles, data, and columns
  Use and distribution

7.0 Test
  Functions to be tested
  Methods to test them:
    Use of drivers, stubs, and hand-generated data
  Input: data, controls
  Output: expected values and acceptance criteria
  Performance: expected and acceptance criteria

8.0 Manual processes
  Procedures
  Manuals
  Training plans
  Resources/supplies and special forms

9.0 Implementation plans
- Data conversion
- Maintenance
- Support
- Facilities: space, power, air conditioning, lighting, and communication lines

Summary

Design has both a technical and a creative aspect. It is not a simple, straightforward process of working backward from the requirements to the design. We must synthesize, or choose and put together components, and then analyze, which involves checking the system's behavior against the specifications. If the specifications are not met, we must make another synthesis, and so on. Good designers, therefore, are able to guess what works so that some iterations can be avoided.

Two common models of design are the top-down design and the design structure system. In the top-down design we draw a design showing the modules, then specify the requirements for the modules, and design the modules. Large modules are designed from smaller modules. The design structure system allows the design of modules with the design of neighboring modules in mind.

The design process follows two general steps: synthesis and analysis. During synthesis we identify the constraints, search for the components that will satisfy our specifications, and then construct a design. The analysis stage involves simulating the design and evaluating the results of the simulation. We may sometimes have to renegotiate the design with the client.

We design an algorithm by breaking down the problem into smaller and smaller solvable problems. A solution tree can be used for representation of the problem-solving process. We carry out the algorithm by hand to indicate what steps, information, and decisions are needed. We develop a variable dictionary and finally we transform the solution tree into a program tree.

Among the useful strategies for designing information systems, backward search helps the designer to work back from the desired outputs to the necessary inputs. Top-down search follows the synthesis and analysis using abstract processes. The data structure first approach derives the program's hierarchical structure from the hierarchical structures of the inputs and outputs.

Structured design tries to organize programs so that other people will be able to make changes later without generating unexpected results. This approach is designed to sustain the integrity of modules, which is called cohesion, and the integrity of data flows, which is tied to coupling. There are five methods for coupling—the best being data coupling—and seven methods for cohesion—the best being functional cohesion.

When we have completed the logical design, we can consider the physical design, which involves allocating the necessary transformations to the software, hardware, and people who will perform them.

From the logical design we see the data structures that will be used. These structures help us to estimate the amount of primary and peripheral memory and input-output processing needed.

Exercises

1. For the CAPERT case study in Appendix A:
   a. Develop a Two-Entity Data Flow Diagram.
   b. Write the annotation forms to support the TE-DFD.
   c. Develop a high-level Tree.
2. Do the following exercises for the design of the IMPERT program described in Appendix B.
   a. Develop a screen and an output report layout for this critical path scheduling program.
   b. Develop the Tree for this program.
   c. Note that if we use the data in this example, when computing the latest finishes we never have to replace a computed value from one successor with a lower value from another successor. Would this example then make a good test case to check that this logic works correctly? Could we develop a better example if we interchanged the labeling and the order in which we read in tasks *A* and *D*?
   d. Let us assume that we want to develop a set of arrays SUCC and SUCCPTR that store the successors just as PRED and PREDPTR store the predecessors. (You might want to do this so you can print out both the predecessors and successors for each task on the report.) How would you develop an algorithm to do this? We will give you hints to represent two ways this might be done. Think about these two methods and consider that we may wish to run this program with up to 3000 tasks with an average of three predecessors each. Compute the number of times the algorithm will read or write data in any of the arrays. Which algorithm would be better? Would the difference be significant even on a modern fast computer? (Such considerations are the concern of courses in the analysis of algorithms. This is just a taste here of a practical example.)

   For Algorithm *A*

   For each task go through the whole list of predecessors finding all the tasks that have this task as a predecessor. Those tasks are then successors that can be entered in the SUCC array. In this way we can build the successors of each task in sequence.

   For Algorithm *B*

   Clear the SUCCPTR array. Then go down through the predecessors and increment the SUCCPTR array to count the number of successors to each task. Then go through the list again, adding these counts so that SUCCPTR now points to the address in SUCC where the first successor of each task will be stored. Now go down through the predecessors and for each predecessor, which implies a predecessor-successor pair, put the successor in the correct place in SUCC and increment SUCCPTR ready for the next successor. This sets up the SUCC array. But when we finish, SUCCPTR is pointing to the position after the last predecessor of each task. We have to make one more pass to put SUCCPTR back to pointing to the first predecessors in SUCC. We do this by moving each element of the array up one position. Remember to put the element at the end of SUCCPTR that points to one position past the end of SUCC. Play with this using the example in Appendix B to see how it works.

3. Complete the design of an algorithm to convert an outline into a Tree, which was begun in this chapter. Develop a solution tree, then convert this solution tree into a program Tree. As you develop the algorithm, think through the following:

   How is the object you are working on represented?
   What data structures do you use for this representation?
   What states are there in the algorithm, and what does the algorithm need to do to produce each state?
   What variables are needed to record these states?
   What decisions are made, and how are the variables computed to make these decisions?
   What is the sequence of operations needed to perform the algorithm?
   What backtracking did you have to do in developing the solution tree—that is, what did you guess that you later had to change?
   What new examples did you have to develop to explore how the algorithm works in special cases?

4. Write and test the program.

References

Alagić, Saud, and Michael A. Arbib. *The Design of Well-Structured and Correct Programs*. New York: Springer-Verlag, 1978.

Bergland, Glen D., and Ronald D. Gordon. *Software Design Strategies,* 2d ed. Los Alamitos, CA.: IEEE Computer Society Press, 1981.

Brookes, Cyril H. P., Phillip J. Grouse, D. Ross Jeffery, and Michael J. Lawrence. *Information Systems Design*. Sydney, Australia: Prentice-Hall of Australia, 1982.

Freeman, Peter, and Anthony I. Wasserman. *Software Design Techniques,* 4th ed. Los Alamitos, CA.: IEEE Computer Society Press, 1983.

Hansen, Kirk. *Data Structured Program Design*. Englewood Cliffs, N.J.: Prentice-Hall, 1986.

Horowitz, Ellis, and Sartaj Sahni. *Fundamentals of Computer Algorithms*. Patomac, MD.: Computer Science Press, 1978.

Jackson, Michael A. *System Development*. Englewood Cliffs, N.J.: Prentice-Hall International, 1983.

Jackson, Michael A. *Principles of Program Design*. Orlando FL: Academic Press, 1975.

Myers, Glenford J. *Composite Structured Design*. New York: Van Nostrand Reinhold, 1978.

Osborn, Alex F. *Applied Imagination: Principles and Procedures of Creative Problem-Solving,* 3d. rev. ed. New York: Charles Scribner's Sons, 1963.

Peters, Lawrence J. *Software Design: Methods & Techniques*. New York: Yourdon Press, 1981.

Polya, George. *How to Solve It*. Garden City, N.Y.: Doubleday, 1957.

Riddle, William, and Jack C. Wileden. *Tutorial on Software System Design: Description and Analysis*. Los Alamitos, CA.: IEEE Computer Society Press, 1980.

Rubinstein, Moshe F. *Tools for Thinking and Problem Solving*. Englewood Cliffs, N.J.: Prentice-Hall, 1986.

Wirth, N. "Program Development by Stepwise Refinement." *Comm. ACM,* 14, 4, April 1971.

Yourdon, Edward N., and Larry L. Constantine. *Structured Design: Fundamentals of a Discipline of Computer Program and Systems Design*. Englewood Cliffs, N.J.: Prentice-Hall, 1979.

16

Implementation

- Implementation involves many types of tasks: programming, developing manual procedures, converting retained data from the old to the new system, training, purchasing hardware and software, setting up facilities, and so on.
- The purchasing process may be of "off the shelf" or specially built items. A bidding process may be required making a list of bidders and sending them requests for proposal (RFPs).
- Among the possible approaches to building a system are thorough requirements, reusable code, versioning or incremental development, rapid prototyping, and application generators.
- Reusable code is created when we store in a library the code generated for each module, then check the library before we consider programming a new module.
- In versioning, or incremental development, we develop a system in increments, using dummy modules for the modules that have not yet been created.
- We test modules as we build them. Systems may be produced in either of two ways—by developing modules starting at the top and working down or starting at the bottom and working up.
- The Mortar First—Bricks Later approach generates the interfaces first and then develops the modules.
- Once a program works we can consider checking its performance to see where it may be improved to run faster or handle larger problems.
- Switch-overs must be carefully planned so the client can meet his or her responsibilities while making the transition from the old system to the new one.

We are sometimes inclined to focus on programming when we think of developing a computer system. But of the total project effort, programming usually represents less than 20 percent, and this percentage can be expected to decrease as systems methods become more mature. Implementation is the major focus of many other books and courses, but here we will spend only a short chapter to show how implementation fits into the total picture.

Appendix D has an Information Systems Analysis and Design Matrix in which the phases of the project appear as columns and the various types of tasks as rows, with specific tasks shown in the intersection of row and column. In this context it can readily be seen that programming is only one of many tasks that must be done to produce a system.

Equipment and software may need to be purchased, which means writing specifications for what is wanted, then shopping for "on the shelf" items or soliciting bids for what must be specially developed to our specifications. Then we must evaluate whether what we get meets these specifications.

If we are replacing an old system that used historical data, that data must be converted to run on the new system. Manuals must be written and people trained to run the new system. Facilities must be arranged with the needed space, electrical power, light, heating, and air conditioning. We must wean our customers away from the old system, making a switch to the new system without the users ever failing to meet their responsibilities to their customers during the change.

If we write our own software, we can take one of several approaches considered in this chapter.

The Purchasing Process

To obtain the necessary program, we can program it ourselves, or we may be able to buy it. If we buy it, we may be able to use it as is, or we may have to modify it to serve our needs. A similar make-or-buy decision must be made for each of the components of the system.

If we decide to buy, we must go through a purchasing process that has many steps very similar to the requirements analysis we go through if we make it ourselves. We must establish very careful requirements so the supplier knows exactly what will satisfy our need and we know what we will be getting.

The usual purchasing process requires that we send a purchase request to our purchasing department specifying exactly what we want. Purchasing draws up a list of vendors who may be able to supply the component and a decision is made whether to buy directly from one vendor (called a sole source purchase), or whether we must allow several potential vendors to bid for it. We may be constrained by our employer's policies to put the purchase out for bid unless we can justify that only one vendor can provide what we need.

If the order is put out for bid, purchasing makes up a bidder's list and each bidder is sent a request for proposal (RFP) specifying as exactly as possible what is wanted. The RFP serves the same function when we buy the product that the requirements specification serves when we build it ourselves. The request for proposal must also state the criteria to be used in selecting the winning bid. The bids must be submitted by a stated time, when they will be opened and compared.

A bid can be rejected if it clearly does not meet the specifications. However, if the RFP does not specify the requirements clearly and unambiguously, and if the lowest bid meets those the specifications as we stated them, we may be required by policy to accept it. Thus, it is in our interest to write the requirements carefully so that we are prepared to accept at the lowest price absolutely anything that satisfies those specifications.

Sometimes different bidders will propose different ways to satisfy the same requirements. Some ways may be better than others. In some cases none of the bidders is able to satisfy all the requirements, or one bidder may not be able to satisfy some minor requirement but has a much lower price. This leaves us with different prices from different bidders representing different proposals of how to meet the requirements. Thus, the bidding is sometimes done in two stages. In the first stage the bidders are requested in the RFP to state how they propose to satisfy the requirements. These suggestions are then used in writing a more exacting set of requirements for a second RFP. The bid on the second RFP is then won by the lowest price of those bids meeting the requirements.

Several Approaches to Building

If we build the system ourselves, we can proceed in one or more of the following ways:

- *Thorough Requirements*—Most of this book has focused on this approach. **Thorough requirements** involves carefully thinking through what is wanted by using abstract representations of the system (see Chapters 3 and 4) to visualize what the final system will look like.
- *Reusable Code*—Every time we program a module, we review it carefully to see if it might be used again in other systems. If there is any chance that it can be reused, we spend the extra care to write it using very general techniques, document it carefully, and make it available in a library. This technique is known as **reusable code**. Thereafter, before we consider writing a module, we check the library to see whether such a module is already available.
- *Versioning or Incremental Development*—We can develop the system as a series of versions, each version containing more modules completed and having more capability than the previous version. Dummy modules must be substituted for the modules that have not yet been developed. This method, called **versioning**, may also be called **incremental development** because the system is developed in successive increments. The key to the success of this technique is very careful

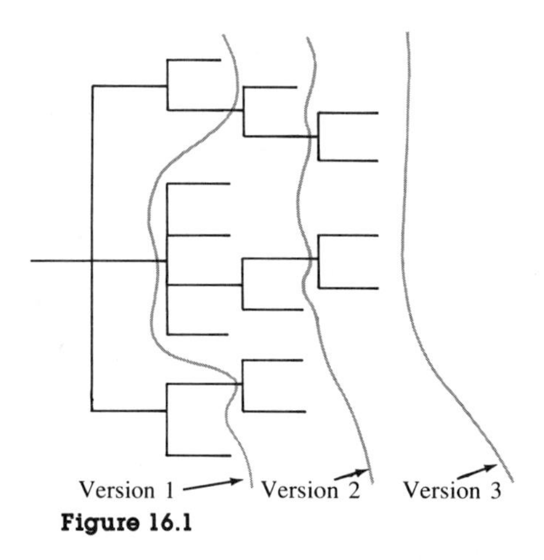

Figure 16.1

planning as to what modules and what dummies are used in each version. The techniques used in this method are related to rapid prototyping.

In Figure 16.1 we have drawn boundaries on a Tree to show how we might represent which parts are to be implemented in each version. Dummy routines are used to simulate the interfaces with the parts not yet implemented.

- *Rapid Prototyping*—Rapid prototyping is a means of getting an early visualization of what certain aspects of the final system will look like, or checking a high-risk aspect of the system by quickly developing and testing certain critical pieces or functions. We discussed rapid prototyping in Chapter 9.
- *Application Generators*—We discussed this approach in Chapter 8. It can be used as part of rapid prototyping, quickly providing a system showing the user formatted screens and reports and performing the basic computations and data base processing. An analysis of performance may show that the system produced with the application generator is adequate, or may indicate that the system must be rewritten in conventional programming language to perform faster or handle more data.

Integration of Implementation and Testing

We test as we build. Otherwise, if we test only once when we are finished building the whole system, there can be so many problems we will be unable to sort them all out. The changes required to fix them can be monstrous. Thus, our implementation strategy must consider how we plan to test the components and the subsystems at each step as we build them and integrate them into larger systems. We must plan how control and data are to be supplied to the modules or subsystems so we may test them. Top-down, bottom-up, and Mortar First—Bricks Later are approaches to how this can be done.

Top-Down and Bottom-Up Approaches

If a module is developed before the module that calls it, we write a dummy program to act as the calling module. This dummy is called a **driver**. The driver must pass fabricated data to the tested module. Then it must either check any data that are returned against precomputed results or print out the results so they can be checked by someone reading the output.

On the other hand, if the module to be developed calls a module that has not yet been programmed, we can write a dummy module to substitute for the module called. This dummy is called a **stub**. The stub will check or print the data sent to it and return prefabricated data to the tested module.

Drivers and stubs may be used to substitute for the parts not yet written in a rapid prototype or incremental version. Stubs can be used to simulate the use of resources by containing arrays to take up space and loops to use up time.

If we write the modules toward the leaves of the Tree first and use drivers to test them, this system is said to be developed bottom-up. If we write the top modules first and use stubs, then this system is said to be developed top-down. The more general term *incremental development* can be used to refer to either of these approaches, or various mixed approaches where some parts are developed top-down and others developed bottom-up.

Rapid prototyping and thorough requirements can be used hand in hand by carefully laying out requirements for the whole system and prototyping the high risk aspects to determine feasibility. The requirements may be modified after the prototype tests.

Rapid prototyping or incremental development can often provide clients with a system they can use temporarily to meet their needs, relieving some of the pressure to have an operating product while the developer produces a more carefully crafted and comprehensive version.

Mortar First—Bricks Later Approach

In the **Mortar First—Bricks Later approach** we develop first the interfaces that hold the modules together (the mortar), then we develop the modules (the bricks). We develop a general program that we can use to write files containing our hand-generated test data. Once these files are written, the modules can be written and tested against these interface files. Thus, many people can be writing and testing many different modules at the same time. Because the people developing modules that must work together are testing them against the same test files, we have confidence that when the modules are written they will work together. We avoid the problems that arise when we try to rewrite modules to fit with other modules, which can cause a succession of rewriting module after module.

Two Views—Control and Data Flow

In design, implementation, and testing we can adopt one of two points of view: (1) the control structure view, and (2) the data flow structure view. When we discuss the top-down and bottom-up approach, we use trees and take the control structure view. When we talk about the Mortar First—Bricks Later approach, we use data flow diagrams and take the data flow point of view. Both the top-down and Mortar First—Bricks Later views emphasize implementing the interfaces early in the development. Generally if the interfaces are correct, the modules will fit into place without trouble. But if different modules are developed assuming different interfaces, then all these interfaces and the modules may have to be redone before we can get the system to work.

Performance Analysis

The first consideration in implementing a program is that it works; that is, it satisfies the functional requirements. A program that is fast but doesn't work is of no value. Once it works, we can consider how fast it performs. It is necessary to make this rather obvious point because programmers often play a game of finding ways to shave microseconds off the performance of the program. In the process, though, they use tricks that are hard to document and maintain, leading to programs that run fast but do not correctly perform their proper functions.

Once we have a program that properly fulfills its functions, we can run experiments to see if its speed is satisfactory. If it is, the program can be left at that. Massaging the program to get more speed may not be worth the cost of depriving other projects of this programming effort.

If the speed is not satisfactory, we need to determine how much and where the speed needs to be improved. Pareto's law is our friend in this process. Usually the greatest amount of the execution time for most of the cases occurs in only a small part of the code. Rewriting only this small part of the code, maybe less than 5 percent, can often improve the speed of the whole system significantly, sometimes even an order of magnitude or more. Often these sections of the code can be redone in the same high-level language used for the rest of the code by using a new algorithm, or by carefully analyzing what is done inside loops that could be done outside, adding buffering of the I/O to the disk, and so on. Using the high-level language facilitates making such changes to the methods used, which can result in significant performance improvements. But sometimes it may be necessary to recode just these vital parts of the program in a lower level language.

The key is to find those critical areas of the code that take most of the execution time. These critical areas are often not the ones we expect. Thus we may wish to add code that will monitor how often the program enters certain parts of the program. By multiplying by an estimate of the execution time of the part, we can develop an estimate of how long the program spends in each part. This can sometimes be done with a **profiler**, a program that analyzes branches in the source code and inserts monitoring code. Then, as the program is run, data are collected on a file, which is analyzed by another program later to produce the necessary statistics.

But we should add a note of caution. Often the speed or volume of data the program can handle is dictated by the data structure. If a change to the data structure is required, this change usually means completely redesigning and reimplementing the whole program. A good designer will recognize the need for speed and the amount of data to be handled, and will initially design the data structure of the program accordingly.

The choice of data structure is particularly important when deciding whether the program is to handle all its data in memory or to use disk access during the processing. This can be a hard decision to change later.

Preparing for the New System

Although computer programs have been a major concern of our book, we must also design and implement the people and data components of the system.

We don't lock the design and implementation of all the aspects of the system together so that all the designs are done in one time phase and all the implementations done in another. Instead we break the design and implementation of each of these aspects into tasks and schedule them using a critical path network. It is only as a matter of convenience that we choose to discuss design implementation and testing in separate chapters.

Procedures

Just as we developed programs to tell the computers in the system how to work, so we must develop procedures that show the people in the system how to do their work. The people who will operate the system must be chosen (hired or transferred) and trained. Before they can be trained, manuals must be written and training plans developed.

When we developed the logical design for the system, we decided what information transformations were to be made. In moving to the physical design, we determined which of these transformations were to be done by hardware, software, or people. This is the starting point for developing both programs and manual procedures.

People

We have to decide what tasks the people need to do, then consider the skills they will need to do them. Can we acquire people who already have these skills, or do we acquire people without the skills and plan to teach them? We must consider where we are going to get these people; by hiring or by transferring them within the organization. What responsibilities are they to have and who is to manage them? If we transfer from within, we must consider how and when these people can be released from their current jobs. How do we schedule their work hours to cover the time the system must operate? This is all part of our personnel planning.

We may need to coordinate training with the acquisition of the hardware or software on which that training will be done, or find some way to simulate the operation of the system on an available machine so we can start training before the final hardware or software is available.

Data

The old system may have records that must be carried over to the new system. If the media on which this information has been stored or its format is not compatible with the new system, this data must be converted. We may find mechanical ways of doing this, or the information may have to be manually reentered. This conversion may begin once the new media and format are known.

Site Preparation

If new equipment is to be purchased, the site must be prepared for its operation. That usually requires space, whether it be a room for a minicomputer or mainframe, or a desk for a personal computer. Mainframe and minicomputers may require special provisions for power, air conditioning, wiring between cabinets, false floors under which to run the wiring, and so on.

Switch-Over

One of the most crucial processes in the systems development is the switch-over from the old system to the new. Supposedly the client had a way of meeting obligations to his or her customers using the old system, although the client probably was not completely satisfied with it. And the client will again have a means of meeting obligations when the new system is fully implemented. But in the early days of getting the new system running properly, there is a great danger that difficulties will arise because of circumstances not properly anticipated in the requirements, errors in the design specifications so that the requirements are not met, or errors due to not implementing the system to the design specifications. If these problems do not become evident until the system is running, the user may not be able to fulfill his or her obligations using this system until the problems are fixed. Some companies have been forced out of business because they were not able to meet their obligations or collect what was owed them during the switch-over period.

The obvious solution is to run the old and new systems in parallel, relying on the old system and checking the results for the new system until all difficulties with the new system have been resolved. But parallel operation can be extremely expensive. During this time the client is paying the staff and expenses for (1) running the old system; (2) running the new system; and (3) comparing and evaluating the outputs of the new system, diagnosing the causes of any difficulties, and making the necessary fixes. Usually the people who run the old system will be running and getting training on the new system at the same time. They are running the new system under the most adverse circumstances because they are trying to learn it before it is working properly. And they will not be up to speed with the new system because they are still gaining experience with it. They are working long hours under great pressure. It is worth a great deal of testing by the system developers before the switch-over to shorten this expensive but necessary parallel operation.

Some systems have gotten into trouble because an early schedule committed the old equipment to be returned or sold when it was thought that the parallel operations would be completed. Then due to schedule slippages the old equipment became unavailable before the new system was ready to assume its responsibilities. The switch-over was premature and had to be made cold turkey. The result: chaos and cost.

Summary

Implementing a system involves making all of the changes necessary to make the new system work. It includes acquiring hardware and software, programming, putting into place manual procedures, training, data conversion, site preparation, and switching over operations from the old to the new system.

We may implement a system in a number of ways. We can buy and use a program as it is, buy it and make modifications, or build it wholly ourselves. The purchasing process usually follows certain established steps. A product may be bought from a sole vendor or a list of bidders may be sent a request for proposal, with the purchase going to the vendor with the best bid.

There are several approaches to building a system. Thorough requirements uses abstract representations to visualize what the system will look like. The reusable code method keeps codes for developing each module in a library so they can be applied later in another system. A versioning, or incremental development, approach builds the system in increments using dummy modules for those that have not yet been created. Rapid prototyping offers advance looks at certain aspects of the system to make an early resolution of questions about requirements or design. Application generators show formatted screens and perform the basic computations and data accesses.

Testing takes place during building. We carry out testing using dummy modules. A driver is a dummy that substitutes for a calling module; a stub substitutes for a called module. Bottom-up development is writing modules toward the leaves first using drivers to test them. Top-down development writes the top modules first and uses stubs to test. The Mortar First—Bricks Later approach designs the interfaces between modules, then develops the modules themselves.

The first priority is that the system works according to the functional specifications. Once that has been ensured, we can check to see that it performs according to performance specifications. In attempting to speed up the program, it is important to find the critical areas of code that take the most execution time.

Switch-overs from an old system to a new system are tricky transitions. The old system must be kept long enough so that the client can meet his or her responsibilities during the switch-over.

Exercises

1. Write an implementation plan for the new CAPERT system.
2. For the IMPERT program from Appendix B consider how you would plan the writing of the program to do incremental implementation.
3. Write a simple version of the IMPERT program.

References

Gilberg, Philip. *Software Design and Development*. Chicago: Science Research Associates, 1983.

Weinberg, Gerald M. *The Psychology of Computer Programming*. New York: Van Nostrand Reinhold, 1971.

17

Testing and Maintenance

- The best method of preventing errors in a system is to write careful specifications, create a good design, and conduct frequent reviews so that errors will not be introduced and later need to be found and fixed.
- Corrective maintenance includes testing, diagnosis, and changes aimed at getting the system to do what it was expected to do. Perfective maintenance is making changes to get the system to do something beyond what it was originally expected to do.
- The number of combinations of conditions that must be checked in a program may be so great that we cannot guarantee by testing that it will meet the specifications.
- The quantity of test cases can be reduced if the system can be broken into independent parts.
- Test cases must be monitored so we know what parts of a program have been tested and what parts haven't.
- Testing includes design testing, module testing, integration testing, system testing, and acceptance testing.
- After a developer has tested the program, a professional tester should search for remaining faults.
- Regression tests make sure that tests that had previously run correctly still do.
- Errors are documented in problem reports.
- A test plan shows how to verify the system is implemented according to the design.

A software error exists whenever a program does not do what the client can reasonably expect it to do. This could be because of a misunderstanding between the client and the developer when the requirements were written, because of a poor decision on how to design it, or because of a mistake in implementing it.

As we have pointed out before, the more work that is built upon a faulty decision or piece of work, the more expensive it is to fix it when the resulting error is found. A large proportion of software errors found in testing were made in the requirements specification, and because requirements are done so early and so much is built upon them before testing occurs, mistakes in the requirements specification can require much work to be redone and can be spectacularly expensive to fix.

How easy it is to find errors depends on the complexity of the program. Complexity implies interaction. By designing the system in modules with a minimum of coupling between modules (that is, good structured design), we can more easily isolate where the mistake is and fix it without introducing other errors. Thus the way we design the program is very important to how expensive it will be to find and fix errors.

The costs of software errors are due to:

- the costs of the testing required to find them
- the costs of diagnosis to find their cause
- the costs of fixing them, which may require redoing much of the work done since the errors were introduced
- the costs clients incur because they relied on bad results or did not have use of the system

This latter cost may include the costs of running backup systems when the primary system is not working correctly and the costs of lost time, opportunity, credibility with customers with a resulting loss of business, legal suits, and possibly even loss of life. As we rely more heavily on software in transportation, factories, and medical systems, the costs of computer failures in our society become less acceptable.

In this chapter we'll see how we contend with the mistakes that appear in a system. We'll describe the types of tests that can be run to uncover errors and the types of maintenance to keep the system accomplishing its goals. Finally, once we fix the faults, we also need to document them and reestablish confidence in the product.

Modes of Failure

Hardware usually fails because some part wears out with use and time. It can be fixed by replacing the part with an identical one put aside when it was made and thus not subjected to any wear. Software generally does not wear out, although the recording of the software may. Software fails because it is exposed to new input and situations that reveal faults that already existed but had not been previously uncovered. Such failures can occur with no warning and at the most unexpected times.

High reliability in hardware, such as required in a manned space mission, can be achieved by using three redundant computers. The system chooses the answer it gets from any two of the three. One cannot achieve software reliability by this means unless the three programs are written by different groups of people. The system could still fail if there was a fault in the requirements used by all these three groups.

A major consideration in software design is to protect against input data that the program is not intended to handle. A requirement of some systems—which probably should be a requirement of most systems—is that it will handle any input, either by producing the correct results or rejecting the input. More than half the code of some programs may be devoted to checking against invalid input, even though this input may occur only infrequently. But when such an input does occur, if it is not caught and people rely on the results, the consequences can be very costly.

As we discussed earlier, our best assurance against errors is careful requirements specifications (which may be based upon rapid prototyping), careful design, and frequent reviews at each stage to catch faults as soon as possible. Above all, we should proceed with the conviction that we are not supposed to make errors.

The only sure way not to have errors in the delivered system is not to make any. Particularly in complex systems, the extra effort needed so as not to make errors usually pays off handsomely because of the very large cost of finding and fixing errors. In principle we could push past this point of diminishing returns by applying too much effort to avoid errors that can easily be caught. But this mistake is not commonly made. We will see shortly why finding and fixing errors can cost so much.

Testing and maintenance can account for a significant part of the system's cost over its full lifetime. For some systems 40 percent of the development time and cost is devoted to testing and changes to get the system to work as originally intended. And once the system has been developed and is working correctly, as much time and effort as was expended in the development may then be spent making changes to enhance its capabilities beyond those originally intended.

Thus, a major part of the system's cost over its lifetime is devoted to changing it rather than developing it. Much of our attention in earlier chapters was devoted to careful development of the system to make these changes easier. This added effort in the development of the system repays itself when making these changes.

Maintenance

There is often confusion in defining what is and is not considered part of maintenance. We consider maintenance in two general categories:

1. **Corrective maintenance**—testing and diagnosis, then designing and making changes to get the system to do what it was expected to do.
2. **Perfective maintenance**—designing and making changes to get the system to do something beyond what it was originally expected to do.

We use the phrase "expected to do" to include the possibility that the correction may be due to errors in the requirements specification, leading to differences between what clients thought they were getting and what the requirements specification called for.

A subset of perfective maintenance concerns those changes that pertain to running the system on new hardware. This category is sufficiently important that it is given its own term, **adaptive maintenance**.

Maintenance can be as much as two thirds of the total cost of the system over its full lifetime. Of this two thirds, roughly one fourth is corrective, one fourth adaptive, and one half other perfective maintenance (Pressman, 1979).

If the perfective maintenance involves significant change, we are facing a whole new project. Even small changes require that we go through the same steps we go through in a full systems project: analyzing the need, determining the requirements, designing the change, implementing the change, and maintaining configuration control for the revision.

Testing—The Battle with Combinatorics

To test a program thoroughly we must check that it does all the arithmetic and logic correctly. Testing the arithmetic is usually fairly easy. We compute values by hand to compare with all the logical possibilities inherent in the arithmetic—for example, for positive, zero, and negative values of parameters that may produce logically different results. There usually are not too many of these arithmetic combinations. But ensuring that the program takes all the right paths is another problem altogether.

In a continuous system we can test that it behaves properly at certain points and know that it will behave properly between those points. When a system makes discontinuous jumps, it has to be tested at each jump. Hardware tends to be continuous or at least have few jumps and is thus relatively easy to test. Software tends to be discontinuous with many decisions and conditions responsible for many jumps, which makes software much more difficult to test (Parnas, 1985, 1326–1335).

The number of combinations of conditions that must be checked in a program of any significant size can be so great as to make it impossible to test the program to ensure that the requirements are completely satisfied. To guarantee that every branch is properly taken and every loop is initialized and exited properly requires not only that we have one test case to test the extreme value immediately on each side of every branch or loop exit, but also that we test every possible path representing all possible combinations for these branches and loops.

We can analyze the source code to map each set of values at the branches back to the input required to produce these values at the branch. Programs can be written to help in this source code analysis. But then we must develop the answers for these cases by some independent means and compare them with the results from running the program to validate the program's correctness.

If we have n branch points in the program, and we wish to test each branch with data for the two extreme conditions on either side of the branch—the greatest (or least) value for which it branches, and the least (or greatest) value for which it does not branch—then the number of possible combinations is 2^n. For a simple module with 24 branches, 2^{24} is 16,777,216 test cases. Even if we could run these cases on the computer, for each test case we must independently develop the correct answer and validate that the module gets that answer. This is impractical for large modules.

Independently developing the answer to validate the module's result often means computing it by hand. Sometimes we are concerned only that the new program gets the same answer as is generated by an existing and already tested program. However, that other program or some program that came before it had to have been validated by hand.

Some types of problems are such that it is easy to make a calculation proving we have the right answer. The solution of a system of equations is an example. We can determine quickly whether the answer is right by plugging it in, even though getting it may have taken a tremendous amount of computation. For these types of problems we can validate each test case by writing a program that checks that the answer satisfies the required conditions. But even in the simple case of solving equations, round-off errors can raise ambiguity about whether the answer is correct, or whether a logical error may be hidden by the round-off.

Combinatorics concerns the problems that result from all the combinations with which we have to work. How can we fight this battle of combinatorics? Primarily we can test in small pieces. For example, if we took our module with 24 branches and broke it into 3 modules of 8 branches each, we would get 3×2^8 different test cases, which would give us only 768 tests. Table 17.1 shows the number of test cases as a function of how many equal-sized modules we break it into. By this reasoning a module with n branches broken into m equal parts would require $m \times 2^{n/m}$ power.

These numbers are very impressive, particularly since a module of 24 branches is still not very large. Another assumption needs to be stated. If you divide a module into smaller parts and test the parts, you must take the responsibility for the assertion that if each of the parts works, then the parts when put together will work. So as we break the module into smaller parts to test it, we are replacing brute force with the need to apply our own powers of reasoning.

We can greatly reduce the number of test cases if we can partition the system into independent parts so that the various combinations can be generated for each

Table 17.1 Number of tests required for a module with twenty-four branches.

Number of Parts	Branches per Part	Tests per Part	Total Tests
1	24	16,777,216	16,777,216
2	12	4,096	8,192
3	8	256	768
4	6	64	256
6	4	16	96
8	3	8	64
12	2	4	48
24	1	1	24

part alone, and each combination in one part does not have to be tested in conjunction with all the possible combinations in the other. To do this requires that we make a careful analysis of the program.

When we get down to testing modules of one branch each, we might as well test each branch as we specify it in the design. Thus our final protection that the system does what the requirements have specified rests with the design, not with tests made by running the program.

Although testing may be of limited use for telling us that the system meets the requirements specification, it may be useful for telling us that it was implemented according to the design specifications.

Test Coverage Monitoring

When we run our test cases, we want to keep track of what has been tested and what has not been tested. Has every part of the program been executed during at least one of the tests? It can be embarrassing if after the program has been turned over to the clients, they run a case that does not work because some part of the program had never been tested.

Statements can be inserted into the source code to collect data on a file counting how many times each side of each branch was taken, and for each piece of code whether and how many times it was executed. Computer programs have been developed to insert these statements automatically. These monitoring statements need be inserted only at the branches and where the branches join since it can be assumed that if a branch is taken, all the code up to the next branch must be executed. Another program can analyze the statistics collected on this file to produce a report showing how well the tests have covered all possibilities. Sometimes it is useful to draw a graph called a **control graph**, which shows as nodes the branches and as arcs the sequence of code between the branches (see Figure 17.1). These graphs can be used to analyze programs to help us determine how and where to test them.

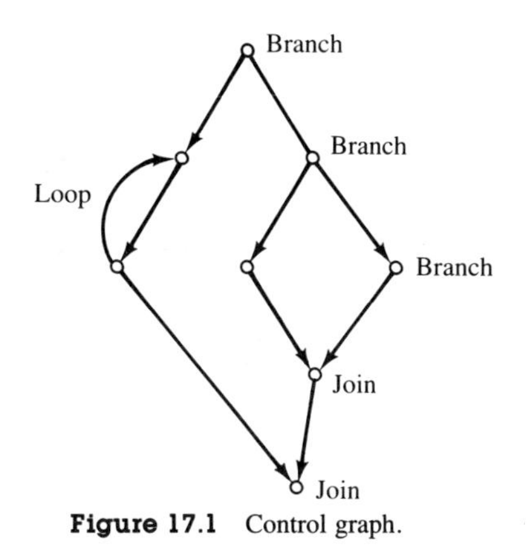

Figure 17.1 Control graph.

When to Test

The earlier we can test the better. It is costly to build components from designs that don't work, or systems from components that don't work. When a faulty design or part is finally found, all the work built with that design or part has to be redone. Thus, we want to test the design before it is implemented by simulating in our minds to determine whether it will work as intended (design testing). As we build modules, we test them (module testing) before they are integrated with other modules into subsystems (integration testing), and then when these subsystems are integrated and tested as a whole system (system testing). Finally, we want to test the system as it will be delivered to the client so we can determine whether he or she is likely to accept it (acceptance testing).

Developing Test Cases

A successful test is one that fails. If any fault exists, the purpose of a test is to find it. If a test case apparently runs correctly, the result can mean either (1) there is no fault, or (2) the test case was inadequate to find it.

As the builders, we'll naturally want our product to work. If we do not watch ourselves very carefully, therefore, we are likely to design our tests to demonstrate the program's strengths instead of seeking out its weaknesses. We must continually remind ourselves of the acute embarrassment that awaits us if someone else finds the faults we missed. After we have made our own tests, we should give the final testing to someone else who is not afraid to reveal any additional faults as efficiently as possible. This person should approach testing with the destructive vengeance of a maniacal fiend. Our testing will be more thorough because of our embarrassment at having someone else find our mistakes. It is best that mistakes not be left to be found by the client, or worse yet, to be found by his or her customers.

As we have seen by looking at the problems of combinatorics, it is not likely that we can ever run sufficient test cases to cover all conditions to guarantee correctness. We had best analyze the program to determine what might be its greatest weaknesses and concentrate our tests on them first.

The more faults we find in the initial testing, the longer it is likely to take to get the remaining faults worked out. If we find an unexpectedly high number of faults, we might begin to wonder whether it will ever be possible to achieve a high degree of reliability by testing and fixing this program. It may be better to throw the program out and start over from whatever point we feel it first began to go wrong. This could be all the way back to the requirements specification.

Clients may not intend to be fiends, but very often, in all their innocence, they will find errors we never thought to test for. Their approach to the program's use does not include our prejudices about how the program should be treated. They also are the ones most interested in seeing that the program works correctly because they must assume the risk to their business or job if it does not.

Diagnosing Errors to Find Faults

Once an error is found, we must diagnose the error by tracing its cause to the responsible fault in the program and then fix it without introducing new faults.

If we are told there is an error, we may wish to run our own test case to confirm and study for ourselves the circumstances under which the error occurs. We try to isolate the fault to as small a part in the system as possible. The approach here is to find a part of the program where correct data goes in and incorrect data comes out.

Next we may be able to develop an hypothesis of what may be causing the problem and run tests to see if we get the errors we would expect if this hypothesis were true. We may make tests to rule out some hypotheses so we can proceed in confidence to other hypotheses.

Finally we may go through the code hand-simulating the execution or use a trace on the running code to dump the intermediate results until we can see something happen that does not look right. Or we may look at a dump of memory to see if we can figure out what happened to produce those values.

Fixing Faults

Once we have found the fault, we must determine how to fix it. In fact it may be easy to come up with a fix that will cause the program to perform correctly for the specific case that revealed the fault. It is much more difficult, though, to come up with a fix that does not introduce errors into other cases.

One of the largest single sources of faults in programs is fixing other faults. Some systems have never worked because fixing the old faults has introduced still more new faults.

Given a proposed fix, we must carefully consider what its effects will be for all legitimate input. After the fix is made, regression tests should be run. **Regression tests** check that cases that previously had run correctly still do.

The fix should be carefully documented through the configuration control process so that the records to be used later will show the correct configuration of the system.

Problem Reporting

When an error is found, we write a **problem report** with the following type of information:

Problem report identification—perhaps a sequential number
Circumstances of the error
- System and operating environment at the time of the error
- Test case(s) that revealed the error
- How should the system behave and how does it in fact behave?
- What are the consequences?

Who is authorizing a fix?
Who needs the fix and when?

Once the fault is found and fixed, we write a **change report**, which includes the problem report and adds the following information:

Was the error confirmed as valid?
What was the cause of the error?
What changes were made?
- Code before the changes
- Code after the changes

What tests were run to confirm that:
- The changes fixed the reported error?
- The changes did not introduce any new errors?

How Do We Establish Confidence in the Tested Product?

Tests may find errors, but they cannot prove correctness. How do we gain confidence that a program is working correctly? One approach is to have someone other than the person doing the testing seed the program with known errors. The assumption is that the proportion of these known errors that are found is similar to the proportion of unknown errors found. But we must be sure to keep good records of the seeded errors so that if they are not found, we can remove them before the program is delivered.

If our professional fiend tests the program and does not find errors, and if he or she has a reputation that few errors are ever found after his or her testing, then this check may give us some confidence.

Some current research is applying the methods of mathematical proof to ensure that a program must work correctly for all valid input. However, these techniques have as yet been successfully applied only to very simple programs. As programs become more complex, the proof process becomes even more complex.

Test Plan and Test Documentation

The **test plan** should be developed at the same time as the specifications the test is intended to verify. When the system is designed, the test plan should be written showing how to verify that the system is implemented according to the design; and when the modules are designed, the test plan should be written showing how to verify that they were implemented according to their design.

A test plan for a module or system should include:

Identification of the module or system to be tested
Identification of test plan
Functions to be tested
For each function:
- What approach is to be used to test this function?

For each test case:
- Test case identification
- Environment for the test:
  - Equipment to be run on
  - Operating procedures to be used in the test
  - Drivers, stubs, and other modules to be used to produce the necessary interfaces
- Input data
- Output expected and acceptable ranges
- Performance expected and acceptable ranges

When the test is run, its documentation should show:
- Identification of test plan and test cases
- Any variations from the plan in how the test was run
- Actual input used
- Output analysis:
  - Expected output for this input
  - Actual output obtained
  - Comparison of expected and actual output

Analysis of comparison
Is output acceptable, and if not, why?
Performance analysis:
Expected performance
Actual performance
Comparison of expected and actual performance
Analysis of comparison
Is performance acceptable, and if not, why?
Summary of errors found
Suggestions for the cause and further tests
Suggestions for fix if this were the cause

As these tests are made, we may wish to retain in a library the test data, which can later be used for regression tests after subsequent changes have been made.

Summary

The best way to avoid errors in a system is not to introduce them. Initial protections include writing careful specifications, making a good design, and conducting frequent reviews. The cost of these precautions pays off in savings when the large expenses of finding and fixing errors can be reduced. Programs should be written so that they protect themselves against input data they cannot handle.

Systems must be maintained over their operating lifetimes to ensure they meet certain goals. Corrective maintenance is testing, designing, and making changes to get the system to do what it's supposed to do. Perfective maintenance aims to get the system to do something beyond what it was originally expected to do. Adaptive maintenance helps to make the changes necessary for the system to run on new hardware. Maintenance can amount to as much as two thirds of the system cost over its lifetime.

Testing involves checking that the system does all the arithmetic and logic correctly. The number of combinations of conditions that must be checked may be so great that we cannot test all of them. One way to reduce the number of conditions needing testing is to divide the system into independent parts so that the various combinations can be generated for each part alone.

As we test the system, we have to keep track of what has and has not been tested. The earlier we test the better, because later fixes are more expensive. We do design testing, module testing, integration testing, system testing, and acceptance testing.

Those who build systems may not be the most capable of spotting errors in their own systems. Professional testers may be used to find as many faults as they can. This step may avoid errors slipping through to be discovered by clients.

When fixing faults, we must guard against introducing new errors. Regression tests check that cases that had previously run correctly still do. Discovered errors are documented in problem reports, and the change is recorded in a change report. To gain confidence in a system we may seed the program with known errors and see if they're found.

A test plan, which is written at the same time as the specifications, shows how to verify that the program is implemented according to the design.

Exercises

1. Write a test plan for the IMPERT program discussed in Appendix B.
2. Write test cases for the IMPERT program.
3. If you wrote the IMPERT program as an exercise in Chapter 16, now you have a chance to test it.

References

Beizer, Boris. *Software System Testing and Quality Assurance*. New York: Van Nostrand Reinhold, 1984.

Glass, Robert L., and Ronald A. Noiseux. *Software Maintenance Guidebook*. Englewood Cliffs, N.J.: Prentice-Hall, 1981.

Lewis, T. G. *Software Engineering: Analysis & Verification*. Reston Va.: Reston Publishing Co., 1982.

Lientz, Bennet P., and E. Burton Swanson. *Software Maintenance and Management: A Study of the Maintenance of Computer Application Software in 487 Data Processing Organizations*. Reading, Mass.: Addison-Wesley, 1980.

Miller, Edward, and William E. Howden, eds. *Software Testing and Validation Techniques,* 2d rev. ed. IEEE Catalog No. EHO 180-0, Los Alamitos, CA.: Computer Society Press, 1981.

Myers, Glenford J. *Software Reliability: Principles and Practices*. New York: Wiley Interscience, 1976.

Myers, Glenford J. *The Art of Software Testing*. New York: Wiley Interscience, 1979.

Parikh, Girsh., ed. *Techniques of Program and System Maintenance*. Cambridge, Mass.: Winthrop Publishers, 1981.

Parikh, Girsh, and N. Zvegintzov. *Tutorial on Software Maintenance*. Los Alamitos, CA.: IEEE Computer Society Press, 1983.

Parnas, David L. "Software Aspects of Strategic Defense Systems." *Comm. ACM,* 28, 12, December 1985, 1326–1335.

Pressman, Roger S. *Software Engineering—A Practitioner's Approach*. New York: McGraw-Hill, 1979, Chapter 12.

Shooman, Martin L. *Software Engineering*. New York: McGraw-Hill, 1983, Chapters 4 and 5.

Appendix A Case Study: CAPERT—A Wholesale Auto Parts Distributor

Donald H. Chambers
Donald V. Steward

The Uses of This Case

This case study accompanies Part 5—Chapters 10 through 17. It is referred to occasionally in text examples and can be used as the basis for exercises and a continuing case to be done by the student.

The CAPERT company has gone through several generations of data processing. They now face problems that will give the student an opportunity to consider possible new approaches as CAPERT moves into a new generation. The student may consider proposed solutions involving distributed systems, portable personal computers, communications, personal computer workstations, local area networks, office information systems, and mainframes.

The case should be of value no matter what the student's area of interest—whether it be applications programming or systems programming. In the newer generation of software development the distinctions between these areas are fuzzy. We know that systems programmers are concerned with operating systems and compilers and drivers for local area networks. But who is concerned with developing the systems that underlie the operation of an electronic office?

In a real-world project a lot of information needs to be organized before it can be understood clearly enough for people to deal with it. We have attempted to make this case study realistic in this sense.

The case presents background information about the development of the CAPERT company and its business. It also describes some of the efforts by the people in the case to study the problem. This description of their efforts is only a way of introducing information students need to work with. It should not restrict how students choose to proceed. They should be encouraged to use their own approaches and to criticize the methods and procedures used by the people described in the case study. Instructors may wish to elaborate or modify this case, or to substitute a case of their own.

A Brief History of the CAPERT Company

The CAPERT company was founded in 1947 by Henry W. Thompson as a manufacturer and distributor of auto parts used by garage mechanics. (The name CAPERT was derived from the maiden name of Thompson's wife; after the fact someone worked it out as the acronym for Car Accessory Parts, Engines, and Repair Tools.) Thompson's small operation in Trenton, New Jersey, grew rapidly during the early 1950s as the U.S. market in automobiles and auto parts mushroomed. CAPERT manufactured some of its own parts, mostly special engine components not generally distributed by the manufacturers, and also acted as distributor for General Motors parts in central and southern New Jersey, as well as areas of eastern Pennsylvania. The business grew so rapidly that Thompson took on a partner, Bill Hatley, who owned a line of auto repair shops in Connecticut. With the new capital generated by this merger, CAPERT opened two warehouses (one in Trenton and the other in Hartford, Connecticut) and expanded manufacturing operations at the Trenton plant. By 1956 CAPERT was one of the larger distributors of auto replacement parts in the northeastern United States.

Thompson and Hatley incorporated in 1959 with Thompson serving as board chairman and Hatley in charge of operations as president. At that time a decision was made to concentrate corporate efforts on the distribution aspect of the business, which was making the best return on investment. As a result, the manufacturing business was sold and new distribution centers were opened in Chicago, Kansas City, and Los Angeles. CAPERT purchased parts from manufacturers in each area and distributed them under their own label. They also continued to distribute GM replacement parts. By 1963 their customers numbered over 1500 jobbers. A jobber is a retailer or garage that buys parts wholesale.

The problems of managing such a widespread operation soon resulted in decentralizing the company and putting more responsibility on the five regional vice presidents. Reporting to each regional vice president were directors of finance, purchasing, inventory control, and distribution services. Personnel, planning, and marketing departments, each headed by a corporate vice president, remained centralized at corporate headquarters, which were moved to Kansas City following Henry Thompson's retirement in 1964. Bill Hatley assumed the board chairman position, and the company promoted John Ruthouse from regional vice president to take on the responsibilities as president of CAPERT. Another regional vice president, Lewis Musgrove was appointed to a newly created position of executive vice president and general manager, whose primary responsibility was to coordinate the efforts of the regional vice presidents reporting to him.

Under John Ruthouse's leadership CAPERT put in their first data processing equipment in 1968. This system consisted of five CPUs, one at each distribution center, with accompanying printers, keypunches, card readers, and tape drives. Since the primary function of these systems was to handle the financial operations (billing, accounts receivable, accounts payable, payroll, and so on), their operation was placed under the director of finance in each distribution center. Regular weekly reports were then sent on to corporate headquarters.

As early as 1973 it became apparent that in order to remain competitive, CAPERT would have to update its order entry and inventory control systems. A consulting firm was called in to make a study of their systems and recommend changes. As a result of these recommendations, the development of the new order entry and inventory control systems was assigned to a new department of information systems development (ISD), whose director reports to the vice president for planning at corporate headquarters. Hired for this newly created department were three systems analysts and four programmer/analysts, some of whom were promoted from the distribution centers.

The data processing personnel at each of the distribution centers were reorganized into new departments of data processing. At each center the department was headed by a director of data processing who reported directly to the regional vice president. The purpose of these data processing departments was to implement at the center the new order entry and inventory management systems designed by ISD, and assume the data processing functions of the finance department. The data processing departments provided all processing services needed by the centers and maintained the programs required. These departments consisted of a manager of programming and a manager of operations. Two or three programmers reported to each programming manager. Three computer operators, eight to twelve processing clerks, four or five data entry operators, and a hardware maintenance person reported to each operations manager.

Because of the individual differences in hardware and personnel at each distribution center, implementation of the order entry system and inventory control system was considerably more difficult than anticipated and very much behind schedule. Eventually it did become operational, due in large part to compromises made in the design. As a result of the constraints placed on the project (severe limitations on new hardware expenditures, for example) and the difficulties faced in the implementation phase, the director of information systems development resigned during the project and two of the systems analysts left shortly after the system hobbled into operation. These departures put an added burden on the data processing departments at each distribution center to fix up the system as best they could. The whole experience left bad feelings all through the ranks. Despite those initial difficulties, each data processing group did an excellent job in adapting the order entry and inventory control systems to their own operations, resulting in a substantial improvement in customer service and overall efficiency of the distribution system for a number of years.

CAPERT's Current Problems and Opportunities

From the seventies through the early eighties, the auto parts industry faced challenges of major proportions. First, a decrease in the availability of oil products during the oil embargo and increase in governmental regulations regarding safety and air pollution control encouraged development of innovations in technology by the auto manufacturers. As a result, CAPERT increased its stock of domestic parts from 100,000 items in 1978 to 120,000 items by 1981.

Second, the demand for replacement parts for foreign-made automobiles mushroomed while demand for parts for U.S. cars slacked off. Up to that time the market for these foreign car parts had been largely dominated by the foreign auto manufacturers themselves. However, CAPERT's top management recognized the need to enter this market in order to provide for the needs of the increasing number of foreign car owners in the United States. This move meant potentially doubling the current number of inventory items, creating major adjustments in the distribution and purchasing processes, and major capital investment.

Third, a general downturn in the U.S. economy created long-term changes in the auto parts market. Many of the repair garages and small auto parts stores serviced by CAPERT went out of business, while "do-it-yourselfers" increased the demand for auto parts in larger chain auto parts stores and department stores. At this point 38 percent of CAPERT's market was composed of these large retail stores, and this percentage was expected to increase in the following few years. In addition, the capital available for new investment had been extremely hard to come by due to high interest rates. Although more recently absolute interest rates have come down, *real interests rates* (that is, interest rates minus the index of inflation) have stayed high or even increased. This real interest rate governs how much the loan costs when it comes time to pay it off. CAPERT's corporate management had started a campaign to streamline operations so that waste and needless duplication could be eliminated. Along these lines, consideration was being given to centralization of as many services as possible at the corporate headquarters, without sacrificing customer service.

With the rapidly changing conditions affecting CAPERT's market and its internal operations, the company experienced serious problems. Salespersons for CAPERT had been complaining about the slow response time in filling orders at several distribution centers. Directors of inventory control reported that traditional methods of determining reorder points were ineffective due to the unpredictable market. Worst of all, many large retail outlets protested the amount of paperwork necessary to place an order with CAPERT. One of the largest of these informed CAPERT that they were taking their auto parts business elsewhere. In a letter to John Ruthouse, they cited delays and errors in filling orders, excessive paperwork in placing orders, and excessively high prices as their main reasons for going to one of CAPERT's competitors.

This letter dramatized a situation that had been developing for some time. Mr. Ruthouse called an urgent meeting of all corporate and regional vice presidents to discuss these complaints, uncover other apparent problem areas, and propose corrective action. He solicited observations from each participant in the meeting concerning the areas they felt needed improvement. Mel Franciscus, regional vice president for the Trenton distribution center, with which this customer did most of its business, defended the methods his people were using. He blamed manufacturers for not supplying needed parts by the dates promised, and the customer's management for unrealistic expansion that had put their operation in jeopardy. Marilyn Morris, regional vice president from Los Angeles, felt that this customer may have a valid point about the paperwork overload. With their nationwide operation, they must place orders at several of CAPERT's distribution centers. She suggested that other large customers have similar complaints, referring to a recent phone call she had received from a top executive at a large discount chain headquartered in the Los Angeles area who is a major customer. Other regional vice presidents had similar comments about the order entry system as it applies to larger retail customers.

Lewis Musgrove acknowledged the problems in service to retail outlets. He suggested that a study be made into the order entry system to get to the heart of the problem and propose revisions in current procedures. Mr. Musgrove also indicated two other problem areas—inventory levels and the need for better operational and management information.

Inventory levels in all distribution centers were excessively high, which tied up capital badly needed elsewhere in the company. This commitment was very costly, particularly during periods of high interest rates. Mr. Musgrove also observed that, due to shifting demand in the market, many centers have experienced shortages of specific items that are surplus stock at other centers. The types of parts in demand may be different in different parts of the country. For example, foreign parts were in higher demand in California than in the Midwest.

Information on customer orders, receipts of goods from suppliers, and inventory levels could not be compiled at corporate headquarters in time to control effectively shortages and delays in shipments to customers. These problems had been traditionally handled by salespersons ordering from various distribution centers in order to obtain sufficient supply for a customer, or by one distribution center ordering from another to fill an order. Both of these methods require additional paperwork and time.

Vice president for marketing, Jason Goldberg, agreed with Mr. Musgrove that the ordering system was causing his sales staff a great deal of trouble. He presented statistics on order processing time for the previous quarter that showed, among other things, that 62 percent of the sales force's time is being spent on processing orders for the large retail stores, whose business accounts for only 38 percent of CAPERT's sales volume. Also, salespersons reported delays of up to three days in placing orders because of the inability of some distribution centers to have up-to-date inventory level information.

Problems have not been limited to large-volume customers, however. Mr. Goldberg also mentioned that small jobbers had increasing difficulty in making inquiries about availability of parts and that short shipments increased 15 percent over the same quarter last year. He pointed out that small repair facilities cut back their shelf inventory in recent months because of interest rates and expected CAPERT to provide their needed parts on demand. Mr. Goldberg noted that other auto parts distributors were likely to increase their share of the small-volume customers by providing faster turnaround time. During the high interest rate periods CAPERT and its competitors learned how to offer faster response to their customers so that the customers could maintain lower inventories. The customers got used to that mode of operation and continued to expect it when interest rates later dropped. He felt that CAPERT's consistent quality of products and his sales staff's constant efforts in maintaining goodwill with customers has held off any mass desertion to competitors, but they would not be able to hold off indefinitely.

President Ruthouse summarized the points that had been raised up to this time and asked Elizabeth Schultz, vice president for planning, to outline corporate plans that were in the works. Ms. Schultz reported on two major efforts by her departments. The first was a feasibility study under way to determine possible benefits and costs involved in entering the foreign auto parts market. She had just returned from a trip to Japan, which resulted in beginning negotiations between a Japanese automaker and CAPERT on potential supply of their parts for U.S. distribution. The second effort was in response to a need expressed by the board of directors for reducing wasted time, manpower, and materials expense in CAPERT's regional operations. This effort was being carried out, in part, by a time-motion study being done at the Chicago distribution center.

The group then turned to proposals for attacking the problems described thus far in the meeting. It was generally agreed that the areas of primary concern were order entry and inventory control systems. Mr. Ruthouse suggested that efforts by the planning division could be focused on these areas, and that the current time-motion study at Chicago could be incorporated. He directed Ms. Schultz to begin such a study immediately, and to report initial findings within two months.

The Systems Project Request

The *systems project request* is the document that starts the systems development process. Anyone with proper authorization in the company can submit such a request. At CAPERT, the vice president for planning must authorize work to be done on any project request. Following such authorization, the initial study phase begins. The systems project request specifies the objectives of the project and describes the problem to be solved or the opportunity to be taken. Once the authorization is given by the vice president for planning, the project request becomes the project directive. Figure A.1 shows the request made for this project written on CAPERT's project request form.

Getting Started

Systems development at CAPERT was handled by the department of information systems development (ISD), then headed by Jack Marcini. At the time ISD

CAPERT

SYSTEMS PROJECT REQUEST/DIRECTIVE Pg. 1 of 1

Request Date: 9/15/87 Required Date: 11/15/87

Requested by: J. W. Ruthouse Subject: order entry—invent. control

Objective/expectations:

1) Improve turnaround time on large volume orders
2) Provide more timely information on inventory levels
3) Increase the efficiency of the order system and reduce costs

Brief description of the problem/opportunity:

I would like to request an investigation of CAPERT's order entry and inventory control systems to suggest feasible alternatives to alleviate several recurring problems. Chief among these problems are apparent delays in processing large volume orders and subsequent delays in shipping. Inventory levels seem to be out of line with demand and poorly distributed among the various distribution centers. Information flow concerning inventory levels seems to be too slow to respond to this problem adequately. In addition to these problems, new opportunities in the foreign auto parts market will potentially double the number of items of inventory stocked. Overall efficiency in the order system is of utmost importance in reducing the current costs of operation. Top management is placing an emphasis on centralization of operations in order to improve control and eliminate duplication. Initial investigation and recommendations are expected by November 15.

Requestor:	Department	Title	Telephone
J. W. Ruthouse	5000	President	X288

Authorization:

Labor		Materials
Hours	Amount	Amount
150	$3300	$800

Authorized by:	Department	Title	Telephone
E. M. Schutz	4010	VP for Planning	X267

Figure A.1 CAPERT's system project request.

received the systems project request concerning the order entry and inventory control systems, they were already behind schedule on a project for the finance department dealing with accounts receivable. Mr. Marcini wasn't optimistic about the chances of getting this new project off the ground within the given time limits. When he and Ms. Schultz met to discuss details of the latest request, he was hoping to postpone this new project for a few months. However, Ms. Schultz communicated the urgency of this new project and its importance to top management. She was insistent that this project become top priority and that staff be taken *off* the accounts receivable project in order to get started. Marcini stated a week would be needed to document where they stood on the accounts receivable project so they could pick up on it later without lost effort. He asked about the possibility of hiring additional staff for this project. Schultz responded that she was not budgeted for added personnel at this time, but, staffing needs for the project should certainly be considered in the initial investigation. She agreed to three man-days to document the status of the accounts receivable work and filled him in on the background of the project proposal as it had been discussed at the meeting of managers in Mr. Ruthouse's office.

With a very broadly stated project request, limited information on the expectations for such a project, and a definite problem of staffing, Jack Marcini went back to his office to sort things out. The first question to answer was: *Who* are the people involved in the proposed project? He prepared a list of individuals and their potential roles in the project development and implementation. This list would give Marcini sources of information that could clarify: (1) the actual *need* involved, (2) the *expectations,* and (3) the general *environment*. Mr. Ruthouse was clearly personally concerned with this project and certainly should be on the "who's who" list. Also included were Ms. Schultz, Jason Goldberg in marketing, directors of inventory control, and directors of data processing operations for the distribution centers.

Getting Oriented

After considerable shuffling of schedules, Jack Marcini assigned two of his staff to the new project, giving one of them three days to document the accounts receivable status. He shared the information he had up to this time and urged them to focus the project on a manageable scope as quickly as possible by identifying the actual needs and devising a plan for meeting those needs.

The next step was to review all documentation available on the present order entry and inventory control systems and logs of the daily operation of these systems. This information would provide them with a base of knowledge when they interviewed personnel and observed the actual process. The following, in narrative form, is what they found.

The Order Cycle

The order cycle begins with the placement of an order by a customer. Orders are placed in two ways: first, customers can order directly by returning appropriate order forms with the truck on the delivery route; second, the customer can place an order through sales staff in the field. Statistics available to the project team indicate that approximately 80 percent of the orders are now received on the delivery routes, and 20 percent are placed through salespersons. All phone orders are placed through a salesperson. The majority of large-volume customers place orders through the sales staff.

Each of CAPERT's distribution center warehouses is divided into sectors (generally five sectors), each of which houses particular types of parts. For example, one sector houses regular maintenance parts (which have high turnover), another houses mechanical engine parts, a third houses large parts (over 10 lb.), and so on. Orders for each type of part are written on a multiple-copy order form, which is color-coded according to the sector in which it is located. This method speeds processing at the warehouse, where each order can be picked from a localized area of the building.

The jobbers send their orders back with the delivery truck and generally get their orders filled within two days. Most of them have a small, stable staff that has been working with CAPERT for many years. They are quite used to filling out a different order form for each warehouse section. Rush orders are phoned in and can usually be filled the next day.

When CAPERT started selling through the retail auto parts chain stores, they met opposition to this practice because of the large number of different people at the chains who would be making out the orders. The chains felt it would be a formidable problem to train so many people, particularly because of their high employee turnover. To get their business CAPERT offered to process these orders through their own salespersons, who would take care of making out the different forms for different warehouse areas.

Receiving Orders

Orders received on the truck route are returned to a basket in the shipping area when the truck returns from deliveries. The basket is emptied four times each day by a messenger who transports them to the data processing office.

Orders placed through salespersons are delivered directly to data processing. On large orders for a particular item, the salesperson usually checks availability of the item at the warehouse by calling a clerk at the distribution center, who checks a computer printout updated twice a week. If sufficient supply will not be available at the local warehouse by the desired shipment date, the salesperson may order from more than one warehouse. All inquiries about order status from customers are channeled through the salesperson responsible for the customer's geographical area (district).

Processing Orders

Once in the data processing office, the order forms are date-stamped and given an order number. A data entry operator keys in the customer identification information and the items ordered. This order is stored in an open order tape file. One copy of the multiple order form is kept in the data processing office; another goes to accounting for a credit check, and the remaining two copies are sent to the appropriate warehouse sector. If the credit status is *not* satisfactory, a credit clerk determines the reason for the credit check failure. As a result of this investigation, the credit clerk can change the credit status on file and allow the order to be processed, or cancel the order processing and return a form letter to the customer explaining why the order is not accepted.

In the warehouse each sector is headed by the sector foreman who assigns a picker to each order. Pickers pick the items that are in sufficient supply and note on the order form those items on which they are short without picking them. Orders fully complete are sent to the shipping department for packing. Orders not completely filled are placed in a holding area and the paperwork is returned to the sector foreman. If the customer has given prior approval for a short shipment it will be indicated on the order form. If not, the sector foreman sends the order form to a warehouse clerk, who contacts the salesperson for the appropriate sales district. It is the salesperson's responsibility to instruct the warehouse on the desired processing of short shipped orders. Once the paperwork is cleared, or the order is cancelled, the partial shipment leaves the holding area.

The shipping department packs orders for shipment and manages the loading of trucks for their regular delivery routes. Large orders are delivered, as much as possible, on a regular weekly or biweekly basis. Small orders go out on a daily basis. Occasionally, orders requiring special handling are made. These orders require considerably more time to process, since special equipment or personnel may be required. Customers are made aware of such potential delays at the time the order is placed.

When the order is packed, the second copy of the order form is used as a bill of lading. The last copy is returned to the data processing office for billing and inventory purposes.

Inventory Processing

Data entry operators enter on key-to-tape items actually shipped using the updated order form sent back from shipping and insert the shipment date. This record is then placed in a shipment file. Prices and shipping charges are added by the billing program using the shipment file record, and an invoice is prepared and printed. Invoices are delivered to accounts receivable for further processing. The order form then goes to an inventory control clerk, who updates the record on a tape file for each item shipped. The order form is signed off by inventory control and filed with the original order form. This completes the order cycle.

Expectations for the Project

The review of available documentation of the current system provided the project team with a base of information from which to define a reasonable scope for the project. They arranged to meet with Jack Marcini and John Ruthouse to discuss expectations and constraints related to the improved order-entry and inventory control systems. Ruthouse detailed plans for consolidation of company functions and expansion into new markets. In particular, he emphasized the following:

1. *Centralization of data processing*—Data processing functions are to be centralized at Kansas City, including financial systems, order and inventory systems, as well as distribution management. This, Ruthouse realized, would necessitate investment in hardware, communication facilities, and new application software. He estimated, however, that $1,500,000 could be saved each year by centralization in these areas and that any new system that could be developed within a 3-year payback period (based on $1.5 million/year) would be acceptable.
2. *Management information*—One reason for the emphasis on centralization, according to Ruthouse, is the need for accurate daily information on inventory, order processing, accounts payable, and accounts receivable so that timely management decisions, influencing distribution and financing, can be made.
3. *Large-volume customers*—Mr. Ruthouse noted CAPERT's trend in recent years toward the large-volume retail market. This, he said, will be accelerated in the near future in order to keep up with the demand from the do-it-yourself mechanic. He challenged the project team to develop a system that could better serve the large-volume customer, and yet, would still provide for the small-volume customer (who is currently reducing his own shelf inventory and requiring a shorter turnaround time).
4. *Imported parts market*—Efforts are currently under way to introduce foreign auto parts into CAPERT's line of products. This will be done on a trial basis in the region served by the L.A. Distribution Center. Managing this new inventory may create challenges to the capability of the present system since the number of stocked items will be substantially increased. Any new order entry and inventory control system must be able to handle large increases of this nature.

Mr. Ruthouse pledged his full support, as well as that of other top management personnel, to the project team in meeting the challenges inherent in this project. He indicated that other pending projects should be shelved if at all possible to make available maximum staff time. He also indicated a willingness to hire additional systems analysts as necessary to speed the process, and that such requests should be incorporated in the initial investigation report, due by November 15. Finally, he indicated the urgency of the entire project by stating that CAPERT must implement changes in the order entry and inventory control systems within eighteen months to remain competitive.

Groundwork Laying

It is important to keep users and management informed of pending action by the task force, providing some persuasive arguments to justify the course of action to be taken. In other words, it is important to lay the groundwork for the proposed courses of action *before* starting out, so that key individuals will not be taken by surprise. Surprises will invariably work to the disadvantage of all concerned with the systems development process. Letters and memos to appropriate personnel are the prime method of establishing this communication. Follow-up by phone conversation and personal contact is desirable, as well. Any means of making the individual feel that his or her opinion is important will help in the success of any project.

Information Gathering by Interview

Communication must continue throughout the whole system development process. Generally it begins during the initial investigation with interviews.

Higher management is interviewed for the purpose of defining scope, constraints, and expectations. As the detailed investigation progresses, operations staff are interviewed to get an idea of what is really going on or where the problems and potential solutions really lie. The ongoing interview process therefore is top-down in the organization and also general-to-specific in the nature of the information acquired.

Preparing for the Interview

The project team at CAPERT, consisting of analysts Helen Windsor and Jess Andaya, initiated the interview process by listing, in order of scheduled interviews, the personnel they needed to contact. A memo was prepared for employees in all areas relating to order entry and inventory control informing them of the investigation under way. High on the list of interviewees was Jason Goldberg, vice president for marketing. Preparing for this interview, Jess Andaya listed a number of questions he hoped Goldberg could answer:

1. How is the sales organization structured?
2. How is information transferred within this organization?
3. What is the role of the salesperson in processing orders from large-volume (small-volume) customers?
4. What are the major problems encountered by the sales staff related to order entry and inventory control?
5. What suggestions does he have for reducing the apparent paperwork overload on the sales staff?
6. What methods for order entry have been tried within the past five years?
7. What lies ahead for the sales organization as market conditions and financial conditions change?

The Interview

Since Jess Andaya had arrived early for his interview with Mr. Goldberg, he had an opportunity to review his notes and to reflect on what he knew of the man he was about to interview:

> Jason Goldberg is an extremely cordial individual who thrives on the personal contacts he makes with both employees and customers. He is good at working with people, and it is readily apparent that he is an extremely capable salesman. He enjoys great influence with the board of directors and has an outstanding reputation among Madison Avenue advertising executives for his marketing savvy. Some Wall Street analysts have credited Goldberg with holding CAPERT's position in the auto parts market despite its occasional financial troubles. Obviously, then, Goldberg has a great deal of vital information concerning order entry from the marketing perspective.

What follows is an excerpt from the Goldberg interview:

Goldberg: Come on in, Jess, and make yourself comfortable.

Andaya: I'm certainly glad you could find time to fit me into your busy schedule today.

Goldberg: Well, I understand you're doing important business for us and I wanted to arrange this get-together before the marketing convention next week in Philadelphia. Can I offer you a cup of coffee?

Andaya: Yes, thank you.

Goldberg: Now, how can I help you out?

Andaya: As I indicated on the phone the other day, I was hoping you could shed some light on the problems being experienced with the order entry and inventory control systems.

Goldberg: Yes, I am aware that John (Ruthouse) has got you working on an updated order system. I'm sure you'll be doing a thorough job on this one, judging from the excellent work your department has put out for finance over the past few months.

Frankly, Jess, I think we're in trouble if we don't update our approach to order processing. I've got a team of highly proficient sales people out there using the latest techniques marketing research has to offer, and they have to deal with an antiquated order system that forces them to put out twice as much time as they should in processing orders.

Andaya: In order to get a handle on the problem, we want to determine exactly *what* is currently being done in order processing, what needs are not now being met adequately.

So, if you could start by giving me some particulars about the sales staff it would help us define our current status a little better. For example, how is the sales organization structured?

Goldberg: The sales network is centralized here in Kansas City under Jeff McNulty, director of sales personnel. Five regional directors, one for each distribution center, report to him. Then each region is divided into six to ten districts, each covered by one sales person.

We've tried to keep a pretty close-knit family here. Daily phone contact is maintained among the staff, and monthly the entire sales staff meets to discuss current issues and conduct department business. In my opinion this structure has provided us with the control necessary to stay on top of things in the marketplace.

Andaya: What is your current staff size and what do you spend annually for staffing?

Goldberg: You'd have to get the exact figures from McNulty, but salesmen number about forty to forty-five, and that translates to an annual investment in the neighborhood of one and a quarter million dollars. The overall sales budget usually runs around $3 million per year.

Andaya: In what ways do you feel that the centralization of your sales department has contributed to its success. *And,* how would you apply it to other areas of company operations—order entry, for example?

Goldberg: Back in 1964, the sales department resisted efforts of Bill Hatley to decentralize all operations. I believe the centralized control we've maintained has helped us in two major ways: First, and most importantly, information flow is quicker and more effective. Managers are in nearly constant contact with men and women in the field, and have the latest information at their disposal for effective decision-making. Second, sales can be a very lonely job. We've provided the support for our people in the field that has provided a very stable and productive organization. Our personnel turnover is the lowest in the industry.

As for centralization in other areas of operation, I'm all for it. Especially in order entry and inventory control we need up-to-date information on order status and inventory levels in all regions. Many of our sales staff's problems could be resolved simply by having a centralized bank of reliable and current information.

Andaya: What do you consider the major problems encountered by the sales staff related to order entry or inventory control?

Goldberg: One of the problem areas is in placement of orders. In the days when ordering was almost entirely done on the truck routes, the customer completed the appropriate order forms. With the increase in large-volume customers, the sales staff became increasingly burdened with the paperwork of order forms. We had to take on this responsibility for two reasons: first, the large retailers wouldn't hire additional clerical staff just to handle our order form system. And, second, large orders often require placement of orders with several distribution centers in order to achieve sufficient supply. Essentially, our staff provided a buffer between the old order system, which was designed for the small jobber market, and the large-volume customer. We've kept that old system working for a long time when it should have been scrapped. Within the past two years the portion of orders handled through our salesmen rather than through the jobber has increased from 5 to 20 percent of our business and the sales staff is swamped with the necessary paperwork. We've had to hire additional clerical help, but even with that the present system is at its limits.

The other problem is the availability of up-to-date information on order status and inventory levels. This affects the salesperson in two ways: in making inquiries about availability of items to fill orders, and, to respond to inquiries from customers about the status of their orders. Rarely is this information immediately available when a salesperson calls. Usually it is ten to thirty minutes later that the salesperson gets a call back from the distribution center answering his or her question. Even then the salesperson can't be completely sure the information is up to date. The order may come up short when it finally is processed.

Andaya: What other order entry methods have been tried within the past five years? What were the outcomes?

Goldberg: One experiment comes to mind immediately. About two years ago we tried a phone order system in Chicago, hoping we could improve our turnaround time. At that time the vast majority of our orders were coming in by truck. The goal was to provide phone ordering for the most urgent orders, which we were led to believe amount to about 20 to 30 percent of the order volume. Order clerks were hired in data processing to receive phone orders and fill out order forms, and the system got under way without a hitch. After about two weeks, however, the idea of phone ordering was getting so popular that the clerks couldn't keep up with the orders. Customers began complaining about busy phone lines, and the rate of order errors substantially increased. And, surprisingly, the overall response time was not significantly improved after six months. We had to make a decision to increase the staff of order clerks or to go back to the old truck route system. We just couldn't see that type of phone system as being cost effective.

Excuse me, Jess, there's an urgent call I have to take.

Appendix B
Case Study: IMPERT—A Critical Path Scheduling Program

Uses of This Case

This case can be used in conjunction with Chapter 15 as an exercise in program design. It illustrates the Tree method shown in Chapter 4, demonstrates some concepts of algorithm design and analysis, and raises questions about the design of tests (Chapter 17). We should already have a good understanding of what a critical path scheduling program does from reading Chapter 7.

We start by providing the data structure to be used in this design. This shows the importance of developing the data design before the program design and indicates how the program design can follow once the data design has been established. It also gets us over the data design hurdle so we can get to use a Tree to design a program.

This exercise illustrates the point that if you understand the structure of the data and how that data is used, you pretty well understand the program. This also shows why it is so important to document the data structure and its use. "Index" is just the index of the elements in each of the arrays.

DESC is the array giving the descriptive names of the tasks. Internally the tasks are referred to by the order they appear in this array (for example, Task *D* is number 4). These numbers are shown at the top of the boxes in Figure B-1.

DUR is an array giving the durations of the tasks. The duration is shown below each box in Figure B.1.

PRED lists the predecessors of each task in sequence. In Table B.1 the underlines show where the predecessors for each task end.

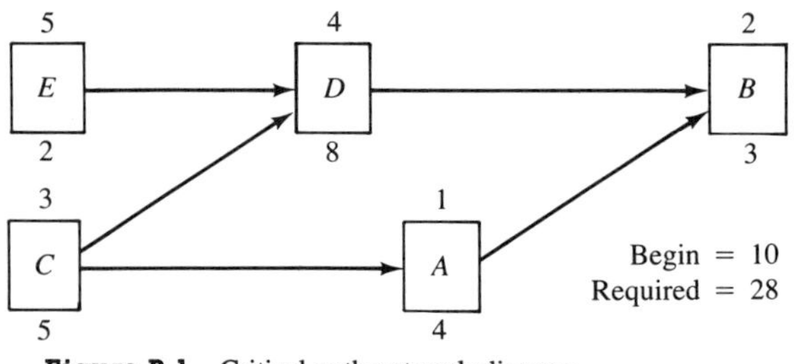

Figure B.1 Critical path network diagram.

Table B.1 Critical path network table.

Index	DESC	TOPPTR	PREDPTR	PRED	DUR	EF	LF
1	*A*	3	1	3	4	19	25
2	*B*	5	2	1	3	26	28
3	*C*	1	4	4	5	15	17
4	*D*	4	5	0	8	23	25
5	*E*	2	7	3	2	12	17
6			8	5			
7				0			

PREDPTR points to the position of the first predecessor for each task. PREDPTR separates the predecessors for each task in the PRED array, playing the role for the computer of the underlines in Table B.1. For example, Task *D* is task number 4. PREDPTR[4] gives the position in PRED where we find the first predecessor to *D*. PREDPTR[5] gives the first predecessor of the next task. Following this logic, the predecessors for any task *I* appear in PRED at positions PREDPTR[I] to PREDPTR[$I + 1$] $-$ 1. Why is there a value in the position after the end of PREDPTR pointing to the position after the end of PRED?

Arrays DESC, DUR, PRED, and PREDPTR are set up while reading the input. The tasks might not be read in the order they can be processed to compute their begin and finish times. When computing earliest times, we want to process each task only after the earliest finish time for each of its predecessors have been computed. Such an order is called a topological order.

TOPPTR is an array that points to the tasks in topological order. Thus, when computing earliest times, we go down the TOPPTR array picking up the tasks to be computed in the correct order. When computing latest times, we go through this list in reverse order.

We compute this topological order by going down through the tasks and assigning the next order number to any task that has no predecessor or all its predecessors have already been ordered. We choose a convenient array that we have not already saved information in, say EF. We clear EF. Then, as we add a task to the order, we place a 1 in that task position in EF to mark that this task has been ordered. We may have to go through all the tasks several times before they are all ordered. We should zero out the EF array after we have finished.

EF (Earliest Finish) is computed picking up the task numbers in the order given by TOPPTR. Assume that the next task for which we will compute EF is task *I*. We find the predecessors of task *I* using PREDPTR and PRED. If it has no predecessor, then its earliest beginning is the project BEGIN. If there are predecessors, then the earliest beginning is the largest value of EF for these predecessors. Now we add the duration to this beginning to get the earliest finish of task *I* and store it in EF[*I*]. Note that we don't have to store the beginnings. We can get them at any time by subtracting the duration from the finish.

If we had a list of the successors, we could compute the latest times in the same way but going backward in the topological order. Let's use a trick that allows us to use the predecessor list to compute the latest times, thus saving us from having to compute a successor list.

To start, we put the value of the project required finish in LF for every task. Then we go through the tasks in the reverse order given by TOPPRR. We pick up the LF for this task and subtract the duration to get the latest beginning. Call it LB. Now for each predecessor, we replace the current value of LF with LB if LB is smaller. Thus by the time we get to a task we will have in LF the smallest value of the latest beginnings of all its successor tasks. This is its latest finish. Note again that we don't need to save an array of latest beginnings, only the latest finishes.

Note that we have a much better understanding of how this program will work if we first go through the algorithm with a hand computation using a simple example.

Appendix C
Computer Systems for Software Development

Those of us who develop computer systems to help other people in their work often overlook how the computer can help us in our own work. It is as though we are willing to inflict computers on others but don't really believe in them ourselves. As we review the previous chapters we recognize that we must deal with a tremendous amount of data, representing what we learn about the old system and the requirements and design for what we want in the new system. We should certainly look to see whether the computer can be of any help to us in developing this data, storing it for later retrieval, and processing it for guidance during design and testing.

We must be able to get sufficient value out of the machine to make it worth the effort to enter the data. The key to this is to represent the data in a consistent and usable way so that it can be used over and over through the various phases of the project. To do this requires a consistent method of representation. We have set the basis for such a representation in Chapters 3 and 4.

Two fundamental domains of representation must be tied together. One is the tree domain, which shows control; the other is the data structure domain with its data flow diagrams and matrices, hierarchical data structures, and the data element dictionary.

The tree domain can be handled using various available outline editors (such as PC Outline), which represent the tree structure on the screen in outline form. Special postprocessors can be used to print the outline on the printer in the form of a tree and to translate the outline into a structured language. The data base domain can be implemented in a data base with the ability to link files by using a field in one record as a key to access a record in another file. DBASE II and III have these capabilities.

A system of personal computer workstations connected together in a local area network, communicating through word processing and electronic mail, and sharing data and a common set of tools constitutes an electronic office for software development. In this environment we can consider tools to do the following:

1. Assist in editing:
   a. To enter and maintain the tree and data structure information
   b. To produce documents and reports
2. Prompt us for further information when the information is not complete
3. Analyze for:
   a. Errors
   b. Omissions
   c. Ambiguity
   d. Inappropriate redundancy
   e. Completeness
   f. Consistency
   g. Adherence to standards and procedures
4. Assist in task management by helping:
   a. Develop breakdowns of:
      1. Requirements
      2. Modules
      3. Tasks
   b. Develop a task list:
      1. From these breakdowns
      2. From an analysis of the tree and data structure matrix
      3. By selecting tasks applicable to the current project from a list of typical tasks such as the information systems analysis and design matrix in Appendix D
   c. Develop a schedule from the task list, predecessors, and estimates of durations
   d. Collect progress data from:
      1. The tree and data structure matrix being developed in the system
      2. Employee input of hours applied
      3. Monitoring lapse time with a clock and calendar
   e. Maintain an accounting of resources used
   f. Help in allocation of resources: Suggest good allocations
   g. Assign tasks to persons and:
      1. Announce who is assigned to do what
      2. Record these assignments to be used for impact notification
5. When a proposed change is considered, an approved change is made, make an impact analysis of the effects of the change to determine:
   a. What else needs to be changed as a consequence
   b. What needs to be verified to see that it is consistent with this change
   c. Who should be informed
   d. How making the change will affect their work and schedule

6. Configuration control:
   a. Receive a request for change
   b. Make an impact analysis (see 5)
   c. Make an assessment of effects of proposed change
   d. Solicit from the people whose work would be affected by a change their assessment of the effect
   e. Notify the affected people of approved changes
7. Produce printed reports
8. Produce notifications using electronic mail
9. Compute and retain metrics to:
   a. Measure and record progress as a function of time, resources expended, and effort expended so we can:
      1. Know what we are spending
      2. Know how we are progressing
      3. Have the data available when needed to make estimates for future projects
   b. Compute the coupling metric
   c. Compute the complexity metric

We would like the following characteristics of such a software development system:

1. most useful results out for the least effort to put data in
2. easy-to-use and easy-to-interpret results
3. accessible to all who should use it.

Electronic Office for Software Development

Here is how an electronic office for software development might work. The needed hardware and software, which can be adapted to this application, already exist in the form of local area networks, outline editors, electronic mail, and data base management systems. Some time and effort would be required to set up the system for a given environment, make the needed adaptations, and train the people who will use it.

The tree editor and the data base that stores the data structure matrix represent the latest state of the system while it is being developed. Programs are compiled from the finished tree. Progress during the whole development process from requirements specification to code generation can be measured and reported by having the computer count branches. Lapse time and effort expended can be monitored and related to progress. This information can be very valuable when estimating projects in the future. Since the information is maintained in the computer system, retaining it for future use can be automatic rather than burdensome.

The Tree, the data structure matrix, and annotations (Chapters 3 and 4) could be developed on the system using special editors. As they are developed, anyone can interrogate their latest status at any time. Reports are written in the system, and people on the distribution list are informed through electronic mail when the report becomes available so they can retrieve the report at their leisure. Only one copy of these files needs to be kept and it can be maintained so it is always up to date.

The plan and schedule may be set up using this electronic office. The information systems analysis and design matrix (Appendix D) may be stored on file and could be used to select an initial set of tasks pertinent to the project at hand. Other tasks unique to the project may be obtained from the work breakdown (Chapter 6). Design and implementation tasks are defined from the tree and the data flow matrix (Chapters 3 and 4). The predecessor relations may be added to develop a precedence table and matrix, which is analyzed to plan how the tasks are to be iterated. Durations may be added and a critical path schedule developed (Chapter 7). Task responsibilities may be assigned to specific people and expectations defined, communicated, and updated by electronic mail (Chapter 8). Effort and lapse time can be monitored and reported against the number of branches developed. People may be notified as the predecessors to tasks for which they are responsible are completed. Management can be kept apprised of the progress and costs.

When a change is made that affects a task, you can go down the column for that task in the task precedence matrix to find all the other tasks that are directly affected (Chapter 7). One has to select only the affected tasks from this list because not all the tasks may be affected. Going down the columns for each of these affected tasks will find the tasks they affect in turn.

When it is proposed that a data element be changed, you can use the data structure matrix to find all the data flows, files, and processes that this data element appears in that may need to be changed. A file of assignments and responsibilities shows who is responsible for these affected flows, files, and processes, so they can be consulted or informed of the change. Other changes of elements appearing in the data structure matrix can similarly have their effects traced.

Thus using the task precedence matrix and the data structure matrix, we can get a picture of all the consequences of a change. These consequences can be factored into the schedule and an estimate made of the time and costs of the change. When the change is contemplated, the system can request information from the people whose work will be affected to evaluate the effects of the change. If we decide to make the change, the system can notify these people of the approved change and resulting work to be done.

In Chapter 4 we discussed several metrics that can be computed by the system from the tree that is maintained in the system. Branches can be counted to measure progress. Another metric we discussed measures the data flows as an indication of the coupling and cohesion, which is an indicator of the goodness of structure of the design. And yet another metric measures the complexity. Thus we can try different variations on the design and get the system to produce goodness measures of structure and complexity for each variation.

In Chapters 16 and 17 we discussed analyzers that would read a source program and insert statements to collect data about performance or test coverage when the program runs. Such analyzers can be added to our repertoire of tools.

Although we may develop the Tree interactively on the screen in outline form, we may wish to print it out in the form of a tree to take away from our workstation to the office of a client, or for presentation in a meeting. A postprocessor may be used to print the tree.

A postprocessor is required to take the tree developed with a tree editor (actually in outline form on the screen) and convert it into source language for a structured language such as Pascal, Modula 2, or ADA. This postprocessor inserts BEGINs and ENDs and interprets double limbs as in IF-THEN-ELSE and CASE statements. The emulation of the UNTIL event in Pascal requires the use of labels and GOTOs, but no other labels or GOTOs should be required.

One program converts from a tree format to a data flow matrix format, or from the data flow matrix to the tree. Another is used to develop the data descriptions for a structured language from the data element matrix and hierarchical data structure.

Other Software Development Systems

Software development systems such as we have described are sometimes referred to as environments. UNIX is an environment, and ADA is an environment as well as a language.

A number of programs or environments handle various aspects of the system we have discussed here. Many of these tools have been developed in industry and some cost as much as $40,000. Unfortunately this places them out of reach for most schools. The student may wish to read the literature on these various methods and report on their advantages and disadvantages as related to the tools developed in this book.

Student Term Projects

Most of the capabilities of an electronic software development office discussed here can be developed using available outline editors and data base management systems. Specifying and/or developing various parts of this system could make useful student term projects.

References

Case, Albert F., Jr. *Information Systems Development: Principles of Computer-aided Software Engineering*. Englewood Cliffs, N.J.: Prentice-Hall, 1986.

Davis, C., and C. Vick. "The Software Development System." *IEEE Trans. Software Eng.*, SE-3, 1, January 1977, 69–84.

IEEE Transactions on Software Engineering, Special Collection on Requirements Analysis, SE-3, 1, January 1977, 2–84.

Miller, Edward. *Automated Tools for Software Engineering*. Los Alamitos, CA.: IEEE Computer Society Press, 1979.

Riddle, William E., and Richard E. Fairley, eds. *Software Development Tools*. Berlin: Springer-Verlag, 1980.

Teichrow, Daniel, and Ernest A. Hershey III. "PSL/PSA: A Computer Aided Technique for Structured Documentation and Analysis of Information Processing Systems." *IEEE Trans. Software Eng.*, SE-3, 1, January 1977, 41–48.

Appendix D Information System Development Tasks

In this appendix we show a list of tasks that are characteristic of the development of an information system. We can consider these tasks when we develop a plan and schedule.

We have organized the list of tasks in the form of a matrix. The rows show the components being analyzed or designed, while the columns show the phases each component may go through. The entries in the matrix show the tasks.

Not all components have a task for every phase, and not all projects will have all these tasks. This is a generic matrix. For a specific project we can pick and choose the tasks from this general matrix that pertain to our specific project. Tasks that apply to this project but don't occur in the matrix must be added.

Time will tend to proceed from left to right. However, it is not necessary that the tasks for all components in one phase must be finished before any tasks in the next phase are started. At any time the tasks for different components might be in different phases.

The matrix is laid out as follows:

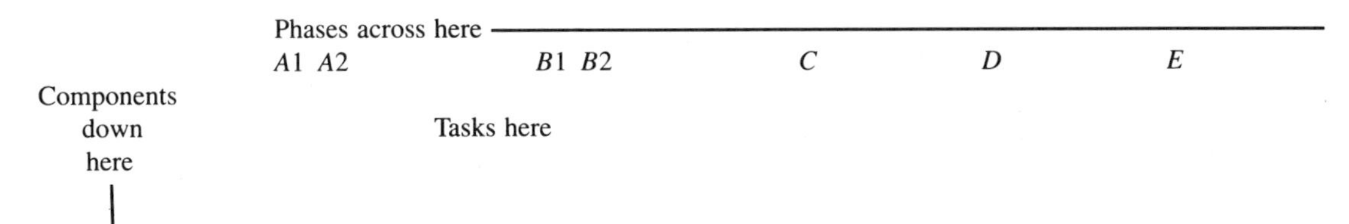

The phases defined for the purpose of this matrix are:

A Fact-finding / systems study

*A*1 Study and evaluate the existing system

*A*2 Study and evaluate alternative/optional systems (These two subphases are shown together in the same column as the same tasks apply to both.)

- *B* Design
  - *B*1 Logical design
  - *B*2 Physical design
- *C* Implementation
- *D* Test
- *E* Operation and maintenance

A short list of the components is shown next to the rows of the matrix. A more detailed list of components follows:

- System
  - Functions
  - Performance
    - Volumes
    - Frequencies
    - Distributions
- Data
  - Output
    - Requirements
    - Layout/screens
    - Distributions
    - Control
  - Input
    - Source
    - Forms
    - Data entry
    - Errors
    - Controls
    - Formats
  - Data elements
    - Sizes
    - Formats
  - Files
  - Data bases
  - Flows
- Processes and modules
  - Function
  - Interfaces
  - Data structures
  - Transaction generation
  - Input/output
  - Decisions
  - Transformations
  - Logic
  - Errors

- Controls
- Test data

Acquisition
- Hardware
- Facilities
- Software packages
- Supplies

Procedures and manuals
- Management overview
- User's
- Operations
- Maintenance
  - Hardware
  - Software

People
- Skills
- Organization
- Responsibilities
- Assignment or hiring
- Staffing schedule
- Training
- Operations scheduling and supervision

Communications
- Distribution of functions / computations
- Distribution of data residence
- Network configuration
- Data flow
- Protocols / disciplines

Controls, backup and security
- Handling of input errors
- Handling of operations errors
- Handling of communication errors
- Handling of equipment malfunction and downtime
- Handling of software errors
- Backup of files and data bases
- Restart procedures
- Physical protection
- Access security

Some of the tasks in the matrix need to be further defined as follows:

Testing
- Of:
  - Modules
  - Integration

System
Acceptance
Tasks
Plan
Data preparation
Test and log
Evaluate and diagnose
Change
Conversion
Plan
Data
Operations
People
Purchasing
Vendor search
Request for proposal/quotation
Evaluation
Installation
Test

During the whole process we must maintain change/configuration control showing the effects on:

User's expectations
Software
Hardware
Procedures
Files
Manuals and documentation

Some tasks in the matrix are shown with an /L or /P, which refer to logical or physical, respectively. The "User" and "Software Engineer" above the columns identify the phases in which these people are involved.

INFORMATION SYSTEMS ANALYSIS AND DESIGN MATRIX

User ______ User reviews ______ User ______

Systems Analyst ______

COMPONENT		A FACT-FINDING A1: REFERENCE A2: OPTIONS	B1 LOGICAL DESIGN	B2 PHYSICAL DESIGN	C IMPLEMENTA- TION	D TEST	E OPERATION
SYSTEM Requirements, Performance, and Test	1	Review or develop documents for A1: Reference or A2: Optional systems using categories under B1.	Requirements specifications Performance vol, freq, distributions I/O, source, destination	Conversion plans	System test plan	Acceptance test	Performance evaluation
DATA Files and Data Bases	2a		Data flow diagram/L	Data flow diagram/P			
	2b		Data element dictionary	Data layout on input, output and files	Data conversion	Verify data	Data entry
	2c		Test data/L	Test data/P			
PROCESSES and MODULES Software	3a		Functions				
	3b		Methods/algorithms		Module test	Integrate test	Software changes
	3c		Tree/L	Tree/P	Write programs		
ACQUISITION Hardware, Facilities, Software	4		Specify/L	Select/P	Build/acquire/install	Hardware test	Run
PROCEDURES and MANUALS	5a		Overview manual	Operations manual	Maintenance manual	Revise manuals	Update manuals
	5b		User's manual				Log operation
PEOPLE	6a		Define changes to organization		Set up new organization		
	6b		Personnel requirements	Assign/Hire	Training	Parallel operation	Schedule and supervise
COMMUNICATIONS	7a		Define location of data & processing				
	7b		Network	Protocols			
CONTROLS, BACKUP and SECURITY	8a		Define/L	Define/P	Implement in components	Backup and security test	Audit operation
	8b			Test data			

Glossary

Abstraction Something for which the requirements have been defined but which has not been designed or built.

Abstract process A process for which we know the properties but which we have not yet built.

Adaptive maintenance Perfective maintenance that is concerned with getting a system to run on different hardware than originally intended.

Agenda of concerns The list of items that currently can be afforded management attention.

Agent The person or machine that performs a process.

Allocation The assignment of functions to be satisfied by hardware, software, or people, or the assignment of performance resources, such as execution time, memory space, and accuracy to individual components of the system.

Analysis The study of a system to determine its expected behavior and the consequences of that behavior.

Annotation A supplement to the data flow diagram that describes a process or a data flow or store in more detail.

Application generator A high-level system for adapting existing procedures to satisfy new applications needs.

Arc A directed line between nodes in a graph or between entities in a data flow diagram.

Artificial intelligence Computer systems that perform feats that would be considered intelligent if done by humans and that are unexpected (at least by some people) when done by computers.

Asynchronous processes Processes that can be run either simultaneously (in parallel) or one at a time in an arbitrary order.

Authority The right and power to do something, particularly to enforce obedience.

Availability The probability that a system will be accessible to perform when needed.

Backward search A design methodology in which the design proceeds from the output back to the input.

Bar chart A chart with bars showing each task and the interval of time on a calendar during which work on that task is scheduled. Also called ***Gantt chart.***

Batch processing Processing in which all the input for a job is accumulated before the job is run.

Behavior What a system does as a consequence of the input or stimulus it receives.

Branch The horizontal line in a Tree.

Capacity The size of a problem that can be handled.

Cause and effect graph A graph showing cause and effect, which may be used to determine what we want to change to obtain the desired effects.

Change control The management process that evaluates proposed changes, determines their consequences on the product, cost, and schedule, approves appropriate changes, ensures that documents and the schedule reflect the changes, and notifies the people whose work is affected. The same as ***configuration control.***

Change proposal A proposal describing a change, why it is needed, and its effects on the product, its cost, and schedule.

Change report A report on a change made as a consequence of fixing a fault.

Chief programmer team A programming organization in which everyone works around a chief programmer who is usually responsible for the system and interface design.

Circuit A path that returns to its starting node.

Clone model A management process whereby the manager evaluates the work done by a subordinate based not on explicit expectations, but on whether the subordinate did it exactly as the manager would have done it.

Cohesion The integrity of the module, that is, how easy it is to comprehend the function of the module.

Complexity The difficulty involved in using the relationships among the parts to infer the behavior of the whole.

Condition A logical statement of the circumstances under which an output is generated by a process.

Configuration The current best description of the product, its parts, their assembly, and any assumptions made about the product.

Configuration control Same as ***change control.***

Configuration management plan Description of the procedures to be used for considering and making a change to the configuration of the product.

Configuration tree Same as a ***parts breakdown tree.***

Conflict Behavior that interferes with another's attaining his or her goals.

Constraint A boundary that delineates what can and cannot be done, or what is and is not possible.

COnstructive COst MOdel (COCOMO) A model for estimating the lapse time and man-months to complete a software project based on a prior estimate of KDSI, the number of thousands of delivered source instructions.

Context diagram The highest level in a data flow diagram, showing the whole system as one process with the immediate interfaces that connect it with other systems.

Control Making sure that work is done how and when the plan calls for and, if it isn't, to understand why and take corrective action.

Control graph A graph of a program, showing branches as nodes and arcs as sequences of code between branches, used to analyze how a program is to be tested.

Controlling document A document constraining what the product is or how the project is managed.

Corrective maintenance Maintenance to get a system to do what it was originally intended to do.

Coupling The integrity of the data flow between modules, that is, how easy it is to comprehend the effect of a change in one module on another module; also, the amount of data flow.

Critical path The longest path in a scheduling network. The tasks on this path have the smallest slack in the network, which is equal to the required project finish minus the computed project finish.

Data base management system A system for managing data storage and access.

Data element dictionary A dictionary showing detailed information about each of the data elements.

Data element matrix A matrix that shows which data elements appear in each data flow, store, or process.

Data file Same as ***data store.***

Data flow The movement of data between processes or between a process and a data store.

Data flow diagram (DFD) A diagram showing the relationships among data flows, processes, and stores.

Data flow matrix A matrix representation of the information in a data flow diagram. The rows are processes or stores, the columns are data flows.

Data store Data waiting to be read by a process. Same as ***data file.***

Data structure matrix The combination of the data flow matrix and the data element matrix in which the data flows (columns) in each of the matrices align.

Decision support system Computer systems used to help people make decisions. This term has been used by some as a replacement for management information systems. Others, including the author of this book, prefer to use it to represent systems in which the decision process is distributed between the computer and the human. In such systems it is not practical initially to give the computer all the information needed to solve the problem. What information is needed is worked out as the problem solving develops.

Delegation Giving someone a task and expectations as to how it is to be done.

Deliverable Anything that is delivered to the client, including hardware, software, documentation, and service.

DeMarco Data Flow Diagram (D–DFD) The original form of a data flow diagram in which processes are shown as bubbles, data flows are written on the arcs, and stores appear as straight lines. No conditions or control information are allowed.

Design The synthesis (putting together of parts) to obtain a system or subsystem and the analysis of the results to see if it satisfies the requirements.

Design specification A specification describing how the system is to be built.

Design structure system A planning and scheduling method for design in which we work with a precedence graph that may contain circuits. We break circuits by showing where to use estimates to obtain a critical path schedule.

Directed graph A graph in which the lines (called arcs) have direction.

Distributed thinking The coordination of decisions made by several people.

Driver A dummy module used to make a call, with fabricated data, to another module.

Dummy task A task with zero duration, inserted in an IJ network to represent the precedences correctly.

Edge An undirected line in a graph.

Electronic mail The transmission of addressed messages from computer to computer.

Electronic office An office using computers to help with the creation, storage, and transmittal of information and the coordination of work.

Embedded computer A computer that is part of a larger system.

Environment That which is outside the system and interacts with it; that is, other systems that interface or people who work with the system.

Event An I or J node in an IJ network.

Events list A list of occurrences that are important to a study and when they occurred.

Expectations list A list showing who has what expectation for the system.

Expert system A computer system that, through the debriefing of an expert, emulates the advice that may be given by that expert.

External dimension The descriptors used for external specifications (requirements). The principal external dimensions are: environment, function, performance, and logistics.

External specification The specifications of how a system should behave as can be determined from the outside. Requirements is an external specification.

Feasible Satisfies all the constraints.

Feasibility study A study to determine the likelihood that a proposed system will be possible, affordable, and acceptable.

File Same as ***data store*** or ***data file***.

Fired The starting of a process when the necessary circumstances occur.

Formal management review A well-prepared review, presented to management to help them make decisions.

Function What a system does or should do, those behaviors that justify a system's existence.

Gantt chart Same as ***bar chart.***

Graph A set of points, called nodes, and lines between certain pairs of points.

Grief–joy chart A bar chart highlighting by exception the work that is behind schedule (grief) and the work ahead of schedule (joy).

Hierarchical method A method of representation that can be read by moving through a hierarchy while digesting only a small amount of information at a time.

Hierarchical organization An organization with the structure of a tree, in which each person reports to one superior.

Hierarchy An organizational structure that may be described either as boxes within boxes, or as parts that have parts that have parts, or as a tree.

IJ network A scheduling network showing the precedence relationships among tasks in which each task is given an I event at its beginning and a J event at its end and precedences are shown by making the J event of one task the same as the I event of the next. May require dummy tasks to show the precedences correctly.

Incremental development Same as ***versioning.***

Inspection A peer review, more formal than a walkthrough, conducted by a moderator, with formal documentation of the results.

Internal rate of return The interest rate that is equivalent to our investment, that is, the interest rate to which we compare our investment such that the cumulative present value at the end of a specified period is zero.

Internal specification The specification of what a product looks like inside, that is, what makes it behave as called for in the external specification. Design is an internal specification.

Iteration The process of repeating, or a repetition after the first.

Lano matrix A representation of the same information that occurs in a data flow diagram. The rows are the processes from and the columns are the processes to and the cells at the intersections show the data flows. Named after its developer, R. J. Lano. Also called Lano N-squared matrix.

Leaf The right end of a branch that has no limb.

Limb The vertical line in a Tree, having one entering branch on the left and one or more exiting branches on the right.

Local area network (LAN) The connection of computers for sharing information, processing capabilities, and peripherals within a local area.

Logical data flow diagram A data flow diagram showing just a logical description, that is, just the information and the changes made to it.

Logical description The description of a system including the information in the system, its source, flow, transformation, storage, and use but not the hardware, software, or people who transmit, transform, or store the information.

Logistics That which is needed to keep a system operating correctly, such as upkeep and maintenance, and issues such as reliability and availability.

Maintenance The testing and diagnosis, then designing and making changes, that occur after a program has been delivered.

Management by objective A management process whereby management and subordinates negotiate mutually agreeable expectations.

Matrix The organization of information as an array of rows and columns.

Matrix, or network, organization An organization in which one may report to several people, each with a different concern, such as a project manager and a programming manager.

Mortar First—Bricks Later approach A development method in which the interface files between modules and their test data are developed before the modules themselves.

Multiprocessing The use of more than one computing processor to work on various parts of the same job simultaneously.

Multiprogramming The switching of computing and IO processors among several programs to obtain more efficient use of the machine.

Network A term sometimes used for a graph, particularly a graph used for describing the precedences among taks in a schedule.

Node A point in a graph.

Objective A specific thing to be accomplished.

On-line processing Processing in which the computer operates as the user process occurs, but the speed of the user process may be affected by the response time of the computer.

Open questions list A list of questions and issues that have not yet been resolved.

Operand Something to which a process is performed, or something produced by a process. May be thought of as a noun.

Operator A process that is performed. May be thought of as a verb.

Opportunity cost The loss of opportunity to do something else with the same resources.

Optimum A solution such that no feasible solution has a better payoff.

Organization A management-subordinate structure that tells individuals whose expectations they are to meet and by whom they will be evaluated.

Organization chart A diagram showing what position reports to what other positions.

Outline processor An editing program that helps one in working with hierarchical structures. It will allow one to hide certain parts of the outline while revealing others. Also called a ***Tree editor.***

Pareto's law A small fraction of the effort or a small number of the participants account for a major part of the consequences.

Parts breakdown tree A tree that shows the breakdown of a system into the parts that make it up.

Patchwork quilt A set of various methods of representing the system during various aspects of its development, as for example using different representations for requirements, design, and implementation. The changes of representation are seams in this patchwork.

Path A sequence of arcs from node to node. A path may be of one or more arcs.

Payback period The time before the cumulative cash flow crosses over to being positive.

Perfective maintenance Maintenance to get a program to do something beyond what it was originally intended to do.

Performance The extent of how well a system performs its functions, including speed, problem size, accuracy, and use of resources such as memory space.

Performance resources Resources used by the system for its operation, such as execution time, memory space, and accuracy.

Phase The divisions of a project into logically different types of goals. One common scheme breaks a project into the following phases: problem definition, requirements analysis, feasibility evaluation, systems design, component design, component implementation, system implementation, operation, and maintenance. To be distinguished from ***stage.***

Physical data flow diagram A data flow diagram to which a physical description has been added, that is, who or what does the transmitting or processing of information and what medium holds the information.

Physical description The addition to the logical description of what hardware, software, or people transmit, transform, or store the information.

Plan The determination of what is to be done, often including by whom, how, and when.

Policy Guidelines showing how to make decisions.

Power The ability to influence or control resources and the actions of other people to achieve goals.

Precedence call A call to a ***precedence procedure*** in a Tree.

Precedence network A network showing the precedence relationships among tasks in which the tasks are shown as nodes and the precedences as arcs.

Precedence procedure A procedure called from several places in a Tree and executed only when all its calls or some combination of its calls have been made.

Predecessor A node that appears before another in a path. The path may be of one or more arcs.

Preliminary study A study made to determine the nature of the problem or opportunity before committing to make a change.

Preliminary study report A report on the preliminary study.

Problem report A report on an apparent problem with a system that is already in operation.

Procedure Step-by-step descriptions of how to perform specific tasks.

Process Something that transforms input into output.

Product The object of a project. It is that which is made, or a change that is produced.

Profiler A system that inserts monitoring statements in a program's source code then, after execution, analyzes the amount of time spent in each section of the code.

Project The process that produces a product.

Project management plan The primary process document. It defines the work of the project and how it is to be done.

Project proposal A proposal to another group that we do work for them.

Project request A request of another group to do work for us.

Prototype A simple system that has some but not all of the properties of the final system, used to explore critical problems in the development of the final system.

Quality management plan A description of the procedures to be used to guarantee that a quality product will be delivered.

Rapid prototyping The early visualization of a product formed by developing a simple version, emphasizing certain aspects that involve a risk that needs to be resolved early.

Rationalization Reasons developed to justify action, or the process of resolving which of several models to use for making a decision.

Rayleigh curve, or Putnam-Norton-Rayleigh curve A model for the number of people needed as a function of the time into the project. Uses a curve originally derived by the physicist Rayleigh for an entirely different process.

Real-time processing Processing in which the computer must respond fast enough that the user process with which it is working is not delayed.

Recursive A process that contains a process like itself.

Regression test Tests run after a change to determine whether cases that had run correctly previously still do.

Reliability The probability that a system will perform correctly for a given period of time.

Request for proposal (RFP) A request made to potential bidders stating what their proposals should cover.

Requirements breakdown tree A tree that shows the breakdown of the requirements into more detailed requirements.

Requirements specification The interface between the client and developer stating the characteristics of the product that would satisfy the client's need.

Responsibility The requirement to be accountable for something done.

Reusable code Code developed with such care in one application that it can be used in another application.

Root The first branch of a Tree that has no limb on its left. Execution of the Tree begins and ends at the root.

Sandwich rules Rules for a Two-Entity Data Flow Diagram that state that between entities of one type there is an entity of the other type, that is, between processes or between a process and a store there is a data flow, and between data flows there is a process.

Satisficing The process of producing a satisfactory answer such that no further work to produce a better answer is likely to be worth the effort.

Scenario A play-by-play description of how a system may be used and how it would behave.

Scope What is and is not included in a system.

Seamless Refers to having no changes in method of representation when going from one aspect of the systems development to another, as for example when going from requirements to design.

Semantics The information about a system that cannot be shown by a graph, for example, the behavior of the parts and what the arcs mean.

Slack The amount of time a task in a schedule can be delayed without delaying the completion of the project, computed as the latest begin minus the earliest begin time.

Software engineer One who integrates the various computer science technologies and people skills to build a computing system to satisfy defined expectations for how well the product meets the client's needs, how much it costs, and when it is delivered.

Software engineering The management of expectations, computer technologies, people and their skills, time, cost, and other resources to create a product that meets the expectations of the client with a process that meets the expectations of the producer or the techniques used to produce a high-quality software product with a high-productivity process.

Software development project The effort resulting from the producer's promise to create a software product in exchange for payment by the client.

Software product A system of computer programs and everything needed to run them not already available to the client.

Stage A division of the project for the purpose of reviewing progress to date and preparing for the next stage of work. To be distinguished from ***phase.***

Standards and procedures manual A manual that describes how an organization expects things to be done.

Steering committee A committee of management who oversees and guides the work of a project.

Structure The information about a system that can be shown in a graph, that is, what is preceded or caused by what.

Structure chart A tree showing the breakdown of a program into modules and the modules into smaller modules. It may show what parameters are passed between calling and called modules. It can be used to analyze for coupling and cohesion.

Structured design A method of designing programs so they are easy to change and maintain, which arranges modules so there is high cohesion within them and low coupling between them.

Stub A dummy module called by another module or a short limb added to a tree to show the data flow.

Subtree A branch and the Tree to its right.

Successor A node that appears after another in a path.

Synthesis The putting together of components to produce a system or subsystem.

System A set of parts that interact in such a way that the behavior of the whole depends on the interactions among the parts as well as on the behaviors of the parts.

System life cycle The pattern of phases used to describe the project from its inception to its completion. Also used interchangeably with system development process. See ***phase.***

Systems analyst The person who does the analysis of the problem or opportunity and helps the client develop the requirements specification. This role may be assumed by the software engineer.

Systems engineer The person responsible for development of the whole system, including both hardware and software aspects.

Task force A group of people concerned with or working on a specific task, often made up of the client and various people representing the various functions to be applied to the task.

Tear A break of an arc in a graph. It may show where an assumption is to be made about a predecessor.

Test plan A plan for how a system is to be tested, showing what is to be tested, with what inputs, using what procedures, and how to evaluate the results.

Thorough requirements The systems development approach involving a careful and complete development of requirements before the system is designed and built, using an abstract representation.

Timesharing Several people simultaneously running a common computer by remote access.

Top-down design A method of design in which the system is first designed using modules, then the modules are designed using smaller modules, and so on. Also called ***top-down search.***

Top-down method A method that considers the whole before it considers the parts, then considers these parts before considering the parts that make them up, and so on.

Top-down search Same as ***top-down design.***

Topological order A numbering of nodes in a graph for which each node has a number higher than those that precede it.

Trade-off Change of the allocation of performance resources.

Transaction The processing that occurs for a particular set of input.

tree A representation of a hierarchical structure in the form of a graph with a root, nodes, and branches such that each node except the root has exactly one entering branch and zero or more exiting branches.

Tree A tree-like structure alternating between branches (horizontal) and limbs (vertical) extending from the root on the left out to right. The Tree can be built successively to represent requirements, design, and finally the program itself. Spelled with a capital T to distinguish this particular type of tree.

Tree growth model A model for the number of people needed as a function of the time into the project. Assumes a project grows like a tree then shrinks.

Trigger An event that plays a role in starting a process.

Two-Entity Data Flow Diagram (TE-DFD) A form of data flow diagram using two general types of entities: processes and stores together represent one type of entity, and data flows represent the other. Additional information such as control conditions and triggers can be added to the diagram.

Undirected graph A graph in which the lines (called edges) do not have direction.

User's manual A manual describing to the user how the system is to be operated and what it will do.

Validation A review to establish that what was specified is what was wanted. It is an external review of specifications against need. It asks: "Are we doing or did we do the right thing?"

Verification A review to establish that what was produced satisfies the specifications. It is an internal review of what was built against specifications. It asks: "Are we doing or did we do the thing right?"

Versioning The development of a system as a series of versions, each version having more modules completed and having more capability than previous versions. Also called ***incremental development.***

Walkthrough A review by peers looking for errors or potential errors in the system.

Warnier-Orr diagram A diagram showing the hierarchical construction of a program or data using braces.

Work breakdown tree A tree that shows the breakdown of work into smaller work units.

Index